SYSTEMS ENGINEERING

SYSTEMS ENGINEERING

Erik Aslaksen
&
Rod Belcher

PRENTICE HALL
New York   London   Toronto   Sydney   Tokyo   Singapore

Acquisition Editor: Andrew Binnie
Production Editor: Gillian Gillett
Typesetter: Monoset Typesetters, Strathpine, Queensland.
Cover design: Max Peatman

Printed in Australia by Macarthur Press, Parramatta, NSW.

ISBN 0 13 880402 8

National Library of Australia
Cataloguing-in-Publication Data

Aslaksen, E.
    Systems Engineering.

    Includes index.
    ISBN 0 13 880402 8.

I. Systems engineering. I. Belcher, W. II. Title.

620 .001171

Library of Congress
Cataloging-in-Publication Data

Aslaksen, E. (Erik)
    Systems engineering / E. Aslaksen, W. Belcher.
        p.   cm.
    Includes bibliographical references and index.
    ISBN 0-13-880402-8
    1. Systems engineering. I. Belcher, W. II. Title.
TA168.A73 1991
620'.001'171--dc20   91-22330
                                                      CIP

Prentice Hall, Inc., *Englewood Cliffs, New Jersey*
Prentice Hall Canada, Inc., *Toronto*
Prentice Halll Hispanomericana, SA, *Mexico*
Prentice Hall of India Private Ltd, *New Delhi*
Prentice Hall International, Inc., *London*
Prentice Hall of Japan, Inc., *Tokyo*
Prentice Hall of Southeast Asia Pty Ltd, *Singapore*
Editora Prentice Hall do Brasil Ltda, *Rio de Janeiro*

PRENTICE HALL

A division of Simon & Schuster

Contents

PART B New concepts 55

PART C Methodology 119

Preface

This textbook has resulted from the development over eight years of a comprehensive set of lecture notes for an introductory subject in systems engineering, given to final-year students in the School of Electrical Engineering at the University of Technology, Sydney. As a result, the case studies and most of the examples are taken from electrical engineering, and in order to fit the material into a one-semester course there is a special focus on the design aspect of large systems; otherwise, the concepts introduced and the methodology expounded are of quite general validity and may be applied to a wide range of problems.

The lecture notes were developed in response to a simple and practical need. From his experience with a major project in Australian industry, it had become apparent to one of us (WRB) that although Australian engineers were both inventive and knowledgeable in engineering science, they lacked a methodology which would allow them to apply their skills to complex problems in a systematic fashion. Therefore, upon becoming Head of the School of Electrical Engineering of what was then the NSW Institute of Technology in 1982, he immediately set about finding the means of introducing a systems engineering subject into the undergraduate curriculum. As the Faculty of Engineering was always based on very strong ties to industry through its cooperative education program, it was natural to seek an industrial involvement in this endeavor, and through the efforts of the Dean, Professor P.J. Parr, the other half of the team (EWA) was recruited.

The introduction of Systems Engineering in the final year of an undergraduate course has turned out to be a great deal more difficult than first envisaged, and it took us a while to realise just how foreign some of the concepts in systems engineering are to students who have experienced nothing but the traditional bottom-up approach to problem-solving; whose first reaction to a problem is to cast about for a suitable formula; who expect the answer to be unique; and who consider any non-technical considerations (particularly economic considerations) to be distractions. Consequently, the lecture notes and the format of the lectures themselves have undergone a considerable development over the eight years of development, and the structure of this textbook reflects our current approach to overcoming these difficulties.

Part A introduces the subject-matter and a number of case studies, and places systems engineering in the context of engineering projects and project management. Particular emphasis is given to two aspects, namely, the general nature of the system concept, and systems engineering as a methodology rather than a collection of solutions.

Part B discusses some of the concepts that are basic to systems engineering, but which are generally new to students. The emphasis here is on understanding; there is little that can be directly applied.

Part C is the central part of the text in that it treats, in as much detail as a one-semester subject will allow, the methodology and some of the main techniques of which it is composed. This is where the essential reorientation towards the top-down approach takes place.

Part D discusses a number of system characteristics and their integration into the design process. As a result of being a part of that process, each one is developed in a top-down fashion, but they are also all developed concurrently, and interact strongly.

From our teaching of systems engineering to undergraduates it has become clear that the aspect causing students more difficulties than anything else is the role played by abstraction and the abstract nature of the systems concept itself. And yet, without a thorough understanding of that aspect, systems engineering would be reduced from a distinct discipline with its own general methodology to nothing more than a set of rules (albeit very useful rules) for how to carry out the engineering of complex entities. Consequently, the role of abstraction is emphasized throughout the book.

It is assumed that the students are familiar with the basic elements of engineering practice, such as writing a specification and a report, and that they understand the fundamentals of economic analysis as a basis for decision-making. Furthermore, a reasonable understanding of statistics and probability theory is essential.

It is a pleasure to acknowledge the contributions of a number of organizations to the discipline of systems engineering and, directly or indirectly, to the development of this book. One of the authors (EWA) was introduced to systems engineering thirty years ago through his work at Bell Telephone Laboratories, an organization which pioneered many of the ideas that make up the subject today. The development of the present undergraduate subject was encouraged and supported by Brown Boveri, in particular the head of the Electrical Division in Australia, Mr Gordon Howard. The US DoD has been a leader in systems engineering for more than twenty years, and the various standards and handbooks produced by it provide invaluable source material. The interest shown by the Australian DoD has been very encouraging to one of the authors (EWA).

Finally, we would like to thank Mrs Rosa Tay for her patience and dedication in typing the text through a number of versions.

Erik Aslaksen
Rod Belcher

Introduction

CHAPTER 1

Systems

1.1 Concept of a system

1.1.1 Importance of systems

The word "system" is used more and more frequently in almost every area of human endeavor; we have, for example, control systems, weapons systems, ecosystems, transport systems and life support systems. This is a reflection of several related factors:

- the increasing complexity of society in general, due to the tighter coupling of its components through the increased mobility of goods and information;
- an increasing capacity for abstract thought and the realization of complex relations and dependencies, due to a higher average level of education;
- more immediate and serious effects of interactions that were negligible on a small scale of operations;
- better means of describing and handling complex relations.

The driving force in this development is technology. In particular, microelectronics is enabling major development in all sorts of systems related to data processing and transmission. This is bound, through its wider interactions, to have a profound effect on society. Thus, systems per se are certain to be of increasing importance.

1.1.2 Features of systems

While everybody uses the word "system" with impunity, relatively few have a clear understanding of what it means. It is easy to find examples of systems and explain exactly what they are and how they work, but the meaning of the word "system" by itself appears elusive. However, looking at such examples, it becomes apparent that one central feature of any system is its composite nature; it consists of and is described in terms of two or more **elements**. For example, a control system may be considered to consist of process interface equipment, a central processor and man-machine interface equipment. A particular weapons system may be considered to consist of a detection subsystem, missile-launching facilities, a guidance subsystem and control facilities. In both these cases the elements are distinguished by their **functionality**; the overall function of the system is described in terms of its external behavior (i.e. as seen from the outside, without any reference to the internal workings) of its composite parts. Such functional elements are also sometimes called "black boxes"; a circuit consisting of more than 1000 transistors may be treated as a single element with a simple function (e.g. the multiplication of two eight-bit binary numbers).

It is not necessary for a system to be composed of functionally different elements; a number of microcomputers linked by a local area network (LAN) is an example of a system where the elements are all functionally equal but spatially (geographically) separated. Moreover, the elements do not have to consist of hardware; they can consist of organizational units or simply people. For example, a company may consist of sales, production and administration departments, each again consisting of a number of persons. The elements can also be nonphysical, such as the activities required by a task or the software modules making up a program. In all of these cases a main purpose is to describe the system in terms of elements, each of which is distinguished by its reduced **complexity**.

This leads immediately to another important aspect of the system concept. If a complex object is simply divided into n parts, the way a cake is cut, the single complex object is just changed into n interacting objects, and it is not clear that anything has been gained. The desired reduction in complexity results only if the choice of elements, the **partitioning**, is made in such a way that each element can be considered largely (but not completely) on its own. In other words, complexity is reduced if the element can usefully be considered in a zero-order approximation as isolated or independent, with the system behavior emerging as a result of first-order perturbations caused by the interactions between the elements. This may be illustrated by the following simple example.

1.1.3 Example: voltage monitoring

A system for the metering and monitoring of an a.c. voltage is shown in Figure 1.1, together with the natural definition of its elements, namely, the four functions of signal conditioning, A/D conversion, display and alarm. For this system it would be completely nonsensical to combine (say) R7, C1 and Q8 into an element, as this group of components would have no function that could be described simply. Essentially, there would be no gain, as far as the ease of description or analysis is concerned, in such a partitioning. (Indeed, partitioning in terms of individual components of the circuit would seem more beneficial than the trial combination.)

The example shown in Figure 1.1 also illustrates the distinction between internal and external interactions. The whole system also has a simple functional description and may serve as an element of a larger system, as shown in Figure 1.2.

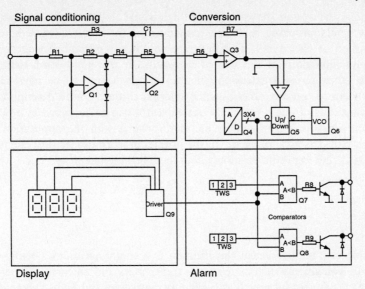

Signal conditioning   Conversion

Display   Alarm

Figure 1.1 System for the metering and monitoring of an a.c. voltage. A partitioning into four functional elements is shown by the four large boxes.

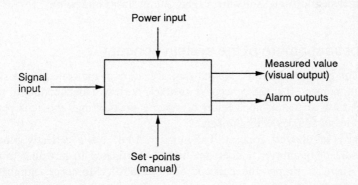

Figure 1.2 System of Figure 1.1 as a single element, with clearly defined external interactions (inputs and outputs).

1.1.4 Partitioning and size

The example described above can be used to illustrate two further aspects of systems.

Firstly, the choice of partitioning (i.e. the way the system is represented) is a major factor in the efficacy of any system description, even though there is no unique representation. **Different partitionings can be beneficial, depending on the particular interest**. For example, it is possible to represent the system in Figure 1.1 by means of only two elements, say, signal processing and man-machine interface. This could be the best partitioning for the purpose of writing a user's manual. However, for the purpose of circuit development the earlier partitioning would seem a better choice.

Secondly, the system shown in Figure 1.1 would not normally be described as a system; it would probably be called a module or (possibly) a piece of equipment (depending on its packaging). This illustrates another important fact about a system. **There is no limitation on the size of a system, in the number of elements, their complexity, their physical extension** (they do not even have to be physical) **or their monetary value, or on any other quantity relevant to the system's description**. An RC network consisting of one resistor and one capacitor can be described as a two-element system, and there is a considerable literature dealing with the application of graph theory to circuit analysis. However, in the normal practice of electrical engineering the word "system" has implied a position within the following hierarchy:

system
equipment
module
device.

An example of this hierarchical classification is a communications **system**, where the elements are **equipments** such as receivers, transmitters and antennas. The elements of the equipments are **modules** such as oscillators, amplifiers and filters. These in turn are achieved with elements that are devices, namely, transistors, capacitors, resistors, inductors and so on. While this applies directly only to hardware, it is generally true that the benefits of systems engineering are most obvious for "large" systems. These can of course include hardware, software, organizations, tasks and so on.

1.1.5 Abstract nature of the system concept

From the foregoing discussion a second central feature of the concept of a system may be apparent, namely, that the concept is **abstract**. A clear understanding of this feature is essential to a successful application of systems engineering, and it is perhaps best introduced by an analogy with numbers.

Number is an abstract concept. The number "four" has a perfectly good meaning (as a measure of quantity). It does not have to be related to anything to retain this meaning. There is simply the concept of "fourness". Moreover, numbers can be manipulated as abstract quantities; for example, the relation $4 + 2 = 6$ is well understood and applies universally.

Systems also have abstract properties, notwithstanding that systems are much more complex than numbers (actually, more like functions of many variables). The usefulness of systems engineering comes from recognizing and exploiting these properties. However, as systems can be so complex, these properties are not self-evident.

All systems share certain properties. These may be regarded as totally abstract and at this point undefined, but they provide the foundations of systems engineering. Large classes of systems have further properties in common, recognizable from experience. Within the confines of such a class the systems engineering methodology becomes in turn more powerful. This leads to an interesting aspect of systems engineering: the desire to find classes that are large enough to give the benefit of wide applicability, but restrictive enough to be rich in abstract properties. This aspect wil be considered further in Section 1.3.

Standing back for a moment and considering the discussion of systems to this point, it would be excusable to feel somewhat confused by a potential anomaly. In the opening paragraph it was said that the word "system" is becoming more frequently used; yet now it is stated that the word describes a highly abstract concept. Does this mean that people are expressing themselves increasingly in abstract terms? The answer is no, but the word is being used (often unconsciously) in two different contexts and has a different meaning in each context. In speaking about a system in daily conversation (e.g. about the education system), the entity that produces education as a service is meant. Normally the speaker is not thinking about it in terms of elements. However, in speaking of representing something in terms of a system, the speaker is thinking of "system" as a way of **describing** something physical, and it is this description or **model** that is abstract and now needs to be defined. This is the subject of the following section.

1.2 Formal definition

1.2.1 Previous definitions

Before presenting the definition of a system to be used in this text, it is interesting to see how other people have tackled the problem. Five definitions, from well-known authors in the field of systems and systems engineering, are presented in Section 1.8.

On review of the definitions, it is immediately apparent that they differ greatly in their precision and general applicability. If their precision is plotted against their degree of general applicability (without being too specific about the scales along the axes), a picture of the type shown in Figure 1.3 is obtained. This figure is meant to convey but one message: the greater the precision of the definition, the less the degree of general applicability.

The several definitions offer another insight relevant to the discussion in the preceding section. Precision makes the concept of a system more **useful**, to the extent that it can be manipulated mathematically, analogous to the way algebra manipulates numbers.

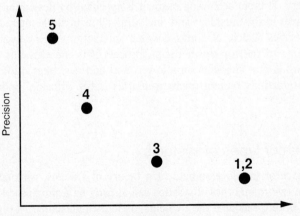

Figure 1.3 Characteristics of the five definitions contained in Section 1.5.4, showing the correlation between precision and general applicability.

1.2.2 Definition applicable to this text

The definition of a system that applies throughout this text is the following: A system consists of *three related sets*:

> *a set of* **elements**
> *a set of* **interactions** *between the elements*
> *a set of* **boundary conditions**.

The elements may be of any type (e.g. hardware, software, activities, concepts, people), but they must form a set; that is, there must exist a rule that allows a decision to be made about whether an object is a member or not of the set.

The interactions also may be of any type; and while they will often take the form of an interchange of energy or information, they are by no means restricted to having a physical nature. They can be purely logical (e.g. an ordering of the elements).

The boundary conditions are the interactions between the elements and all other objects (i.e. the outside world), to the extent that they are chosen to be considered in any given case. The set of boundary conditions is often divided into subsets that are treated in basically different ways as far as the functions of the system are concerned, such as inputs/outputs and environmental conditions.

In the application of this definition to practical cases, a point that sometimes causes confusion is the distinction between what is an element and what is an interaction. Take, for example, the case of a power distribution system, consisting of transformer and/or switching stations and the lines and/or cables between them. Then, for doing power flow analysis, the stations would be chosen to be the elements and the lines to be the interactions. However, for doing reliability analysis, both the stations and the lines would be chosen to be elements. In the latter case an interaction is simply a recognition that two elements are interconnected; it has no physical representation.

It may appear from this example that there is little distinction between elements and interactions, but that is not so at all. Whereas there is often a choice, depending on the issues under consideration, of which physical components to classify as elements and which as interactions, once this choice has been made, elements and interactions describe two very different sets. Any element is specified by describing completely how it looks or behaves to the outside world, including all its possible interactions. The set of interactions specifies which of these possible interactions are actually used in that particular system. In the top-down design process it is the elements that are further subdivided, creating new elements on a lower level and new interactions between them. **The original interactions remain unchanged**; this is the principle underlying interface control.

1.2.3 Two major types of systems

At this early stage of the development of a theory of systems, with no more than the above definition two main types of systems can already be distinguished.

A system is **homogeneous** if the elements are indistinguishable. A typical example of this is a system whose elements are water molecules; depending on the temperature, this system is a volume of water vapor, a volume of water or a piece of ice. However, it is by no means necessary that the physical elements be actually identical; it is only important that they be identical in the context of the particular system—or in other words, that the differences not be perceived by the set of interactions being considered.

A system is **heterogeneous** if its set of elements can be subdivided into distinguishable subsets, with the elements within a subset being identical. The limiting case of a heterogeneous system is of course the case where every element forms a distinguishable subset; that is, the elements are all distinguishable from one another.

These two definitions are concerned only with the set of elements; the interactions and boundary conditions do not enter into determining whether a system is homogeneous or heterogeneous.

1.3 Abstraction and modeling

1.3.1 Introduction

The theory of systems developed in the preceding sections offers generally useful principles for understanding and dealing with complex entities or situations. Two of these principles have been emphasized, namely:

1. Complexity can be reduced by appropriate partitioning. The individual elements of any system are less complex than the whole system. In most cases, however, these same elements can also be regarded as (new or different) systems, amenable to further partitioning.
2. Systems share common properties and characteristics. By definition, some properties are common to all systems, while further properties are specific to a class of system.

In this section attention is directed towards the second of these principles.

Properties are found by **abstraction**: by filtering out all the details that pertain only to individual systems or to subclasses of systems and building a more generally valid model system. An important example of this is the transition from the *physical* to the *functional* domain. A system (subsystem, element) may be described in one of two ways: by what it *is* or by what it *does*. The former is the description of a particular physical system (e.g. in terms of dimensions, material characteristics, and so on); the latter is the description of a functional system which may encompass a large class of physically different systems. It may be said to *model* the behavior of a *class* of physical systems.

The process of abstraction and modeling is one of the most fruitful and important activities in systems engineering. Two examples may be helpful in illustrating the principle involved.

1.3.2 Example: buffering and risk

Consider a system consisting of two elements: a charger and a reservoir, as shown in Figure 1.4. The purpose of the system is to supply a commodity to a user, whose demand D is a stochastic variable with values between D_1 and D_2 and is distributed according to some probability density function p(D). The user is supplied from the reservoir, which has a capacity C, whereas the charger, which keeps the reservoir topped up, has a maximum feed rate S. However, the charger is not perfect; it has a failure rate f and a fixed repair time g.

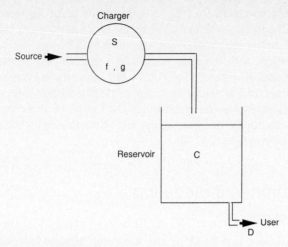

Figure 1.4 System consisting of a charger with maximum feed rate S, failure rate f and repair rate g, a reservoir with capacity C and a user with demand D.

The reservoir serves a twofold purpose. On the one hand, it smooths the demand fluctuations, thereby reducing the requirement placed on S. On the other hand, it serves to bridge supply during a failure of the charger. Nevertheless, there remains a finite **risk** Q of a supply failure, and this risk is a function of the two parameters C and S (for a given p (D)).

The modeling of this system and determination of the optimum values of C and S require more detailed information, thereby narrowing the model's applicability to a smaller class of systems. An example of this will be given later in Section 6.3. However, as defined here this system represents a large class of practical systems, such as:

a water supply system
an uninterruptable power supply
the cash flow of a company
parts or materials supply.

The common system is identified by abstracting from the particular entities of any one system. A dam, a tank, a battery, a bank account and a store all have a capability for "storage". This abstract capability can be expressed more precisely as "the temporal decoupling of supply and demand".

1.3.3 Example: trunk network

A **trunk network** is a transport system consisting of a number of **trunks** linking a number of **nodes**, as shown in Figure 1.5. The nodes fall into two classes: **trunk nodes**, which serve only to interconnect trunks, and **access nodes**, which also allow users access to the network. A user inserts the goods to be transported at one access node (the **source**); it travels through the network and exits at another access node (the **sink**). A route through the network is called a **circuit**. A trunk is simply a transport medium that can support two or more connections simultaneously. A node has the capability to

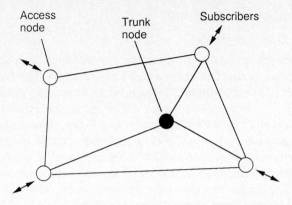

Access node

Trunk node

Subscribers

Figure 1.5 A trunk network, consisting of nodes interconnected by links.

establish connections through it; this activity is called **switching**. The multiple switching involved in establishing a complete circuit is called **routing**.

With the several entities expressed in these abstract (or generalized) terms, the trunk network represents a large class of systems, including, among others:

telephone systems
power transmission systems
water, oil or gas pipeline systems
road transport systems
railroad systems.

For example, the abstract concept of switching, in the case of a road transport system, is realized by the activity of steering trucks from one road to another at the intersection of two roads. In the case of power transmission systems it is realized by opening and closing circuit breakers.

1.4 Case studies

1.4.1 Introduction

To illustrate the application of the systems engineering methodology it becomes necessary to introduce examples of actual systems. However, as the purpose of the methodology is to deal efficiently with complexity, systems of interest must necessarily be complex and thereby require a considerable amount of description and explanation. The three systems chosen here represent a compromise; they can be treated in some detail, but they are still large enough to show many typical system features, the problems arising from their inherent complexity, and how these problems are attacked using the methodology and tools of systems engineering.

The case studies are developed along with the subject; as each aspect of systems engineering is treated in a chapter of this text, it is also illustrated by showing its relevance and application to each of the case studies.

1.4.2 Combustion optimization

1.4.2.1 Background information

Combustion processes are used to produce heat; the combustion product is a gas at a high temperature (i.e. its molecules have a high kinetic energy). This heat can be utilized directly as heat (e.g. to melt iron ore, to burn bricks or ceramic products), directly as mechanical energy (e.g. internal combustion engine, jet engine, rocket), or indirectly to produce mechanical and finally electrical energy (e.g. boiler, steam turbine, generator).

Combustion processes obviously play a vital role in any modern economy; in particular, their efficiency affects the viability of the associated products or processes.

One of the factors affecting the efficiency of any combustion process is the fact that the exhaust gas (i.e. the gas that leaves the complete process after the energy has been extracted) still contains some thermal energy, which is a loss. This loss can be minimized in two ways: by extracting as much heat as possible and lowering the exhaust gas (or flue) temperature as far as possible (an option usually limited by condensation, which produces corrosion); and/or by producing as little exhaust gas per unit energy as possible. It is the latter approach that is the subject of the combustion optimization system to be discussed here, limited to industrial processes.

In any combustion process the fuel is mixed with air at a high temperature, which leads to an exothermic oxidation process. The mixing is carried out in a **burner**, which falls into one of three classes:

- modulating burners, where the combustion rate can be varied over a wide range, typically 10-to-1 turndown;
- two-level burners with only a high and a low rate (typically 3-to-1 turndown);
- simple burners with a single rate, which can only be regulated by turning them on and off.

In all three cases combustion optimization will lead to fuel savings, but the main area of interest is modulating burners.

A typical installation is shown schematically in Figure 1.6. The combustion rate, or **power level**, is set by means of a servomotor operating on the fuel valve, and the appropriate amount of air is provided by making the position of the air valve a particular function of the position of the fuel valve. If the mixing process were ideal, there would be a fixed ratio between the amount of fuel consumed and the amount of air required for perfect combustion, both measured in units of mass per unit of time. This ratio, the so-called **stoichiometric ratio**, depends only on the caloric value of the fuel, as the oxygen content of air is very constant and equal to about 21 per cent.

However, no burner is quite perfect; the mixing of fuel and air is always slightly inhomogeneous. So, to ensure that the air-to-fuel ratio nowhere falls below its stoichiometric value, as this would immediately produce unacceptable levels of pollutants such as carbon monoxide, CO, the average ratio is kept slightly above the stoichiometric value. As a result there is a certain amount of oxygen to be found in the flue, and the volume of flue produced is slightly higher than necessary. Furthermore, it is generally true that the inhomogeneity of the mixing increases with decreasing gas velocity, so the resulting situation is that illustrated by the full curve in Figure 1.7. This curve represents, for a given burner, the minimum value of oxygen (or excess air) in the

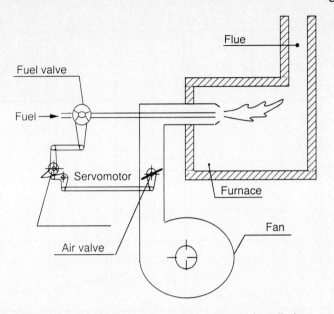

Figure 1.6 Schematic arrangement of a modulating burner installation.

flue for acceptably low levels of pollutants and is the one that an ideal feed-forward control mechanism would want to reproduce.

In reality, variations in the parameters entering into the combustion process, such as the heat of combustion of the fuel, the barometric pressure, the humidity of the air and the ambient temperature, as well as tolerances and wear in the control mechanism itself, cause the actual measured value of the oxygen content to fluctuate around this ideal value. Again, to be on the safe side and always within the law, the control mechanism must be set for a slightly higher average value, as indicated by the broken curve in Figure 1.7. As a result the flue volume and flue loss are somewhat higher than necessary, and it is this excess loss that the introduction of combustion optimization aims to eliminate, or at least reduce significantly.

The basic idea behind the combustion optimization is to use a feed-back loop superimposed on the feed-forward control in order to eliminate the effect of the fluctuations in the process parameters. A high-level view of such an arrangement is shown in Figure 1.8. The details of the optimization system are developed progressively throughout this text. One aspect should be mentioned here, concerning the absence of a normal feed-back control loop for the whole control. The reasons for this are twofold. Firstly, a large number of installations exist that utilize a feed-forward system, and they represent an important part of the market for combustion optimization. Secondly, combustion control systems have to satisfy extremely stringent reliability and safety requirements, and the mechanically coupled feed-forward system is unsurpassed in this regard, in addition to being relatively inexpensive. The add-on feed-back loop, as shown in Figure 1.8, does not have to meet the same safety and reliability requirements, as long as it is designed to have no significant influence once it has failed; that is, it must only be capable of making small corrections to the air/fuel ratio.

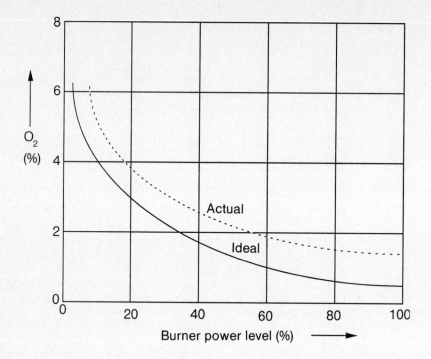

Figure 1.7 Ideal and actual flue oxygen content as a function of the burner power level for a typical modulating oil-fired burner.

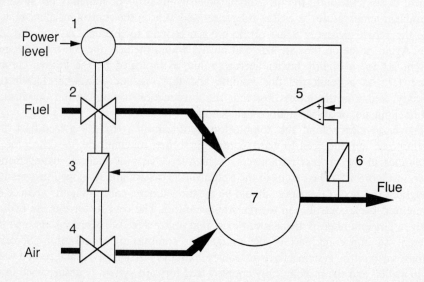

Figure 1.8 Simplified diagram of the feedback loop superimposed on the feed-forward control. 1 = servomotor, 2 = fuel valve, 3 = electromechanical transducer, 4 = air valve, 5 = processor, 6 = electrochemical transducer, 7 = combustion process.

1.4.2.2 Purpose of the optimization system

The very first activity in the system design process is a definition of the **purpose** of the system or, equivalently, its **functionality**. That is, without any thought as to how it will work, what it will consist of or even what type of technology will be used, it must be possible to state unambiguously what the system is required to do. Such a definition does not arise by inspiration; it must be developed in a logical top-down fashion.

At the highest level the purpose of the system might be defined as: "To increase the cost-effectiveness of a variable-rate combustion process". There are many ways in which this objective could be met, such as using a cheaper fuel, reducing the cost of the plant, designing the plant to need less maintenance or reducing heat losses. Of these, only the latter is the subject of this exercise.

Heat losses fall into two main groups: losses due to inadequate insulation (i.e. through conduction), and flue losses. Again, only the latter are considered here, and these can be subdivided into loss due to excessive flue temperature and loss due to excessive flue volume. Reducing the latter could be tackled in different ways; the only one under consideration here is using a feed-back loop to correct the air/fuel ratio, with the oxygen content of the flue as the controlled parameter.

The above top-down development is illustrated in Figure 1.9. As will be discussed in Chapter 2, and again in more detail in Chapter 6, a number of decisions have been taken, and possibly a large amount of work has been carried out, in order to arrive at the definition of the purpose of the system and thus to provide (a part of) the starting point for the system design.

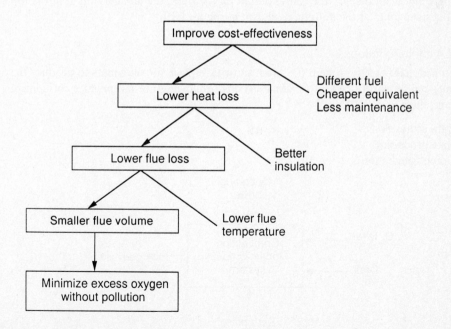

Figure 1.9 Top-down development of the purpose of the combustion optimization system.

1.4.2.3 Identification of inputs and outputs

For a system to achieve its purpose it must influence the external world in some way, and this part of its set of interactions is called its **outputs**. In the present case the main output, namely the correction of the air/fuel ratio, can immediately be identified. At this stage it is also the only output that follows directly from the definition of the purpose; but as the design progresses and more detailed requirements emerge, further outputs may well be necessary.

To produce this output the system needs certain **inputs** from the external world. From the foregoing description of the system the following inputs can be identified:

* the oxygen level in the flue;
* the power level at which the burner is operating;
* the set-point function for the particular burner;
* power to operate the system.

Already in this example it is worthwhile to point out that the inputs can be divided into subsets, which have very different properties. The two first inputs are **dynamic** or **real time** variables; they are functions of time. The set-point function is a **static** variable; it represents knowledge about the outside world (i.e. the particular burner) and is input only once at the beginning of operations. The power needed to operate the system is an **auxiliary** input when the functionality of a system is expressed as relations between inputs and outputs. The auxiliary inputs will not appear in this case study.

At this stage, where the system is still regarded as a single entity without any internal structure, the system can be graphically represented by a single block, with the dynamic inputs on the left, the (only) output on the right, the auxiliary input at the top and the static input at the bottom, as shown in Figure 1.10.

1.4.2.4 Main functions

The functionality of the system (i.e. what must take place for the inputs to produce the output) can be partitioned into **main functions**. Following normal practice for control systems, these are:

> data acquisition
> data processing
> output generation.

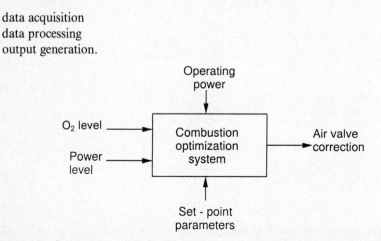

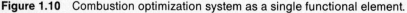

Figure 1.10 Combustion optimization system as a single functional element.

In addition, it may or may not be wished to list explicitly the auxiliary function of the internal power supply:

   power supply.

Consequently, at this stage in its development the system consists of four elements.

1.4.3 Tactical communications system

1.4.3.1 Background information

Trunk systems were introduced in Section 1.3.3. The properties of such systems depend on factors that can roughly be divided into two groups: those which are intrinsic to the elements themselves, and those which arise from the way the elements are interconnected (i.e. from the network **structure**). Consequently, in the design of any normal telecommunications system, such as a common carrier network, considerable effort goes into devising an optimal structure. However, in the special case of a **tactical trunk network** there can be no fixed structure, as the forces to be supported with telecommunications services are constantly moving and changing in size and composition. One is therefore faced with two problems: how to characterize a tactical network, and how to formulate the criterion for an optimal design.

1.4.3.2 Purpose and main functions

The purpose of a trunk network is to transfer information between the members of a set of physically separated **subscribers**. Consequently, a main function of the network must be to overcome distance; this is the **transmission** function. Then, in order to utilize the transmission capacity efficiently it must be possible to establish circuits between pairs of subscribers as required; this is the **switching** function. Finally, the individual subscribers must somehow be able to interface to the multichannel trunk network; this is the **access** function.

   As mentioned in Section 1.3.3, the equipment making up the system is concentrated in **nodes**, and at each node there is equipment belonging to each of the three functions defined above. In the case of the access and switching functions, all the equipment belonging to one of these functions and located at one node is considered to form an **element** of the system; whereas in the case of the transmission function, there is an additional subdivision according to the **link** to which the equipment belongs. This is illustrated in Figure 1.11, where three different symbols have been introduced to represent the three types of elements.

   So far, then, the tactical communications system consists of a set of elements that is subdivided into three subsets. Within each subset the elements are all equivalent at this stage; that is, they are characterized by their functionality alone; any actual physical differences are ignored. The interactions between the elements show their capacity to exchange information; this is described in terms of the structure of the system (see Section 3.5). What remain to be defined are the boundary conditions (i.e. a characterization of the traffic incident on the network).

1.4.3.3 Subscriber population and traffic

To each node there are attached a number of subscribers; together they form the **subscriber population**. Each subscriber generates **calls**. Calls may be generated at

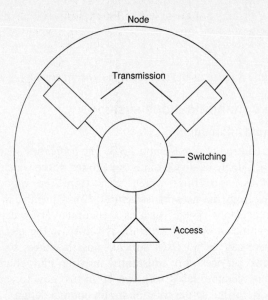

Figure 1.11 Composition of a node in a trunk network, showing the three types of elements: switching, access and transmission.

random times or according to some pattern, and the duration of each call may be fixed or distributed according to some probability distribution. For present purposes assume that calls are generated at random (i.e. according to a Poisson distribution), that each subscriber generates the same average number of calls per unit time and that the duration of a call is fixed. Consequently, the result is a characterization of the traffic volume incident on the network in terms of the average traffic generated at each node; it is the number of subscribers at the node multiplied by the average number of calls generated by a subscriber per unit time multiplied by the duration of a call.

However, the volume alone is not a complete characterization; it is also necessary to specify the destination of the traffic generated at each node. The complete specification of the traffic that the network is required to support (i.e. the boundary condition) is therefore a matrix where the elements are indexed by the nodes and where the values of the elements are the average traffic flows between the corresponding nodes; it is called the **traffic matrix**.

The problem encountered in tactical networks is therefore reflected in a time-dependent traffic matrix. Not only do the values of the matrix elements change, but the number of nodes also changes, with the changes taking place at discrete but random times. This is clearly a complex situation. In later chapters the systems engineering methodology will be applied in order to develop a well-defined characterization of tactical trunk networks.

1.4.4 Electricity supply system and substation automation

1.4.4.1 Background information

A public electricity supply constitutes a very large system. It can be subdivided into three subsystems: generation, transmission, and distribution. In generation and transmission the associated control systems have long played a major role in the operation, ensuring a constant system frequency, voltages within prescribed limits, most economic utilization of plant and so on; but distribution systems have, with some exceptions such as ripple control, remained passive networks in which any switching operations are carried out manually. This situation is in the process of changing, with the advent of microprocessor-based local intelligence and a plummeting price/ performance ratio for electronic equipment. Not only can new functions, such as substation reconfiguration and staggered restoration, be implemented, but also all the existing functions, such as tapchanger control, auto reclosure and power factor correction, can be integrated into one system, which results in significant cost savings. On top of all this, certain useful peripheral functions, such as sequential event recording and telecommunications, can be added at very low incremental cost.

The beginning of this development was the introduction of telemetry systems bringing data to a central location, giving the operators there an overall view of system performance and allowing them to control the system configuration. The next step was to introduce computers at the control center to do the initial data processing, so that the mass of data would be easier for the operators to digest.

As the understanding of power network behavior improved, it became clear that certain conditions demanded defined responses, consisting of predetermined sequences of commands. Thus, the execution of these responses could be delegated to the computer. This alleviated another problem that was becoming increasingly evident as power networks grew in size and complexity, namely, the probability of operator error due to overloading of the operator's capabilities. However, this concentration of control power in one central location involved risk analogous to putting all one's eggs in one basket, while requirements on the data transmission capacity were considerable.

In recent years a technological development has had great impact on control system design. The very large-scale integration of electronic circuits has lowered the cost of both computing power and data storage by many orders of magnitude; so it is now possible to distribute the control functions and to associate partly autonomous process control and data-processing equipment with local sections of the network. The result is a hierarchical structure, with all those functions being performed on a particular level that require knowledge mainly of only one system element on that level. With this design a distribution substation can be looked at as a node in a data-processing network. Before looking further at substation automation, it will be useful to see how it fits into the electricity supply system.

1.4.4.2 Structure of the electricity supply system

It has already been mentioned that the electricity supply system may be considered to consist of three subsystems: generation, transmission and distribution. This partitioning arises from looking at the system as a process, with electricity as the material being processed. The electricity must first be generated by conversion from some other form of energy. The main concern in the **generation** subsystem is the cost of the energy source and the cost-effectiveness of the conversion process. As the energy source is most often located a distance away from locations where the electricity is needed, it becomes necessary to transport the electricity. The **transmission** subsystem is concerned with overcoming long distances in the most cost-effective manner. Finally, the **distribution**

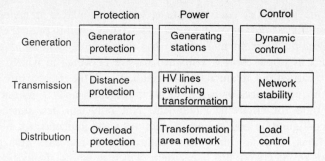

	Protection	Power	Control
Generation	Generator protection	Generating stations	Dynamic control
Transmission	Distance protection	HV lines switching transformation	Network stability
Distribution	Overload protection	Transformation area network	Load control

Figure 1.12 The nine main functional elements of the electricity supply system.

subsystem is concerned with covering an area as efficiently as possible, in order to supply the electricity to a large number of consumers.

However, it is also possible to partition the system from another point of view, by saying that one subsystem, the **power** subsystem, takes care of the physical handling of the electricity, while another subsystem, the **control** subsystem, controls the handling of the electricity, and a third subsystem, the **protection** subsystem, protects the handling of the electricity. Combining the two points of view, it is possible to identify nine main functional elements of the electricity supply system, as shown in Figure 1.12. Substation automation fits into the element in the lower right-hand corner (i.e. distribution control). The labeling of the boxes representing the elements shows just some key features normally associated with the various elements.

The generation subsystem on its own does not form a proper subsystem because its elements, the generators or generating stations, cannot interact without the transmission subsystem. For some purposes it is possible to overcome this problem by assuming ideal interconnections between the elements. This assumption certainly allows the term "generation subsystem" to be used. However, it is often most convenient to combine the generation and transmission subsystems into one true subsystem: the **bulk supply subsystem**. This subsystem then interacts with one or more distribution networks via bulk supply substations. It is one such distribution network that provides the environment for substation automation.

1.4.4.3 Purpose of substation automation

Substation automation will be introduced only if it is seen to bring an economic advantage over the traditional way of controlling substations, taking into account both capital and operating costs. Thus, at the highest level the purpose of substation automation is to increase the cost-effectiveness of the distribution system, and consequently the objective of the substation-automation system design must be to maximize the cost-effectiveness (i.e. to maximize the ratio of the distribution system effectiveness to its lifecycle cost).

Note here the interplay of the two systems. The substation automation system does not bring any benefit on its own, only a cost. The benefit comes only when it is embedded in its host system, the distribution system. Such a situation is quite common; it is also true of the combustion optimization system introduced in Section 1.4.2. It follows that cost-effectiveness models must embrace more than the system itself; how

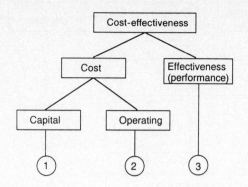

Figure 1.13 The three major components of cost-effectiveness.

much more depends on the particular case and the accuracy desired. For example, should consideration be given to the influence of substation automation on unemployment, which might lead to a higher crime rate, and so on? In the end the whole universe is one big system and every problem infinitely complex; the purpose of systems engineering methodology is to reduce systematically this complexity.

Returning to the question of the cost-effectiveness of introducing substation automation, three major components would immediately be identified: acquisition or capital costs, operating or running costs (including maintenance) and performance (i.e. value of the service rendered to customers, as judged by them and expressed in their willingness to pay for it), as shown in Figure 1.13. Some of the more important possibilities influencing these components are as follows:

1. *Decreased capital cost*:
   - through functional integration (e.g. letting one microprocessor carry out a number of functions);
   - through consistent modularization, thus covering a wide range of substation sizes with a small number of different modules;
   - through better utilization of existing plant, such as transformers, thus deferring the acquisition of new plant.

2. *Operating cost savings:*
   (a) Power losses:
      - lower transformer losses (e.g. by disconnecting nearly idling transformers);
      - minimized power losses by automatic power-factor correction.
   (b) Personnel cost:
      - reduced need for manual operation;
      - improved maintainabilty through better monitoring as input to maintenance planning.
   (c) Spare parts cost:
      - reduced spare parts usage through limiting stresses (e.g. overtemperature) on equipment by such means as gradual restart.

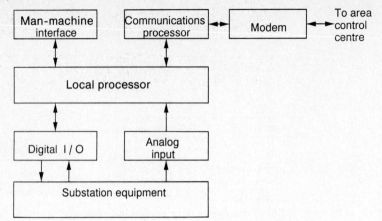

Figure 1.14   Block diagram of a substation automation system.

3. *Performance increase:*

- increased reliability through the early identification of unsafe operating conditions;
- increased availability through automatic restoration;
- improved voltage regulation (e.g. by use of a least squares algorithm).

1.4.4.4 Main functions

The main fuctions of the substation automation system are:

- process interfacing (analog and digital);
- data processing;
- communications, both local (i.e. man–machine) and remote.

As a result a block diagram of the system takes the form shown in Figure 1.14.

The functionality of the process interface is fixed by the type of substation equipment to which the system must interface; only its data-handling capacity (i.e. its size) depends on the size of the substation. The local processing functions, however, may vary considerably depending on such factors as operating philosophy and the type of transmission lines (e.g. cables or overhead lines). In order to accommodate both the range in capacity as well as the range of implemented functions with a minimum of station-specific engineering, the design of hardware and software has to be highly modular. This modularity, together with the requirement that any function must be able to be retrofitted simply by loading the associated software, presents the designer with a considerable challenge.

1.5 Summary

1. The concept of a **system** is a way of describing something. When that something is complex, it is not possible to think of it, nor practical to work with it, as a single entity; and from the natural desire to subdivide it into simpler elements that can be treated at first in isolation and that subsequently form the total entity by interacting, arises the concept of a system as a set of interacting elements.

2. Formally, a system consists of three related sets:

   a set of elements
   a set of interactions between the elements
   a set of boundary conditions.

3. The system concept is only useful if there are certain characteristics that are common to all systems, or to significant classes of systems, and if these characteristics can be used (i.e. manipulated) in the design and optimization of systems. The larger the class, the fewer the applicable characteristics (or system parameters); the process of enlarging the class is called **abstraction**.

4. Because a system describes only the external functional behavior of an entity, it can be said to be a **model** of that entity. Consequently, modeling is a central activity in describing system behavior.

1.6 Short questions

1. What characteristic of an object necessitates the introduction of the system concept, and why has this mode of description become more prevalent in the last 50 years?

2. Define the term "system".

3. Name two common subsets of the set of boundary conditions.

4. What is meant by "abstraction"?

5. Describe the purpose of the combustion optimization system in one paragraph.

1.7 Problems

1. *Home entertainment system*: With the rapid increase in the use of electronic equipment in the home entertainment sector, it becomes necessary to consider home entertainment systems rather than separate pieces of equipment. That is, instead of the TV and the Hi-Fi having their own amplifiers and loudspeakers, they can share the same equipment within an integrated system; and instead of having a programming facility on the video recorder, there can be sequencing operations within the system. Some examples of integrated functions would be telephone answering and wake-up alarm.

   Although such a system is relatively small, it is still quite complex and must be approached in a systematic fashion. The first step is to list all the **functions** one would want such a system to be able to perform. At this stage no thought whatsoever should be given to hardware or to how these functions might be realized.

   (a) List all the functions required of a home entertainment center.

   To perform these functions the system will need a certain number of **inputs and outputs**. These fall into large groups, such as man–machine interactions, signal inputs and outputs, and mass media inputs and outputs. The second step is therefore to produce a list of all inputs to and from the system, still without any reference to hardware.

(b) List all inputs and outputs, using logical grouping.

The third step is to identify the major **functional elements** the system would have to contain in order to process the inputs and produce the outputs. A certain reference to hardware (i.e. power supply, monitor) will be inevitable at this stage, but it is important to keep it as general as possible. Do not make any restrictive assumptions or statements that are not necessary.

(c) List the major functional elements which make up the system.

As a fourth step it is useful to have a clear picture of which functional elements are needed in order to realize each of the functions. This is best done in a matrix form with all the functions down a left-hand vertical column and the elements horizontally along the top. Each matrix element, c_{ij}, will be either one (1) or zero (0) depending on whether the functional element is required for that particular function or not.

(d) Produce a matrix which shows the connection between functions and functional
    elements.

Finally, one would take a step in the direction of detail by considering the question of space/time diversity, that is, the fact that many functions might be required in more than one location, but probably not all at once. This leads to the concepts of signal channels and interface units.

(e) Give a short description of an interface unit and of a signal channel, and
    determine an optimum number for each.

2. *Substation automation:* Consider the substation automation case study. If a utility (e.g. an electricity commission) were to give the job of developing such a system to a firm of consulting engineers, what would be the major objectives contained in the project definition? Pay particular attention to a specific and unambiguous formulation.

1.8 Previous definitions of systems

1. A system is an array of components designed to accomplish a particular objective according to plan (Johnson, Kast and Rosenzweig 1963).

2. A system is defined as a set of concepts and/or elements used to satisfy a need or requirement (Miles 1973).

3. To define a system it is necessary to define the inputs; it is necessary to define the states; it is necessary in some cases to be explicit about the outputs, although this is sometimes arbitrary; and finally, it is necessary to describe how the system changes state in terms of its input and present state. The output of a system is any function of the state of the system. Each state of the system must contain all the information necessary to compute the desired output of the system at any time (Wymore 1976)

4. A system is a set

$$Z = \{S,P,F,M,T,q\}$$

where:

S is a set not empty,
P is a set not empty,
F is an admissible set of input functions with values in P,
M is a set of functions each defined on S with values in S,
T is a subset of R containing 0,
q is a function defined on $F \times T$ with values in M such that q is onto and:

- the identity mapping I is in M, and for every f in F,

$$q(f,0) = I$$

- if f is in F, and s, t and (s + t) are in T, then

$$q(f,t)q(f,s) = q(f,(s + t))$$

- if f and g are in F, and s is in T, and f(t) = g(t) for all t in R(s), then

$$q(f,s) = q(g,s)$$

(Wymore 1967).

5. A dynamical system S is a composite mathematical concept defined by the following axioms:

(a) There is a given time set T, a state set X, a set of input values U, a set of acceptable input functions

$$F = \{f: T \rightarrow U\}$$

a set of output values Y, and a set of output functions

$$G = \{g: T \rightarrow Y\}$$

(b) (Direction of time.) T is an ordered subset of the reals.

(c) The input space F satisfies the following conditions:

- (Nontriviality.) F is not empty.
- (Concatenation of inputs.) An input segment $f(t_1,t_2)$ is f in F restricted to the intersection of (t_1,t_2) with T. If f and f′ are in F, and $t_1 < t_2 < t_3$, there is an f″ in F such that

$$f''(t_1,t_2) = f(t_1,t_2)$$

and

$$f''(t_2,t_3) = (f'(t_2,t_3)$$

(d) There is a state-transition function

$$\{p: T \times T \times X \times F \to X\}$$

whose value is the state

$$x(t) = p(t_0;s,x,f)$$

in X resulting at time t in T from the initial state $x = x(s)$ in X at initial time s in T under the action of the input f in F. p has the following properties:

- (Direction of time.) p is defined for all $t > s$, but not necessarily for all $t < s$.
- (Consistency.)

$$p(t;t,x,f) = x$$

for all t in T, all x in X and all f in F.

- (Composition property.) For any $t_1 < t_2 < t_3$ we have

$$p(t_3;t_1,x,f) = p(t_3;t_2,p(t_2;t_1,x,f),f)$$

for all x in X and f in F.

- (Causality.) If f and f′ are in F, and

$$f(s,t) = f'(s,t)$$

then

$$p(t;s,x,f) = p(t;s,x,f')$$

(e) There is given a **readout map**

$$\{n: T \times X \to Y\}$$

that defines the output

The map

$$y(t) = n(t,x(t))$$

$$(s,t) \to Y$$

given by

$$w \to n(w,p(w;s,x,f))$$

with w in (s,t), is an **output segment** (i.e. the restriction g(s,t) of some g in G to (r,t)).

(Kalman 1969).

1.9 References

Blanchard, B.S. and Fabrycky, W.J., *Systems Engineering and Analysis*, Englewood Cliffs, NJ: Prentice Hall, 1981.

Chestnut, H., *Systems Engineering Methods*, New York: Wiley, 1969.

Johnson, R.A. Kast, F.W. and Rosenzweig, J.E., *The Theory and Management of Systems*, New York: McGraw-Hill, 1963.

Kalman, R.E. et al., *Topics in Mathematical System Theory*, New York: McGraw-Hill, 1969.

Mahelanahis, A., *Introductory Systems Engineering*, New York: Wiley, 1982.

Miles, R.F. (ed.), *System Concepts*, New York: Wiley, 1973.

Sage, A.R., *Methodology for Large-Scale Systems*, New York: McGraw-Hill, 1977.

Wymore, A.W., *A Mathematical Theory of Systems Engineering–the Elements*, New York: Wiley, 1967.

Wymore, A.W., *Systems Engineering Methodology for Interdisciplinary Teams*, New York: Wiley, 1976.

CHAPTER 2

Systems engineering

2.1 Definition of systems engineering

2.1.1 Formal definitions

For the purpose of this text, **systems engineering** is defined as *the art of designing and optimizing systems, starting with an expressed need and ending up with the complete set of specifications for all the system elements.*

This is perhaps a narrower definition than usual, and it is geared specifically towards electrical systems engineering and the design activity. A much more general definition is the following (Sage 1980):

- an understanding of that carefully defined segment of human knowledge denoted **systems science** and operations research, as demonstrated by the ability to develop and use analysis techniques for the resolution of large-scale issues;
- an understanding of **systems methodology** and design, such as to be able to move swiftly from one disciplinary area to another in order to bring many sources of knowledge to bear both on problem-solving efforts and on the design of systemic process adjutants for formulation, analysis and the interpretation of contemporary issues in both the public and private sectors;
- an understanding of **systems management**, to enable the successful interaction of the algorithmic with the behavioral that is needed for the resolution of large-scale issues at the level of symptoms, issues and values.

2.1.2 Discussion

The above formal definitions may appear a little daunting for those with limited or no experience of a typical engineering project. Nonetheless, they ought to convey several insights that are directly relevant to a graduating engineer:

1. Systems engineering provides the basis for a structured and logical approach to projects.
2. Systems engineering is a distinctive discipline. For engineers it typically complements and extends their technical design expertise in a chosen field. (It also offers a natural bridge into management, but it is **not** management in terms of its major responsibilities.)
3. Systems engineering can be regarded as an interdisciplinary "pseudoscience", owing its importance to the complexities of large-scale issues.
4. Systems engineering demands both discernment and judgement. These usually must be culled from a combination of essential and superfluous data. Systems engineering makes extensive use of probabilistic models.
5. The need for systems engineering increases with the size of projects. In large projects, however, its formal application becomes essential.
6. The language of systems engineering can cause difficulties affecting understanding or communication, with the need for abstraction being a major contributing factor.
7. The illustration of abstract concepts with examples is generally essential to an understanding of systems engineering. However, an understanding of the principles (not just the examples) should be the ultimate goal.

It is sometimes also useful to define something by saying what it is **not**. Apart from systems engineering some of the major areas of electrical engineering needed to create systems are device development, equipment design and manufacturing. Device development is concerned with the basic laws of matter and electromagnetism, with the properties of materials, and with the technology used to work with these materials. There is a strong emphasis on mathematics and physics. In equipment design devices are used as building blocks to fulfill certain clearly defined functional requirements—requirements that often result from a marketing activity. The emphasis is on creativity and a good understanding of how devices work. Finally, manufacturing engineering develops the processes that produce the devices and the equipment. There is a strong emphasis on organization and management, on getting the processes to run smoothly and efficiently. (Of course, a production process can be looked upon as a (nonelectrical) system, and the same methodology applied to it.)

It might be argued that systems engineering can be looked upon as the next step beyond equipment design, with items of equipment as the system building blocks. Such an argument would see no underlying difference between systems engineering and the classical forms of engineering, except scale. However, this invalid perspective ignores the inherent "top-down" nature of systems engineering. Engineers and scientists have traditionally been trained to search for the universally valid fundamental laws of nature, and then to build their theories on these basic laws. A significant proportion of scientific research has been aimed at studying molecules, atoms and elementary particles, and then characterizing more complex materials in terms of these building blocks in what is essentially the synthesis of a "bottom-up" approach. Systems engineering starts from the other end of the spectrum, from the often overwhelming complexity of a large-scale problem as a whole, and proceeds downwards through a structured analysis and partitioning process until the questions of interest can be answered. (Depending on these questions, of course, partitioning may cease far above the most detailed level possible.)

Engineers often feel uncomfortable with and even frustrated by such an "unfounded" approach to a problem. A little story may illustrate this attitude. An engineer observes that a tree outside his house is able to lean a little to either of two sides, and that when it leans to the one side the dollar-to-yen ratio will start to rise, whereas when it leans to the other this ratio will start to fall. So what does he do? Does he trade accordingly on the foreign exchange market? No, he spends all his time trying to find out how it is possible for the tree to behave in this manner, and ends up both poor and frustrated.

2.1.3 Main features

The main features of systems engineering, which at the same time distinguish it from traditional engineering activities, may be summarized as follows:

1. The need that is to be fulfilled is most often not expressed in engineering terms. Therefore any work on developing and designing a system must be preceded by work leading to a definition of the objectives of the project.
2. The concepts needed to characterize a system cannot be adequately defined in a universally valid way; it is necessary to develop precise definitions for each particular class of system.
3. Certain parts of a system as well as some of its external interactions cannot be characterized in electrical terms at all; for example, one part of a system may be a human being.
4. Many systems are so complex that just finding a systematic way of handling all the variables and their interactions can become the most important problem.

2.2 Phased approach

2.2.1 Introduction

Engineering does not only consist of factual knowledge about the physical world, about basic laws of nature, materials, devices and processes; it is just as much a way of thinking, a way of attacking a problem. Indeed, with the increasing application of computers and computer-based information systems, such knowledge becomes increasingly accessible and hence of diminishing value, whereas the ability to apply it in a profitable manner becomes more important.

Nowhere is this more evident than in systems engineering, where the complexity of the problem makes a rational and methodical approach a necessity. This approach is not in principle different from that used in attacking any engineering task, but what is often an almost negligible part of a traditional engineering design task becomes the major part of engineering a system. The size of systems and the financial consequences of their characteristics can be immense. Consequently, it is advantageous to spend a considerable amount of time and effort on a careful analysis and a conceptual design phase before proceeding to equipment design.

Clearly, an effective strategy is required to master the high level of complexity inherent in most systems. A phased approach offers the logical basis for such a strategy and is therefore central to the discipline of systems engineering. In other words, **the**

engineering of a system consists of several distinct phases. The relative amount of work involved in the individual phases varies greatly from case to case, but no phase should be left behind without due consideration. The phases are:

definition
analysis
design
implementation
verification.

Before these phases are described in more detail, it is important to recognize that systems engineering is but one of many activities or disciplines present in a project that aims to create a system. The particular structured approach described in this chapter applies to systems engineering only, and the phases introduced here may not coincide with a partitioning of the whole project into parts or stages. The latter is determined by quite different factors, such as fiscal years, budget periods, the decision-making process within the funding organization and the division of project responsibilities between organizational entities. Indeed, it is perfectly possible for the whole of the systems engineering effort to take place in a single stage of a multistage project.

2.2.2 Definition phase

The purpose of the definition phase is to define the **objectives** of the project—that is, to answer the questions: What is the purpose of this project? What is the project expected to have achieved when finished? In defining the objectives care must be taken not to fall into the trap of describing the project itself, instead of its results. It is necessary first to know exactly where one wants to go before starting to plan how to get there or, even worse, actually setting out on the journey. As the saying goes, "If you don't know where you are going, how will you know when you have arrived?"

The objectives can of course only be defined by close cooperation between the user (or client) and the engineer. In this cooperation the user is the primary partner; the engineer plays a secondary role, which consists of questioning and complementing the user's objectives until they form a complete and unambiguous basis for further development. This interactive process can extend over a considerable period; it is not unknown for the definition phase to extend over years and cost millions of dollars. At the other end of the scale it may be completed in a couple of meetings, but in all cases it must be recognized as a formal part of the project on which are placed very definite requirements.

Also, note that so far there is no mention of a system. While the objectives of a project may include explicit reference to a system, it is more common for the system to emerge later as a means of satisfying some or all of the requirements arising from the objectives.

It is usually not adequate just to state the objectives, as these do not exist in a vacuum. They are related to a multitude of other issues, and it becomes necessary to restrict the extent to which these are to be taken into account. This is done by stating the **scope of work** and by making a number of **assumptions** or setting a number of **boundary conditions**. For instance, to reach the objectives it may be necessary to carry out market research, but that task is not included in the scope of work; it will be carried out as part of another project, and the necessary information will be supplied to the present project as a boundary condition. Another type of boundary condition may be exemplified by the case of a project with the objective of determining the most cost-effective way of generating electricity in a remote location, where the possibility of using nuclear power is excluded.

As a result of the above considerations, a **project definition** should normally consist of the following three mandatory (and one optional) sections:

Introduction (optional)
Objectives
Scope of work
Assumptions and boundary conditions

The **introduction** includes background material that focuses the reader's mind on the general subject matter of the project and makes the rest of the definition document easier to read. It may outline the history that led to the project, the persons and organizations involved, and how the project fits into the greater scheme of things.

The nature of the objectives can best be understood by recognizing that the purpose of the project is (roughly) to create a system. This system in turn produces a **service**, and it is this service that satisfies the user's needs. The user's requirements are therefore really requirements on the service; logically, the user should not care that it takes a system to produce the service or how that system is created. One of the most common errors made in formulating objectives is to use the language or concepts of a particular system, thereby restricting or influencing the later development.

An example of this is the case where a community has developed over the years on both banks of a river, and there is a need for better transport services between the two parts. It is possible to define a project to construct a bridge and to write out the objectives in terms of the requirements placed on the bridge. However, this would not be correct; the service required should be formulated independently of whether it is decided to use a bridge, a ferry or a tunnel. Only when the objectives are properly formulated in that manner can a corresponding decision model be developed to determine the optimal solution. The bridge is but one possible system that can deliver the required service.

It may be thought that to formulate such a definition is an almost trivial task; in talking about a project at all the definition is practically self-evident. Experience shows that this is far from being true, and many projects have either come to grief or been seriously disadvantaged by the lack of a proper definition. There are several reasons why it is not so easy to arrive at a proper definition.

Firstly, the objective to be attained by a system must almost always be described in words, perhaps by drawing a parallel to a similar system, perhaps by using an analogy. Only rarely does the objective present itself in strict mathematical terms; and as there are usually several people involved in arriving at a definition, it is difficult to find a wording that is unambiguous to all concerned.

Secondly, it often requires a lot of extra work to quantify an idea or a concept. It is much easier (and involves less personal risk) to state the objectives in qualitative terms, but in some cases the lack of quantification can render the definition useless, as far as providing a point of departure for the systems engineer. Such phrases as "shall provide a significant improvement in performance over the existing system", or "shall minimize the adverse environmental impact", are fine-sounding, but their information content is practically zero. Furthermore, it is very important to include some sort of timescale in the definition, as this can have a great influence on the number of possible solutions. For example, if the objective is stated as "increasing the peak hour capacity of public transport by 100%", a timescale of 12 years would allow a subway system as a solution, whereas a timescale of 4 years would limit the choice to buses on existing roads.

Thirdly, it may happen that the people involved in formulating a project definition pursue at least partially different objectives, some of which are not of a technical nature at all, such as career opportunities, political influence, building a power base within an organization or advancing the interests of a particular manufacturer. It then requires a lot of tact and skill on the part of the project manager or systems engineer to obtain a consensus. It may be necessary to resort to formal methods, such as the Delphi method (see Section 6.2), in which the convergence is achieved in a step-by-step procedure by using questionnaires, processing the answers, issuing new questionnaires incorporating the advances made as a result of the first ones, and so on.

Fourthly, a definition may be unreasonable or unrealistic, stating objectives that either represent an ideal or a wish or are at variance with certain basic facts, such as a physical law. Such a definition is again of little practical value, because what happens in this case is that the engineer is left to "do the best out of the situation", in which case the engineer effectively becomes the one to set the objectives, a role for which he or she may not be suited.

The above problems can be visualized in the following (highly simplified) manner. If a project is defined by specifying the minimum or maximum acceptable value for n parameters, any conforming solution must lie within a volume in n-dimensional hyperspace. For the case of $n = 2$, say performance P and cost C, a project can be represented by a point in the (P,C)-plane, as shown in Figure 2.1, and the project definition would demand $P > P_0$ and $C < C_0$. Conforming solutions lie in the shaded area defined by P_0 and C_0. Assume now that all physically realizable solutions are known; they are represented by the full points in Figure 2.1. Each one defines an area of possible solutions (as the performance can always be made worse or the cost higher), and together they define the **boundary of physical realizability**, drawn as the heavy, stepwise curve. If, as is the case in Figure 2.1, requirement (P_0,C_0) lies outside the boundary, the project definition is unrealistic, and the best that can be achieved is a compromise. Which of the possible solutions is chosen will depend on the relative importance given to performance and cost. If the project definition intersects the boundary, there is one or more conforming solutions.

2.2.3 Analysis phase

The object of the analysis carried out in the analysis phase is the project definition, and the purpose of the analysis is to identify a system and produce a system specification that can serve as the basis for the system design. Extracting a system from the objectives of the project definition is an activity similar to **parsing** (i.e. using the syntactic structure of a command language to identify the elements of a command). Within the statement

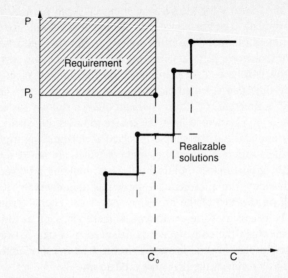

Figure 2.1 A two-dimensional requirements space, with P = performance and C = cost. The requirement is ($P > P_0, C < C_0$); realizable solutions lie to the right of the heavy stepwise curve.

of objectives there will usually be a statement of the form "to [design, develop, manufacture, . . .] something that will produce a [product, service, . . .] that will satisfy [requirements]". The "something" is what becomes the system, and the product, together with the requirements on that product, allows a set of **outputs** to be defined.

To produce the outputs, the system needs certain **inputs**, such as information, power, raw materials, manpower and capital. The system transforms these inputs into outputs, and the corresponding "transfer function" is the **functionality** of the system. The analysis now proceeds to partition that functionality into a set of interacting functions, which become the **elements** of the system.

The considerations that go into the **partitioning**, and thereby the choice of elements, depend both on the particular system and on what aspects of the system it is wished to develop. There are no set rules and procedures that can be automatically applied to all cases, nor is there generally only a single result that is acceptable or useful. The matter of partitioning, which is central to the systems engineering methodology and forms the basis of "top-down" design, is further explored and illustrated throughout this text; here only three very general guidelines will be introduced:

1. It must be possible to describe the functions of the elements completely, relatively simply and unambiguously.
2. The interactions between the elements should be as simple or as few as possible.
3. The elements must be realizable.

As a result, the definition is placed within the context of the real world or environment, and there emerges the distinction between internal and external relations or interactions. One of the three sets defining a system is the set of elements, and the definition of any set must contain a rule that can be applied to determine whether any object is a member of the set or not. The analysis provides this "rule" as a first step in specifying the system and later serves to tell the designer which element characteristics have already been chosen as classification indices; any further partitioning must use characteristics that are compatible with the earlier choice.

The relations between the elements are described and quantified as far as possible without going into the details of how the elements are realized. It is important to remember that at this stage the elements simply represent functions derived from the project definition. Therefore the elements can only be described in terms so general that they do not unnecessarily limit the choice of solutions to be examined during the design phase. However, there are still several general ways to characterize systems, their elements and their interactions, and it is often astonishing how much can be said about a system without going into the design. Such general concepts as information flow, batch or real-time processing, power and/or materials flow, and of course basic boundary conditions imposed by the external world, are usually adequate to provide a clear and detailed system specification.

High level decisions are often made in the analysis phase, such as the choice of technology, basic structure (e.g. centralized or decentralized) and assignment of responsibility. These decisions are usually based on models, and gathering the input data required by these models is another important analysis-phase activity.

As far as the further systems engineering is concerned, the result of the analysis phase is the **system specification**. However, it is important to realize that the scope-of-work section in the project definition may contain a number of other tasks, such as marketing, financing, lobbying, public relations, environmental impact study and land acquisition. These tasks usually have a defined interface to the systems engineering task and provide inputs at certain stages, but how they are conducted is not the concern of systems engineering as it is understood in this text.

2.2.4 Design phase

The design phase takes the system specification and breaks it down, step by step, into specifications for smaller and smaller subsystems until the system has been decomposed into **elements** that are normally recognized as forming design or production units. At each step several activities are carried out to ensure that the whole design activity is proceeding along the **shortest route** towards the fulfilment of the requirements of the system specification. These activities include cost control, performance verification through calculation or simulation, interface control and coordination. Without the latter it can happen that two design teams, working on different sections of the system, are working towards different, but possibly equally good, solutions.

A very important activity that must proceed in parallel with the design of the physical system is the development of the **documentation**. At the end of the design phase it is not enough to have the specifications for every piece of equipment; also needed is all the documentation necessary for operation and maintenance on the system level. This documentation must try to anticipate every possible contingency and must include enough background material to train new personnel and give them a thorough understanding of the relevance of the system.

The result of the design phase is a set of **element specifications** that provides all information necessary for the successful realization of the system.

2.2.5 Implementation phase

In the implementation phase the systems engineer is involved either only peripherally or mainly in a managerial role. No actual systems engineering takes place during this phase. The main activities are equipment design, purchasing, cost and time schedule control, inspection, factory testing and installation, many of which could be carried out by people more oriented towards management, with only limited design experience.

2.2.6 Verification phase

The final phase, verification, depends considerably on the type of project and the complexity of the objectives. Its purpose is always the same: to verify that the requirements of the system specification have been satisfied; but whereas this may take the form of a series of measurements during a day or two of operation for a simple system, some systems may have proving periods of a year or even several years. Moreover, some systems may require elaborate testing facilities that actually are systems in their own right, including the simulation of special environmental conditions, as is the case with systems involving space segments.

2.2.7 Example: weapons system

As a very abbreviated and simplified example of how a project may pass through these various phases, consider an initial objective expressed as: "A weapons system to destroy all enemy aircraft approaching Australia". Clearly, this is a rather sweeping objective that should be clarified somewhat before being used in a project definition. As a result of going through the definition phase, a more realistic objective may be: "Assume that attacking aircraft arrive in waves with N aircraft per wave, and let the frequency of waves be V. Then the probability P that an aircraft will come within 10 km of Australian shores will be a function of N and V. The objective is a weapons system that will result in P being less than a given function." This is illustrated in Figure 2.2.

In the analysis phase the consequences of this definition are looked at. The first consequence is that all aircraft have to be detected. Then there has to be some means for determining whether they are friendly or not. Furthermore, there has to be some graded approach to attacking unidentified planes, as it would not be desirable to use a SAM on an aircraft in peacetime simply because it was unidentified. Then the characteristics of potential enemy aircraft, both present and projected, have to be analyzed and their impact on the objective determined. The same has to be done for the geographical and meteorological data, and as a result requirements on logistics and communications are formulated. Out of all this there emerges a picture of a system consisting of various functional elements (e.g. detection, identification, interception, destruction, supply and maintenance, communication, control, decision) and of the relations between these and between them and the inputs and outputs. For example, it becomes evident that the probability P will be a function of both the probability of

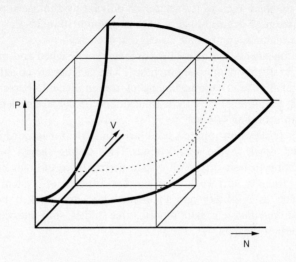

Figure 2.2 Performance requirement for the air defense system, where
P = probability of penetration, N = aircraft per wave and V = waves per
unit time. The performance requirement is that P must lie below the
surface defined by the heavy lines.

detection and the probability of destruction at least, as a plane that is not even detected
will surely reach Australia, while not every plane that is detected will also be destroyed.

Once the system functions and the requirements placed on these by the project
definition have been specified completely and unambiguously, the system design phase
may commence. Somewhat naively expressed, concern now is no longer with the
implications of the project definition or with the what and why; concern now is simply
with the how: how to realize the performance requirements spelled out in the system
specification. The system designer does not work with abstract concepts alone;
progressing through the top-down design, the systems designer works increasingly with
real system components (e.g. radars, radio transmitters and receivers, data-processing
equipment, missiles, aircraft), and these components are now characterized by exact
physical parameters (e.g. output power, sensitivity, bandwidth, radiation resistance,
turning radius, payload). However, the system elements are still **functional** elements.

Whereas the analysis phase is rendered difficult by the non-exact and partly
subjective nature of the data, the major difficulty in the design phase is usually the sheer
mass of data that has to be taken into account. Here the computer becomes invaluable,
and it is only the development of information processing that has made very large
systems, such as the Apollo program and complex weapons systems, possible. The
systems design phase ends with the issuing of specifications for every facility (e.g.
runways, refuelling systems, access roads, personnel accommodations, hospital facilities,
storage facilities) and every piece of equipment, from aircraft and radars down to office
equipment. The operating manuals on a system level consist of thousands of pages
detailing procedures for peacetime and wartime, for conventional and for nuclear
attack, for how to get the best performance out of a partially damaged system, and so

on. The system designer also has to tackle such difficult psychological problems as how to maintain the required degree of alertness throughout a 10 to 20-year period of peace, building into the manuals appropriate checks and exercises.

During the implementation phase the equipment is designed and manufactured, or such items as aircraft may simply be purchased. The facilities are constructed, and then follows the installation and commissioning of the equipment. Personnel (e.g. radar operators, pilots, technicians) have to be trained, and finally all the elements have to be integrated into an operating system.

During the verification phase, tests are carried out according to the test requirements laid down in the specifications. This testing should be as realistic as possible. In addition to tests of overall system performance through simulated attacks (e.g. by some friendly airforce), many other aspects of system or subsystem performance, such as reliability and maintainability, are checked by the extensive logging of operational data (e.g. fault reports, time studies, staff interviews) and the use of models to interpret the data.

2.3 Applicability of systems engineering

2.3.1 Introduction

While it may be generally true that systems engineering methodology brings the greatest benefit when applied to large systems, it must be clearly understood that the methodology applies to **any** engineering task (and many nonengineering ones also); it is simply a question of the most cost-effective level of the effort invested, and whether that level is high enough to consider the systems engineering as a separate activity. As the appropriate amount of effort to be expended on systems engineering, and on the individual phases, varies widely depending on the project, no fixed rule can be given; common sense must be used. A very rough idea of order of magnitude is given by the following distribution of costs, which a large communications system might show:

project management	5%
systems engineering	5%
equipment engineering	15%
manufacturing	50%
training and documentation	10%
testing	15%

The importance of recognizing systems, and of treating them in the proper top-down fashion described in Section 2.2, can best be illustrated by two contrasting examples of very large systems that have been around long enough to show very clearly the effects of systematic and uncontrolled development respectively.

2.3.2 Two contrasting examples

The first example is the telephone system. Very early in its development the telephone system was recognized as a system, and many of the techniques for analyzing, designing and maintaining systems were first developed for use in the telephone system. The need for coordination and organizational unity was taken into account by the formation of one, or at most a few, national companies and by organized international consultation. Extensive national and international standards were developed and maintained in the course of technological innovation. The result of this systematic effort is apparent in overall system performance; the extent and reliability of the telephone service in most Western countries are so good that they are taken completely for granted.

The second example, public transport, stands at the opposite end of the performance scale. Never treated as a system, this area has always been a free fight for all, including such diverse interests as the automobile industry, railroads, the oil industry, the aviation industry, land developers and speculators, all political parties and the steel industry. Whatever little systems-oriented work was done was always lost in this turmoil, and the results are accordingly poor. Cars designed to carry four people at 120 km/h crawl along at an average speed of 20 km/h with one person, very little or no parking is provided at railroad stations, valuable highway space is used as parking space (a few parked cars essentially reduce the width of a highway by one lane), and so on. In a city like Sydney at least 100,000 manhours must be lost due to inefficient transport every working day. And the safety record would never be accepted in any other industry; in the USA alone over 1 million people have been killed by cars since World War II.

2.3.3 Space industry

The space industry is a very visible example of the successful application of systems engineering. Space missions such as the Apollo program or the deep space (interplanetary) probes would not have been possible without a strong commitment to systems engineering methodology. In the case of the deep space probes the systems engineering was carried out by the Jet Propulsion Laboratory (JPL) under contract to NASA, and the following description, condensed from a JPL publicity document, shows how the process outlined in Section 2.2 was followed step by step.

Each project goes through the phases of systems engineering under a management procedure specified in a Project Development Plan, and each project is generally divided into four systems:

- the Spacecraft System, consisting of instrumentation payload, support subsystem, spacecraft frame, attitude control and communications;
- the Launch Vehicle System, consisting of the launch vehicle and its support equipment;
- the Tracking and Data System, which is responsible for the provision and maintenance of the earth-based tracking, telemetry and command stations, for ground communications and for the operational facilities for the mission;
- the Mission Operations System, comprising the management organization responsible for the design and execution of the mission operations.

Further breakdown of the systems into subsystems and components is made to reach a level of complexity where each element can be treated as a single unit. Concurrent with the system breakdown, interfaces between the elements are established that define the functional boundaries of the elements.

This breakdown of a system into functional elements cannot be made arbitrarily. A great amount of managerial and engineering skill is required to select the interface topology, which affects the management control of a project, both administratively and contractually, and the engineering, integration and operation of the system. From a management standpoint, interfaces are defined so as to optimize visibility and control, to isolate independently subcontracted elements, and to delegate authority and responsibility. From an engineering standpoint, interfaces are located to separate independent functions and to facilitate the integration, testing and operation of the overall system.

In the early design stages the exact mode of implementation for a system may not be known. For example, should the spacecraft be spin-stabilized or inertially stabilized? Should power be obtained from solar cells or a nuclear device? Thus, a system is initially defined in terms of performance specifications and constraints for each function, not in terms of implementation schemes.

At this point the design is defined by a set of subsystems, functions and constraints for each subsystem and by an outline of the interface topology. Then the subsystem designs are projected so that the alternative subsystem implementations can be understood. The subsystem performance characteristics and constraints are translated back through the system, and the process is iterated to produce an optimized self-consistent preliminary design.

The preliminary design is specified as a set of functional requirements and interface control documents. These documents define the overall requirements and constraints levied on the spacecraft design by the mission, the major system interfaces, the subsystem interface topology, and the functions and constraints imposed by the system on the subsystem design. They describe the design in sufficient depth to allow the detailed subsystem definition to proceed independently.

2.4 Systems engineering management

2.4.1 Definition

Systems engineering management is *the management of the engineering and technical effort required to transform an operational need into a description of system performance parameters and a preferred system configuration.* The particular management procedures and activities, as well as their extent, depend on the type and size of the project. They should be laid down in a **Systems Engineering Management Plan (SEMP)** to be developed during the analysis phase (MIL-STD-499A, 1974). The plan should be comprehensive and describe how a fully integrated engineering effort will be managed and conducted. The SEMP is usually in three parts:

Part I: Technical program planning and control: This part of the plan is concerned with the management of those design, development, test and evaluation tasks required to progress from an operational need to the deployment and operation of the system by the user. It identifies: organisational responsibilities and authority for system engineering management, including control of subcontracted engineering; levels of control established for performance and design requirements and the control method to be used; technical program assurance methods; plans and schedules for design and technical program reviews; and control of documentation.

Part II: Systems engineering process: This part contains a detailed description of the process to be used, including: specific tailoring of the process to the requirements of the system and project; procedures to be used in implementing the process; inhouse documentation; trade-off study methodology; types of mathematical and/or simulation models to be used for system and cost-effectiveness evaluations; and generation of specifications.

Part III: Engineering specialty integration: This part of the plan is concerned with the integration and coordination of program efforts for engineering specialty areas (e.g. reliability, maintainability, safety, survivability, human engineering) into the total engineering effort, in order to achieve a best mix of the technical/performance values. Detailed specialty program plans, such as:

Reliability Plan
Maintainability Plan
Quality Assurance Plan
Human Engineering Plan
Safety Plan
Logistic, Support Analysis Plan
Electromagnetic Control Plan,

would be summarized or referenced as appropriate. The SEMP must depict the integration of specialty efforts and parameters into one systems engineering process and show their consideration during each iteration of the process. Where specialty programs overlap, the SEMP has to define the responsibilities and authorities of each.

2.4.2 Technical program planning and control

The tasks that fall under the heading of "Technical program planning and control", and their relative importance and extent, vary considerably from project to project. As with all management tasks they must be tailored to meet the needs of any one particular project, and the effort spent on them must be justified in terms of a reduction of the overall project cost. The following is a list of major tasks found as part of the systems engineering management of most systems. The list also indicates the engineering activity to which the management pertains and references the corresponding section in this textbook:

1. *Development of the work breakdown structure*: The work breakdown structure (WBS) forms the basis for the systematic identification of all items within the project and as such is an indispensable management tool. It will be discussed in Section 2.5.2.
2. *Development of a specification tree*: Specifications form the major part of the deliverables resulting from the systems engineering process, and consequently any planning of that process must involve determining the types and numbers of specifications to be produced. In keeping with the top-down approach, the specifications also form a hierarchy and are therefore best planned and listed in the form of a tree. This will be discussed in Section 7.2.4.
3. *Risk management*: Throughout the systems engineering process but particularly in the early phases, every decision must be considered not only for its planned effect but

also for what other effects could possibly result, and for how likely these are. This constitutes the **risk** inherent in any decision, and it must be minimized by proper engineering management actions. These include on the one hand measures to minimize the probability that adverse results of the decision will occur, and on the other hand measures to reduce their effect in the unlikely event that they do occur by having appropriate remedial action planned and prepared. This will be discussed in Section 7.4.4.

4. *Test planning*: No requirement should be formulated without also considering how to test that the requirement has actually been met once the system has been completed. Thus, it is necessary to plan ahead with regard to such issues as where and when testing is to take place, what facilities are needed, what testing is going to cost and how it is going to affect the project schedule. Testing will be treated in Chapter 8.

5. *Cost/schedule control*: The overall cost/schedule control system (C/SCS) is implemented as part of the project management structure, but the engineering activities are often controlled separately by the engineering manager using the C/SCS. Cost/schedule control systems are not treated in this textbook.

6. *Technical performance measurement*: In managing the engineering process the manager needs to know not only how much of the budget has been expended at any one time but also how far the project has progressed. Progress can be measured in two ways: by the basic work packages (tasks) that are **finished**, and by the state of the system performance (i.e. how close the system is to meeting the requirements of the system specification). The latter monitoring activity is called **technical performance measurement** and will be developed in Section 9.2.3.

7. *Technical reviews*: Technical reviews constitute the formal and most visible manifestation of the control of the engineering process, and they also provide an interface between different parts of the total project team. They will be described in more detail in Section 7.3.

8. *Configuration management*: All project documentation that is subject to change must be uniquely identified and placed under configuration control. The planning, implementation and maintenance of a secure and efficient configuration-control system is an important management function; and while it is not specifically treated in this textbook, some of the requirements placed on any such system will be briefly discussed in Sections 7.2.5 and 12.2.4.

2.5 Relationship between systems engineering and project management

2.5.1 Systems engineering as a supporting function

From the previous sections it will have become evident that systems engineering includes some elements of management—in particular, getting a group of people to formulate a common set of objectives. However, further management aspects of systems engineering emerge when it is realized that systems engineering is not a closed process; it can only exist in the context of a project. Its activities are closely interwoven with other project activities; and in particular, management aspects result from the participation in the overall project management.

To a certain extent systems engineering may be seen as a staff function in relation to project management. It provides the project manager with background information, such as the relative merit of competing technologies, and with system performance parameters as seen by the users of the system, not by technologists concerned with the mechanisms that make it work. These performance parameters are of major interest to project managers; with their help project managers can construct meaningful milestones that will allow them to control the progress of projects.

However, systems engineers also usually carry out some project management functions. The extent of this activity varies considerably from project to project (and of course some projects do not utilize any systems engineering at all). Ultimately, the systems engineering role depends on the type of project and on the qualifications of the personnel available for the management functions in its early stages. The reason for this is simply practicality and is shown by the relative effort that goes into project management and into systems engineering during the phases of a project.

Figure 2.3 shows a typical example, and it can be seen that the two activities complement each other. In the early phases almost all the effort is centered on the systems engineering function, answering all the questions about performance, interactions with other systems, economic and political aspects, and so on, and it is reasonable to incorporate the management of this effort directly into the systems engineering. As the project moves into the implementation phase, the systems engineering effort becomes very small or even nonexistent, whereas the project management effort reaches its peak, keeping any number of main contractors and subcontractors working smoothly along the planned path. Nevertheless, even though the project management effort is much greater than the effort expended in managing the systems engineering, many of the techniques developed for project management can be used in managing the systems engineering effort. In particular, activity charts and formalized activity reporting are very useful for controlling the sequence of activities. If the temporal relations between the activities are not clearly understood and controlled, delays and inefficiencies are almost inevitable.

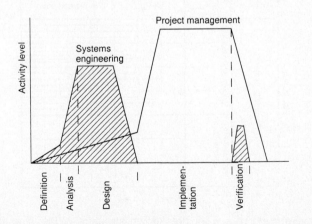

Figure 2.3 Relative efforts going into systems engineering and project management throughout the lifetime of a project.

2.5.2 Work breakdown structure as an interface

A particularly important interface between systems engineering and project management is embodied in what is usually called the **work breakdown structure (WBS)**. To see how this arises, consider the systems engineering process represented in two dimensions. Along the vertical axis are all the elements of the system; along the horizontal axis, the time axis, are the phases of the engineering activity, as shown in Figure 2.4. In the progression from definition through analysis to design, the system is partitioned into finer and finer elements until, at the end of the system design phase, there is a complete list of all the elements making up the system.

In the implementation phase each of these elements goes through a succession of subphases, such as development, design, testing and production, and each of these activities for each individual element represents a certain amount of work: a clearly defined **work package**. The definition of a work package must include:

- a complete statement of the task (i.e. what work has to be accomplished);
- identification of the inputs that must be available before this work can start;
- a detailed description of what the output or result of the work should be, and in what form it is to be presented.

The matrix of work packages constitutes the WBS; and although it is usually presented in a one-dimensional form (i.e. as a list), it is essential never to lose sight of the fact that the underlying structure is two-dimensional. In other words, a work package is bounded on **four** sides.

A WBS is a product-oriented family tree encompassing hardware, software, services and data that results from project engineering efforts during work on a system and completely defines the project. A WBS displays and defines the projects to be developed or produced and relates the elements of work to be accomplished to each other and to the end products.

The WBS occupies a central position in the project (and systems engineering) management scheme; it is the key that allows the different resources and products to be

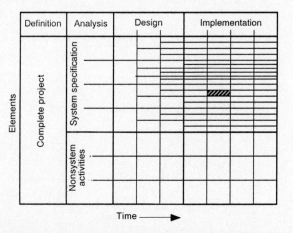

Figure 2.4 Development of a work breakdown structure (WBS). The elements of the WBS are work packages; one of these is indicated by the cross-hatched rectangle.

related to each other, through an appropriate accounting and control system. To each element of the WBS are related budgeted and actual values of manpower, material, facilities, overheads and so on.

Being a family tree, the WBS consists of a number of **levels**, starting with the complete system as level 1 at the top and progressing downward through as many levels as are necessary to obtain elements that can be conveniently managed. The elements on the lowest level are always **work packages**.

A work package is described by a statement of the work to be carried out, the results to be obtained by this work and the prerequisites necessary to commence the work. Elements on higher levels may or may not be work packages, so to distinguish work packages on the lowest level (i.e. the smallest work packages that are individually managed) the work to be carried out within such a work package is called a **task**.

To accomplish a task it is normally necessary to carry out a number of **activities**. An activity is a type or class of work (e.g. producing a draft document, reviewing the results of a previous activity, finalizing data), and these activities are usually described in the engineering procedures that specify how various tasks are to be carried out. For estimating the effort involved in carrying out a task, manhours may be assigned to the various activities making up the task, but these are not controlled (except perhaps by the individual engineer carrying out the task). Also note that the internal structure of a task is always linear; the activities follow one after the other.

Besides simplifying task analysis and cost estimation, the consistent use of activity types allows certain activities to be designated as **overhead activities**. Take, for example, the activity of reviewing ten documents produced within ten different tasks, and assume that this takes 1 hour. It would be highly impractical for a member of the review panel to have to put down ten different job numbers on his or her timesheet and charge 6 minutes to each. Instead, the panel member simply charges 1 hour to the job number corresponding to the overhead activity called "reviews".

What constitutes an element at any level but the lowest depends very much on the type of system under consideration. For example, the case study in Section 2.5 is concerned with analysis only; all the elements are work packages, none are hardware or software products. However, for many projects in the field of electrical engineering, the upper three levels, also called the **summary WBS**, are made up of the following types of elements (see also MIL-STD-881A 1987):

Level 1:
    whole system.
Level 2:
    prime system elements
    training
    peculiar support equipment
    systems test and evaluation
    system/project management
    data
    operational/site activation
    common support equipment
    initial spares and initial repair parts.

1. *Prime system elements* are the functional elements of the system: the ones resulting from the analysis of the project definition in order to satisfy directly the expressed need or to reach the project objectives. Typical level 3 elements are:

   integration and assembly
   sensors
   communications
   automatic data-processing equipment
   computer programs
   data displays
   auxiliary equipment.

   Note the diverse nature of these prime system elements; they can be work packages, devices, services, equipment, software and so on.

2. *Training* encompasses everything connected with providing the human elements of the system, and its elevation to a separate level 2 segment emphasizes the importance of human performance in attaining the required overall system performance. The heavy involvement of humans as system elements is a central feature of systems engineering. (That is, the humans are not "users" of the system; they are part of the system.) The breakdown to level 3 can be as follows:

   equipment
   services
   facilities.

3. *Peculiar support equipment* encompasses any hardware and/or software designed especially to support the operation and maintenance of this particular project or system. Typical items are jigs and fixtures, special tools, special-purpose mobile workshops, special transport containers and so on. Level 3 may be organized according to the maintenance concept (see Chapter 11) into, for example:

   organizational level
   intermediate level
   depot level.

4. *Systems test and evaluation* consists of all activities and equipment needed to investigate any aspect of system behavior throughout the design, implementation and verification phases. The term "trials" is often used instead of "evaluation" when it refers to the verification phase. A possible level 3 breakdown is:

   development test and evaluation
   operational test and evaluation
   markups and simulators
   test and evaluation support
   test facilities.

5. *System/project management* is self-evident, and has the two level 3 components:

   systems engineering
   project management.

6. *Data* essentially means documentation, but the information does not have to be in the form of hard copy. Typical level 3 items are:

> technical publications
> engineering data
> management data
> support data
> data depositor.

7. *Operational/site activation* covers all activities connected with getting the system into its operational form, such as:

> technical support for site establishment
> site construction/conversion
> assembly and installation
> checkout and commissioning.

8. *Common support equipment* encompasses any hardware and/or software that is necessary to support the operation and maintenance of the system but is available as "off-the-shelf" equipment. Level 3 can be as for the "peculiar support equipment".

9. *Initial spare parts and initial repair parts* will be subdivided at level 3 according to the breakdown used for the hardware under "prime system elements" and support items.

2.5.3 Organizational aspects

There is one major difference between project management and the management involved in systems engineering. In the former the people and organizational units being managed are in some way formally responsible to the project manager, be this within a matrix organization or an ad hoc project organization. In systems engineering this is mostly not the case; the people and organizations that should provide the inputs to the systems engineering process have no formal relation to the project. They may not even work in a technical field but may be politicians, lawyers, business people and so on, and the systems engineer's management activities therefore take on the aspect of public relations, persuasion and motivation. The systems engineer must get people to cooperate without having any formal authority over them.

A further delineation of systems engineering can be seen in a typical project organizational chart. (Here it is assumed that the appropriate organizational structure for effective design is a product-oriented team, as opposed to a function-oriented team.) Figure 2.5 shows the key roles and their basic relationships. The **management** dimensions of systems engineering should be carefully noted, both in terms of both responsibilities and interfaces. It should be seen that systems engineering is functionally distinct from two related activities: overall project management (i.e. planning, organizing, coordination, directing) and program control (i.e. overviewing progress and costs). In small projects of course one engineer may be responsible for all three functions, while very large projects may require several persons to cover just one function.

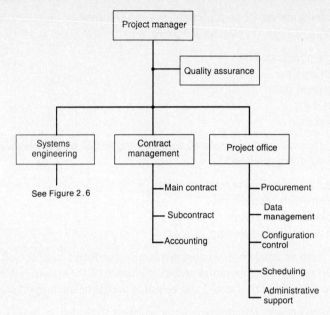

Figure 2.5   Main elements of a systems project management organization.

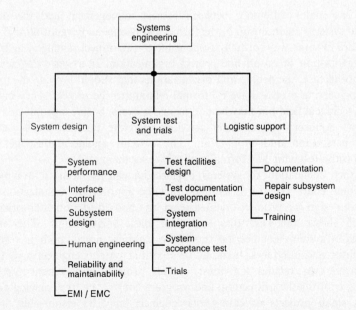

Figure 2.6   Systems engineering functions within the project organization.

A more detailed presentation of the functions controlled by the systems engineering manager is shown in Figure 2.6; they include system analysis and evaluation, system design, interface control, and system integration and test.

2.5.4 Systems and projects

The previous sections examined the relationship between systems engineering and project management (i.e. between two activities). However, it is equally important to have a clear understanding of the similarities and differences between the underlying objects, namely, systems and projects. Of the differences, two stand out immediately.

Firstly, the relative duration of systems and projects is different. A project lasts only until the system has been created and placed in operation, whereas a system has a prolonged **lifecycle** during which it is maintained, revised, extended, modernized and in the end dismantled. Therefore, for example, a reference to lifecycle costing must clearly refer to the system and not the project; and the two have completely different cost structures.

Secondly, in the context of engineering the system is a clearly defined physical entity that produces certain outputs in response to given inputs, whereas the project is an umbrella for a large number of diverse activities and structures that are of only a temporary nature. Very often the elements of a system are physical, such as pieces of equipment or subsystems, whereas the elements of a project are either activities or activity-related (i.e. virtual) organizational units, such as accounting, personnel administration, and shipping and receiving.

However, despite these and other obvious differences, systems and projects have much in common, and much of what is treated in this text is directly applicable to projects. Just as systems are introduced in order to handle complexity, projects are created in order to handle the complexity of a multitude of line organizations interacting for a common purpose. One can even go a step further and say that the project is also a system, with the elements being activities and the interactions being the relationships between them, as expressed in a PERT diagram, for example. Therefore many of the tools used are the same, and this fact is a part of the justification for devoting a considerable part of this text to developing these tools.

The relationship between project and system can be further explored by observing that, while systems engineering is one of the project activities, the project is but one stage in the life of the system. So in one sense the system is embedded in or subordinate to the project, whereas in another, temporal, sense the project is but one part of the system, determined by and subordinate to the overall requirements placed on the system. A similar situation, which may serve as an illustration, is presented by the case of a nation at war. The war is the project; the nation is the system. For the duration of the war the military command (project management) takes on a leading role, displacing many of the civilian functions, but this is (hopefully) a temporary situation. Once the war is over, the military influence is reduced; the war itself is but one phase in the ongoing life of the nation and is subordinate to the overall goals of that society, such as freedom, equality and prosperity.

An insufficient or improper understanding of this dichotomy in the relationship between system and project can sometimes lead to serious difficulties in satisfying the requirements originally placed on the system, or at least to gross inefficiencies, as is the case when the project management becomes incorporated into the system.

2.6 Case study: substation automation

2.6.1 Purpose

Only one of the case studies, namely substation automation, will be used to illustrate some of the material treated in this chapter. By presenting a WBS for the analysis phase the following points should become clear:

- the extent of the analysis phase and the type of issues to be considered in that phase;
- the facts that systems engineering is a part of the project and that its activities must be planned and managed on an equal footing with any other activities in the project;
- the degree of formalism needed to ensure a controllable process—it is not adequate just to produce a system specification at the end of the phase; every step must be documented and this documentation reviewed at several milestones within the phase;
- the use of cost-effectiveness models to narrow the field of possible functions down to a smaller set of probably viable ones, thus reducing the complexity of the design task;
- the importance and extent of data gathering—as much as half of the effort expended in this phase goes into extracting information, in both verbal (i.e. through discussions) and written (either by questionnaires or by locating existing documents) form.

2.6.2 Scenario

The details of the work carried out in the analysis phase depend on how the beginning of the analysis phase was arrived at, how the systems engineering is embedded in the project and what the environment of the project is. All in all, this can be expressed in terms of a **scenario** for the case study at that point in time. (It is analogous to the set of initial conditions needed to continue the integration of a differential equation at a particular point in time.) In the present case the scenario might be as follows:

An electricity distribution authority (e.g. county council) has decided to investigate the possibility of introducing some form of substation automation. To this end the authority has developed a project definition and has now engaged a firm of consulting engineers to carry out the analysis phase.

The main objectives of the project, as far as the analysis phase is concerned, are to determine:

- what functions can sensibly be included in a substation automation system;
- which of these functions are economically viable;
- what requirements must be placed on them in order to optimize their cost-effectiveness;
- what requirements must be placed on them due to the environment in which the automation system is embedded;

and to produce:

● a system specification.

Boundary conditions are:

1. The automation system should be applicable to both existing and new substations.
2. The analysis phase is to be completed within 6 months and within a budget equivalent to 1.5 manyears of a senior engineer.

2.6.3 WBS for the analysis phase

The WBS in Table 2.1 is specific to the substation automation case. However, the structure represented by the first levels of headings is typical of a wide range of systems.

Table 2.1 Work breakdown structure for the analysis phase

---

1 REVIEW PROJECT DEFINITION (PD)
  1.1 *Study existing documentation*
    1.1.1 Check that all documentation pertaining to the PD is available.
    1.1.2 Read PD thoroughly, making notes of any unclear or inadequately formulated points.
    1.1.3 Produce a review report.
  1.2 *Discussions with all parties*
    1.2.1 Discuss all unclear or inadequate points with the author(s) of the PD.
    1.2.2 Check all sources of information used in the PD.
    1.2.3 Amend the review report.
    1.2.4 Discuss report with client to ensure that there is full agreement.

2 OBTAIN DATA AND INFORMATION
  2.1 *Obtain data on the present distribution system*
    2.1.1 Produce a list showing the number and locations of substations.
    2.1.2 Determine the complexity of existing substations (i.e. busbar structure).
    2.1.3 Produce an overview (e.g. by listing feeders and control circuits) of the sizes of the various substations.
    2.1.4 Determine the present level of loading and the protection philosophy.
    2.1.5 Obtain data on operating conditions.
  2.2 *Obtain information on future (planned) distribution system and substation design*
    2.2.1 Determine the system expansion rate.
    2.2.2 Determine the trend in substation size.
    2.2.3 Identify any trends in substation design, complexity or technology.
  2.3 *Obtain system cost data*
    2.3.1 Produce a listing of cost factors (e.g. purchase cost, installation cost) for the main substation elements and for the network.
    2.3.2 Obtain data on operating costs, including losses.
  2.4 *Obtain operating data*
    2.4.1 Collect statistical information (and/or estimates) on the number and duration of outages.
    2.4.2 Determine the present level of manning and operating procedures.
    2.4.3 Conduct interviews with operating personnel to identify any existing problems.

3 ECONOMIC ANALYSIS OF DISTRIBUTION SYSTEM
  3.1 *Identify cost-saving possibilities*
    3.1.1 Produce a list of all distribution system parameters and/or functions that could possibly be influenced by the control system.
    3.1.2 Identify which of the items in 3.1.1 are presently exploited.

3.2 *Quantitative estimate*

    3.2.1   Use the data obtained under 2.3 to estimate the potential savings.

    3.2.2   Produce an ordered (according to importance) list of the items in 3.1.1.

3.3 *Present vs new substations*

    3.3.1   Determine which of the items in 3.2.2 can be realized in present substations (possibly with modifications) and which can be realized only in new designs.

    3.3.2   Correlate the data from 3.3.1 with those from 2.2.1.

    3.3.3   Produce an estimate of actual savings.

FIRST REVIEW

4 COST/BENEFIT ANALYSIS

4.1 *Function analysis*

    4.1.1   Identify all SCADA functions.

    4.1.2   Classify 4.1.1 into existing/new and local/mixed/central.

4.2 *Cost/benefit data*

    4.2.1   Construct a table showing (a rough estimate of) the cost and benefit (by using 3.3.3) associated with each control function.

    4.2.2   Compare cost/benefit for local vs remote (central) system architecture.

    4.2.3   Produce a report identifying the functions to be included in the new automation system.

SECOND REVIEW

5 SYSTEM CHARACTERIZATION

5.1 *Inputs and outputs required to support the functions*

    5.1.1   Determine the types of analog inputs to be accommodated.

    5.1.2   Determine the quantitative range of analog inputs needed (using 2.1.3 and 2.2.2).

    5.1.3   Determine voltage levels for digital inputs.

    5.1.4   Determine the quantitative range of digital inputs needed (using 2.1.3 and 2.2.2).

    5.1.5   Determine voltage levels for digital outputs.

    5.1.6   Determine the quantitative range of digital outputs (using 2.1.3 and 2.2.2).

    5.1.7   Identify requirements on the man-machine interface (e.g. data flow rate, complexity) (using 2.2.3 and 2.4.3).

5.2 *System proposal*

    5.2.1   Identify the main functional elements.

    5.2.2   Identify technologies available.

    5.2.3   Select a system architecture.

    5.2.4   Determine the optimum modularity in order to satisfy size range requirements.

5.3 *System cost estimate*

    5.3.1   Identify elements to be developed and estimate their cost.

    5.3.2   Obtain the cost of off-the-shelf items.

    5.3.3   Estimate system design and manufacturing costs.

    5.3.4   Estimate installation costs.

    5.3.5   Construct a simple cost model, and compare it with 4.2.

    5.3.6   Modify proposed system to optimize cost/benefit.

THIRD REVIEW

6 PRODUCE THE SYSTEM SPECIFICATION

6.1 *Draft specification*

    6.1.1   Set up the specification structure.

    6.1.2   Identify missing information.

    6.1.3   Obtain missing information.

    6.1.4   Write a draft specification.

   6.2   *Specification review*
      6.2.1   Determine who must review the specification.
      6.2.2   Obtain review reports.
      6.2.3   Correlate reports and produce the final manuscript.
   6.3   *Produce the document*
      6.3.1   Have all drawings marked up and drawn for reproduction.
      6.3.2   Arrange the final typing of the text.
      6.3.3   Copy and collate (bind).
      6.3.4   Distribute.

2.7 Summary

1. The starting-point is a **need** or a set of **requirements** that are to be satisfied.
2. The entity that is to provide this satisfaction is the **system**, consisting of a number of **functional elements** (as follows from the requirements) and **relations** between them.
3. To create the system a **project** is defined. The **project definition** contains the objectives, the **scope of work** (as part of the work may be supplied by others) and all relevant **boundary conditions**.
4. The project contains a number of activities, prominent among which are **project management** and **systems engineering**.
5. The project (and thereby the systems engineering as one of its activities) progresses through a number of **phases**, whose **extent and relative importance will vary from project to project**.
6. Because systems engineering is a **methodology**, a way of attacking complex engineering projects, it necessarily has a **strong management component**.
7. The master document for the systems engineering effort within a particular project is the **Systems Engineering Management Plan (SEMP)**.

2.8 Short questions

1. Describe two features of systems engineering that distinguish it from classical equipment or device engineering.
2. What is understood by the "phase" of systems engineering?
3. Give two reasons why it is not a straightforward matter to produce a project definition.
4. What is analyzed in the analysis phase?
5. Mention a few issues or quantities typically found as boundary conditions in a project definition.
6. Both the system definition and the project definition contain boundary conditions. How do they differ?
7. What is a main difference between a project organization and a line organization? Mention a few consequences of this difference.

2.9 Problems

1. *Electricity supply system:* A country has a very rudimentary electricity supply system, and the government has now decided to initiate a project to create a new system. Its very first step is to engage a firm of consulting engineers to write a **project definition**.

(a) What are some of the issues that the engineering firm will have to discuss with the government?

(b) What items of information will the firm have to obtain before writing the project definition?

(c) How should the project definition be structured, and what will be the content of each major section?

2. *Work breakdown structure*: A communications system, such as the Telecom network, offers the transmission of information (e.g. data, voice) to its subscribers. However, this transmission can take place in any one of a number of **modes** (e.g. switched circuit, leased line, packet-switched, but note that the nature of these modes is irrelevant at the level of study considered in this problem), and each of these modes is characterized by values of such technical parameters as bandwidth, delays and error rate and by a tariff structure.

Using these network capabilities, it now becomes possible for third parties to offer subscribers sophisticated services (so-called value-added services) by inserting special equipment (e.g. modems, answering machines, faxes, network control equipment, packetizers) between the network and the user. Indeed there is a whole industry emerging that is based on this idea.

Assume that the requirement for a particular user service has been identified and that a project is to be established to investigate the viability of actually providing this service. Describe how the project should be carried out by giving a logical, relevant and formally correct **work breakdown structure** and by identifying the main issues and parameters to be treated under each work package.

2.10 References

Archibald, R.D., *Managing High-Technology Programs and Projects*, New York: Wiley, 1976.

Blanchard, B.S., *Engineering Organization and Management*, Englewood Cliffs, N.J.: Prentice Hall, 1976.

Chase, W.P., *Management of System Engineering*, New York: Wiley, 1974.

Hajek, V.G., *Management of Engineering Projects*, McGraw-Hill, 1977.

MIL-STD-499A *Engineering Management*, Washington D.C.: US Department of Defense, 1974.

MIL-STD-881A *Work Breakdown Structures for Defense Material Items*, Washington D.C.: US Department of Defense, 1978.

Sage, A.P., "Desiderata for systems engineering education", *IEEE Trans. Systems, Man, and Cybernetics* SMC-10, no. 12, 1980.

Shinners, S.M., *A Guide to Systems Engineering and Management*, Lexington Books, 1976.

System Engineering Management Guide, Defense Systems Management College, Fort Belvoir, VA, 1990.

New concepts

CHAPTER 3

Structure

3.1 Concept of structure

3.1.1 A high-level description of relations

The properties of a system depend on the properties of the elements making up the system and on the way in which these elements **interact**. The same elements interacting in different ways generally result in different systems. The most basic description of these interactions, one applicable to any system, is **structure**.

The structure of a system is the **pattern** of the interactions between the elements. At the risk of being somewhat restrictive, it can be defined as follows. Let the **adjacency matrix** A of a system be defined by:

$$A = [a_{ij}]$$

$$a_{ij} = \begin{cases} 1, & \text{if there is an interaction} \\ & \text{from element i to element j} \\ 0, & \text{otherwise} \end{cases}$$

Then, two systems have the same structure if and only if they have the same adjacency matrix.

It is clear from this definition that structure is defined with respect to a particular interaction or a particular set of interactions. It is perfectly possible for a system to support more than one type of interaction, and in this case each type of interaction may have a different structure. However, it is also possible for two completely different systems—that is, systems with elements that are not of the same general type (e.g. equipment, organizational units, activities) and support quite different interactions (e.g. information flow, power flow, temporal ordering, spatial ordering)—to have the same structure. This case is of great interest, as those properties that depend on structure only must then also be common to the two systems.

The importance of structure is most impressively demonstrated by a class of systems that lies outside electrical engineering but that should nonetheless be familiar, namely molecules—in particular, hydrocarbons. Not only do molecules with different numbers of carbon and hydrogen atoms have completely different properties, but even molecules with the same formula can have different properties depending on how the interactions are arranged (i.e. on their structure). These are the so-called isomers, and Figure 3.1 shows the structures of the three pentane, C_5H_{12}, isomers. As an example of how properties change with structure, their boiling points are:

pentane   36°C
isopentane   28°C
neopentane   9.5°C.

In discussing the concept of structure it cannot be emphasized too strongly that the theory of systems not only is not restricted to electrical systems but also is in no way restricted to systems whose elements are pieces of hardware. Some of the most interesting and powerful applications within systems engineering relate to abstract entities, such as tasks or activities. As a very simple illustration of this, a normal letter (e.g. business letter, letter to a friend) can be considered as a system. The elements of the letter are its various parts, characterized by their purpose or contents, namely:

greeting/introduction
presenting the subject matter
proposing a course of action
conclusion/regards.

The structure is a simple linear one; it is an ordering of the elements on the paper such that the recipient will read first the introduction, then the subject matter and so on. This ordering (or **relation**) of the elements is important. The meaning of one element is influenced by having first read the preceding elements; that is, the interaction between the elements takes place in the brain of the person reading the letter. However, considering the task of composing such a letter as an exercise in systems engineering, it can be seen that the writer proceeds by first determining which elements of information

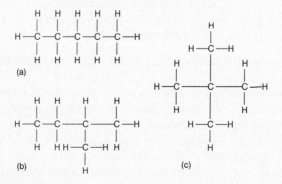

Figure 3.1 Structures of the three pentane isomers: (a) pentane, (b) isopentane, (c) neopentane.

are to be conveyed by the letter and then determining their logical sequence. Thus, the essence of the systems engineering approach is to **structure the task** of writing the letter, and this structure corresponds to the **structure of the physical system** (i.e. the letter).

3.1.2 Interactions as properties of elements

In defining a system one set of elements and two sets of interactions (internal and external) are employed, but it is important to realize that these are not independent; the interactions that are possible between the elements must be a property of the elements themselves. It is which of these that are actually utilized in a particular system that is expressed by the set of interactions. However, the concept of an interaction must not be restricted to a physical interaction; it is a much more general and abstract concept. "Relation" would perhaps be a better word, as it can be anything that relates the elements to each other and thereby creates a structure, and the possible relations depend on what aspects of the elements are of interest.

As an illustration of this, consider a radio receiver. As is shown in Figure 3.2, it can be looked upon as a system consisting of six elements, and the physical real-time interactions between these elements are also shown. They are information and/or power flow between the elements and give the system a particular structure. This would be an appropriate representation for studying the signal processing resulting from the transfer functions of the individual elements; but for studying the reliability of the system as a function of the element reliabilities, the reliability block diagram shown in Figure 3.3 would be more useful. The elements are the same, but the interactions are no longer physical interactions; they represent **logical** interactions. The structure is now quite different and expresses the fact that the system reliability is the product of the element reliabilities; that is, the system fails if any one element fails. In this structure the functionality of the elements plays no role; each element is characterized by its failure rate alone, and even the order of the elements in the chain is irrelevant. However, the requirement that the elements be connected in series is still a property of the elements, of their physical construction. It is not possible to connect two of them in parallel in the reliability block diagram.

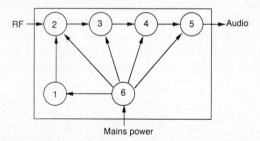

Figure 3.2 A radio receiver as a system consisting of six elements: 1 = oscillator, 2 = mixer, 3 = IF amplifier, 4 = detector, 5 = audio amplifier and 6 = power supply.

Figure 3.3 Reliability block diagram of the radio receiver shown in Figure 3.2.

In the two systems shown in Figures 3.2 and 3.3 the structures are fixed by the functionality of the elements; but many systems can have different structures, even though they consist of the same elements, as was the case with the isomers. Such systems are sometimes called **reconfigurable**. A familiar example is a power distribution network in the form of a ring, where a fault leading to the opening of a circuit breaker changes the structure from a ring into a linear one. Different structures (i.e. different sets of interactions with the same set of elements) are also called different **states** of the system (see Section 5.5 for a discussion of states). As interactions are broken and reformed, the system makes transitions from one state to another. If it is known for certain that the system is in a particular state, this is expressed by saying that the system is in a **pure** state. Most often, however, it is not known exactly which state the system is in as the system makes transitions under the influence of random events (e.g. equipment failures and repairs); then only the probability of its being in a state can be determined. The system is then said to be in a **mixed** state, characterized by a probability distribution over all possible states.

3.1.3 Choice of what is an element and what is an interaction

In Section 1.2.2, when defining a system, it was pointed out that sometimes there is a choice regarding what physical parts of a system to designate as elements and what parts to consider as forming the interactions, and a power system was given as an example. This can sometimes be a little confusing, and it is important to state clearly which choice is being considered at any one time, but it is nevertheless a very useful ability. For example, to investigate how a system changes its state under the influence of certain equipment failures, those equipments must be made part of the interactions; whereas to develop the functionality of the system in terms of its elements, it is necessary to include all the equipment within the elements and to let the interactions be interfaces.

There is, however, an even more extreme choice. For every system it is possible to define the system's **dual**, which is simply the interchange of the two sets, elements and interactions. Whether this is useful or has any physical meaning is a different matter, but it is always possible, and Figure 3.4 illustrates how it is done. The system in (a) has four nodes and six interactions. If the nodes are now considered as interactions and the interactions as nodes, each of the original nodes is transformed into three of the new interactions, and the result is the system in (b), which has six elements and twelve interactions. The system in (b) is clearly a different system from that in (a)—it has a

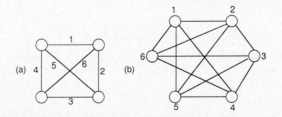

Figure 3.4 (a) A system of four nodes and six interactions and (b) its dual, which has six elements (the interactions of (a)) and twelve interactions.

different set of elements, a different set of interactions, and a different structure—but the two can still both represent the same physical system, only different **aspects** of that system.

3.2 Representations of structure

3.2.1 Graphical representation

3.2.1.1 Nets, relations and digraphs

In illustrating systems and their structure, use has already been made of a graphical representation where the elements are shown as little circles and the interactions as lines between the circles. Such graphs are indeed very useful, and there exists a whole theory of their properties: graph theory. However, before discussing a few useful and important parts of this theory, note that, as with any visual representation, the use of graphs requires caution, as will be illustrated by the following three examples.

As a first example, consider the system depicted in Figure 3.5(a). It contains eight elements, and from the graph it is immediately clear that the system can be viewed as consisting of two loosely coupled subsystems of four elements each. If the same system is depicted by the graph in Figure 3.5(b), this fact is not so immediately obvious (although still true of course), and this latter representation could be said to be misleading.

A second example is given by the two representations of the same system shown in Figure 3.6. The graph in (a) immediately gives the impression of a hierarchically organized system (e.g. with one manager, two departments and three groups), but the graph in (b) no longer makes this type of relationship clear. It is an example of a case where the graph is used to convey something in addition to the formal structure as previously defined.

The third example, shown in Figure 3.7, illustrates a further development of the idea of **levels** within a hierarchy. If the level of an element within a hierarchical structure is defined by the number of links between it and the apex of the structure (e.g. amount of influence with the boss), the element labelled X is one level below the apex. However, if the level is determined by the number of links down to the lowest element linked to the one in question (e.g. a measure of responsibility and number of

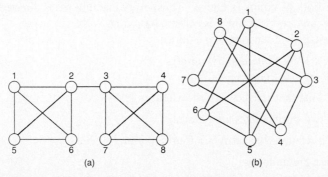

(a)

(b)

Figure 3.5 A system with eight elements, represented by two superficially different graphs.

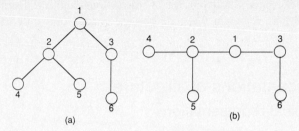

Figure 3.6 A system with six elements, where in (a) a hierarchical relationship is implied but in (b) it is not.

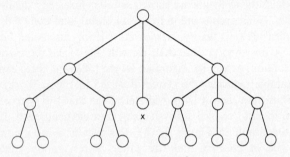

Figure 3.7 Position of an element X within a hierarchical structure.

subordinates), element X is on the lowest level. Finally, if the level is simply that shown in the graph (e.g. if the graph is used to show salary levels or organizational titles in addition to the structure), X is two levels below the apex.

Returning now to the representation of structure by means of graphs, a suitable starting point is the definition of four **primitives**:

P_1: a set V of elements called **points** (or sometimes also **vertices**)
P_2: a set X of elements called **directed lines** or, more briefly, **lines** (or also **edges**)
P_3: a function f whose domain is X and whose range is contained in V
P_4: a function s whose domain is X and whose range is contained in V

For any line x in X, fx is the **first point** of x and sx the **second point** of x. The line x is **incident** with both its points fx and sx. Thus, every line of a net is directed from its first point to its second point. A line x is called a **loop** if fx = sx.

A **net** satisfies the following two axioms:

A_1: The set V is finite and not empty.
A_1: The set X is finite.

A **relation** is a net that satisfies a third axiom:

A_3: No two distinct lines are parallel.

A **digraph** is a relation that satisfies a fourth axiom:

A_4: There are no loops.

Two digraphs are **isomorphic** if there exists a one-to-one correspondence between their points that preserves their directed lines (e.g. as in Figure 3.5).

3.2.1.2 Characteristics of digraphs

A **semipath** is a sequence of lines (and points) between any two points u and v, in which no point occurs more than once, and a **path** is a sequence of lines from v to u (i.e. all lines in the same direction), in which no point occurs more than once. The point u is **reachable** from v if there exists a path from v to u.

A **cycle** is a closed path (i.e. one that starts and ends at the same point). Correspondingly, a **semicycle** is a closed semipath. A **tree** is a connected digraph that contains neither a cycle nor a semicycle.

A digraph that contains no cycle is called **acyclic**, and such digraphs form an important class as far as applications go. In terms of electrical networks (or any control system), the absence of cycles means that there is no feedback; conversely, the presence of cycles may indicate unwanted feedback loops. Activity networks (i.e. PERT or CPM diagrams) must be acyclic; any cycle would imply an erroneous network. The same is true of decision logic, where a cycle would usually imply an improperly structured problem.

The number of lines in a path is called its **length**. A **geodesic** from u to v is a path of minimum length. The length of a geodesic from u to v is the **distance** from u to v, denoted d(u,v). Note that d(u,v) is not necessarily equal to d(v,u). If there is no path from u to v, then d(u,v) is infinite.

A **maximal path** in a semipath L is a path contained in L but is not a subpath of any longer path in L. Two maximal paths in L can have but one point, the **linking point**, in common. In any semipath there are at least two maximal paths and at least one linking point.

Two points u and v are always 0-joined, 1-joined if there is a semipath between them, 2-joined if there is a path from u to v **or** from v to u, and 3-joined if there is a path from u to v **and** a path from v to u.

The digraph D is **strong** (or strongly connected) if every pair of its points is 3-joined; D is **unilateral** if every pair is 2-joined; and D is **weak** if every pair is 1-joined. If D is not even weak, it is **disconnected.**

A **subgraph** S, generated by a set of points S in V, is the graph whose point set is S and whose lines are all those lines of D that join two points of S. A subgraph with a certain property is said to be **maximal** with respect to that property if no larger subgraph contains it as a subgraph and has the property. A **strong component** of a digraph is a maximal strong subgraph; a **unilateral component** is a maximal unilateral subgraph; and a **weak component** is a maximal weak subgraph.

Let D be a digraph. Then every point and every line of D are contained in exactly one weak component. Every point and every line lie in at least one unilateral component. Every point is contained in exactly one strong component; each line is contained in at most one strong component. A line is in a strong component if it is in a cycle.

3.2.1.3 Related digraphs

The **condensation** of a digraph D, denoted by D*, is the partitioning into strong components and the representing of each component by a "point" in D*. Thus, the "points" of D* are subsets $S_1, S_2, \ldots, S_n$ of V.

If D is strong, D* consists of exactly one point. If D is unilateral, D* has a unique complete path (i.e. one linking all points). An acyclic digraph cannot be condensed. That is, for an acyclic digraph D* = D.

To obtain non-overlapping categories for classifying connectedness, it can be said that **two points are i-connected if they are i-joined but not (i + 1)-joined.**

The **reachable set R(v) of a point v** is the collection of all points reachable from v.

$$R(S) = \{UR(v), v \text{ in } S\}$$

A **point basis** of a digraph D is a minimal collection of points of D from which all points are reachable. It follows that all points of indegree 0 (no incoming lines) are in every point basis, and that no two points of a point basis lie in the same strong component. Every acyclic digraph has a unique point basis consisting of all points of indegree 0. It follows that the condensation D* of any digraph D has a unique point basis consisting of all its points of indegree 0. Every point basis of D consists of exactly one point from each of the strong components in the point basis B* of D*.

The **antecedent set Q(v) of a point v** consists of all points of D from which v is reachable.

A **point contrabasis** of a digraph D is a minimal set S of points such that Q(S) contains all points of D; that is, Q(S) = V. It follows that all points of outdegree 0 are in every point contrabasis, and that no two points of a contrabasis lie in the same strong component. Every acyclic digraph has a unique point contrabasis consisting of all points of outdegree 0. It follows that the condensation D* of any digraph D has a unique point contrabasis consisting of all its points of outdegree 0. Every point contrabasis of D consists of exactly one point from each of the strong components in the point contrabasis C* of D*.

A point v is a **source** of D if R(v) = V. Dually, v is a **sink** of D if Q(v) = V. A digraph is strong if every point is both a source and a sink.

To illustrate some of these concepts, consider the two networks shown in Figure 3.8. Digraph (a) is unilateral, transitive and acyclic; (b) is a weak digraph. In the case of (a), D* = D, the unique complete path in D* is (3-1-2-4), the point basis of D is the single point 3, which is the only point of indegree 0, and the point contrabasis is the point 5, which is the only point of outdegree 0. In the case of (b), D* = {1, 2, 3, 5, 8, 9, (4, 6, 7)}, the point basis of D is the single point 1, and the point contrabasis consists of the two points 5 and 9.

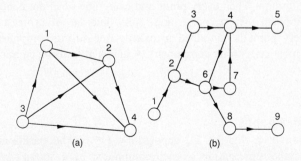

(a)   (b)

Figure 3.8 Two examples of small systems.

3.2.2 Matrix representation

3.2.2.1 Forms of matrices

Digraphs can be described in terms of **matrices**. The **adjacency matrix** is a square matrix with rows and columns labelled by the points v_i of D, as shown in Figure 3.9. If there is a line from v_i to v_j, then $a_{ij} = 1$; otherwise $a_{ij} = 0$. Note that $a_{ii} = 0$ (no loops, axiom A_4). The sum of the ith row is the outdegree of v_i, and the sum of the ith column is the indegree of v_i.

The adjacency matrix can display some of the previously introduced features of a digraph. For example, a digraph is disconnected if and only if there exists some ordering of its points such that the adjacency matrix is of the form

$$A = \begin{bmatrix} A_1 & 0 \\ 0 & A_2 \end{bmatrix}$$

where A_1 and A_2 are square submatrices. A digraph is weakly connected if and only if there exists some ordering of its points such that the adjacency matrix is of the form

$$A = \begin{bmatrix} A_1 & 0 \\ A_{21} & A_2 \end{bmatrix} \quad \text{or} \quad \begin{bmatrix} A_1 & A_{12} \\ 0 & A_2 \end{bmatrix}$$

Conversely, for a strong digraph no ordering of the matrix will result in a square zero submatrix in the adjacency matrix. Finally, a digraph is acyclic if and only if its points can be ordered such that the adjacency matrix is an upper (or lower) triangular matrix.

3.2.2.2 Associated matrices

In the **reachability matrix** $R(D)$, $r_{ij} = 1$ if v_j is reachable from v_i; otherwise $r_{ij} = 0$; that is, $r_{ij} = 1$ if there exists a path from v_i to v_j. As each point is reachable from itself, $r_{ii} = 1$. If p is the number of points in D, then

$$R = (I + A)^{p-1} \quad \text{(Boolean algebra!)}$$

Let v_i be a point of the digraph D, and let R' be the transpose of R. Then the strong component of D containing v_i is given by the entries of 1 in the ith row (or column) of the matrix $R \times R'$.

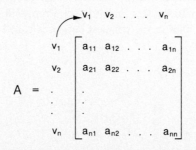

Figure 3.9 Form of the adjacency matrix. Matrix element a_{ij} equals 1 if there is an interaction from element i to element j, and 0 if there is not.

The **connectedness matrix** $C(D)$ has the number $n = 0, 1, 2$ or 3 in its (i, j)-location whenever the points v_i and v_j are n-connected in D. Clearly, $C(D)$ is a symmetric matrix. For any digraph D, the connectedness matrix $C = \{c_{ij}\}$ is obtained from the reachability matrix $R = \{r_{ij}\}$ as follows: If v_i and v_j are in the same weak component,

$$c_{ij} = r_{ij} + r_{ji} + 1$$

otherwise $c_{ij} = 0$. (Note: A weak component may be unilateral or strong, or include a unilateral or strong component.)

In the **distance matrix**, $N(D) = \{d_{ij}\}$, element d_{ij} is the distance from v_i to v_j. (Note: d_{ij} may or may not equal d_{ji}, and $r_{ij} = 0$ implies that d_{ij} is infinite. Also, $a_{ij} = 1$ implies that $d_{ij} = 1$.)

Returning to the two networks in Figure 3.8, the adjacency matrix of (a), shown in its upper triangular form, is

$$
\begin{array}{c|cccc}
 & 3 & 1 & 2 & 4 \\
\hline
3 & 0 & 1 & 1 & 1 \\
1 & 0 & 0 & 1 & 1 \\
2 & 0 & 0 & 0 & 1 \\
4 & 0 & 0 & 0 & 0 \\
\end{array}
$$

The adjacency matrix of (b) exhibits the zero-submatrix typical of weak digraphs; with the ordering $(1, 2, 3, 4, 7, 6, 8, 5, 9)$ there is a 4×4 zero-matrix in the lower left corner.

3.3 Structure-related properties

3.3.1 Connectivity and cut-sets

The usefulness of digraphs is greatly enhanced by introducing a further concept, that of **cut-sets**. A cut-set is a set of lines whose removal from digraph D leaves D disconnected, and no subset of which has the same property. For example, in the digraph shown in Figure 3.10 the sets $\{a, b\}$, $\{a, c\}$, $\{b, c\}$ and $\{d\}$ are cut-sets, but $\{b, c, d\}$ is not. The number of lines in the smallest cut-set is called the **line connectivity** of a digraph.

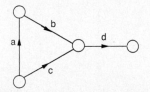

Figure 3.10   A small digraph with a line connectivity of 1.

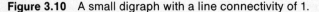

In Section 3.1.3 it was remarked that it is always possible, and often advantageous, formally to interchange elements and interactions. In particular, the physical elements of a system can be represented as points or as lines in a digraph, and a transformation between the two representations is especially common in the case of so-called **transport networks**. A transport network is one in which something (e.g. information, spare parts, manpower, money) flows from a **source** to a **sink**, and a very simple case is shown in Figure 3.11. The network in (a) consists of three elements that take part in transporting something from the source x to sink y. Clearly, for the network to operate, all three elements must be operating; that is, element 1 must carry out the transport from x to intermediate point s, element 2 must carry out the transport from s to intermediate point t, and so on, as shown in (b). However, this fact can be equally well represented by the digraph in (c), where the original elements have become lines.

Consequently, it must be equally possible to have cut-sets that consist of lines and cut-sets that consist of elements. In the latter case the term "line connectivity" cannot be used, but otherwise all aspects remain the same. A very important application of cut-sets, namely to system reliability, will be discussed in Section 10.2.4; here a small example will demonstrate their utility.

In many cases it will be true that the elements in a transport system each have a definite **capacity**; the flow through an element can take on any value up to and including this capacity, but cannot exceed it. Then the following theorem holds (Harary et al. 1965):

> The maximum flow from the source to the sink is equal to the minimum value of the capacities of all the cut-sets that separate the source from the sink.

The capacity of a cut-set is defined as the sum of the capacities of the elements of the cut-set in the direction towards the sink. (Note that in this case the cut-set is defined without regard to the direction of flow within the elements, i.e. as if all elements could sustain flow in both directions.)

Consider a system consisting of six elements, connected as shown in Figure 3.12, where the elements are represented as lines. Part (a) illustrates a particular cut-set,

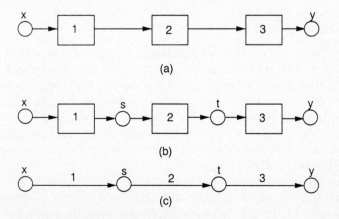

(a)

(b)

(c)

Figure 3.11   Dual representations, (a) and (c), of a transport network, with (b) illustrating the transition from one to the other.

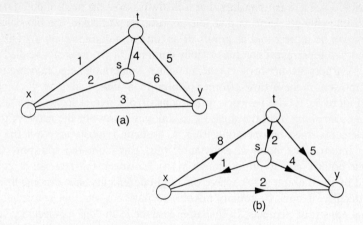

Figure 3.12 A transport network with six elements represented as lines; (a) shows the numbering of the elements, (b) their capacities.

namely {2, 3, 4, 5}, and part (b) gives the capacities, which in this case are all unidirectional. The capacities of the cut-sets are then as follows:

Cut-set	Capacity
1, 2, 3	10
1, 3, 4, 6	14
2, 3, 4, 5	9
3, 5, 6	11

Thus, the maximum flow would be 9 units. Note that the maximum flow value is not just the sum of the outgoing capacities minus the incoming capacities at the source, as might be believed. The capacity of element 2 is irrelevant; it does not enter into any of the cut-set capacities. This arises because the flow pattern (which is not determined by this theorem) at maximum flow will obviously require the flow from s to x to be zero.

3.3.2 Redundancy

An important structural property is **redundancy**. A system is said to have (a degree of) redundancy if it is possible to remove one or more elements without changing its functionality. Redundancy is represented graphically by the connection of elements in **parallel**. However, the converse is not true; the connection of elements in parallel does not necessarily signify redundancy. For example, the connection of two 10 kohm resistors in parallel in order to form a 5 kohm resistor does not constitute redundancy. Neither does the parallel operation of two 10 kW generators in order to supply a 20 kW load, but if the load is only 10 kW, it does.

The most common form of redundancy is where two functionally identical elements are connected (functionally) in parallel, such that if one fails the other can provide the same functionality. The word "functional" is, as always on the system level, important; there is no need for the two elements to be physically identical, as is illustrated by hydraulic and mechanical steering, for example. However, duplicating the functionality can sometimes be unnecessarily expensive, as for example in the case of the two 10 kW generators supplying a 10 kW load. It may be adequate to have three 5 kW generators; it is then necessary for two of the three to be operating in order to supply the load. This reduced type of redundancy is called **n-out-of-m redundancy**; in this case 2-out-of-3.

The opposite situation can also arise, namely that a higher degree of redundancy is required than is afforded by two elements in parallel. If (n + 1) elements are connected in parallel such that only one of them has to be operating in order for the system to function, this is called **n-fold redundancy**.

3.3.3 Vulnerability

However, the case of functionally identical elements connected in parallel is only the simplest form of redundancy. In complex systems the system function or parts thereof may be able to be carried out by two or more subsystems, and each of these may have a complex structure and interact in more ways than just a parallel connection. Classical examples are communications networks and power transmission networks.

Furthermore, in many systems redundancy as described above is not very relevant, as it is not a simple question of how many elements out of a subset can fail before the system fails; it is a question of gradual reduction in performance as elements fail, so-called **graceful degradation**.

This sort of complex behavior cannot be adequately described in terms of such simple concepts as n-out-of-m redundancy. Rather, the **vulnerability** of the system with respect to a particular **threat** or **hazard** is described. Vulnerability must be defined for each particular system; it is the degree to which some measure of performance or security is changed (degraded) as a function of a particular threat or combination of threats.

The threat can be something that is directly describable in terms of the system structure, such as the loss of a line or the loss of a node; but it may also be more indirect, such as "vulnerability to terrorist attack" or "vulnerability to operator error". In the latter case, determining the vulnerability becomes a two-stage process; first it is necessary to determine the probability that the threat will change the system structure (or the system state, see Section 5.2), and then to determine the effect on system performance of such a change in structure.

3.4 Order and entropy

3.4.1 Structure as a form of order

A set of noninteracting and unrelated elements is nothing more than the sum of its elements. The set does not have any properties that are not properties of at least one of the elements, and the elements do not form a system. This changes as soon as **relations** between the elements are allowed. Relations are more general than interactions, or, conversely, interactions are a subset of relations; but without going into this aspect any further, it can be simply said that a relation expresses an element's "position" with regard to some parameter. For example, if the parameter is size, and the elements are characterized by the classifications small, medium and large, then any two elements x and y are **related** by one of the three statements "x is the same size as y", "x is larger than y" or "x is smaller than y".

The existence of a relation implies a degee of **order**, where order is a measure of the extent to which a fixed relation (or, perhaps better, a fixed "value" of a relation) is valid. If a relation is completely fixed and immutable, there exists perfect order (with respect to the corresponding parameter). In the case of structure, which expresses a direct relation between the elements, it follows that **any** fixed structure represents perfect order. However, if the system is not restricted to one structure but has a set of possible structures, the degree of order is a measure of the degree of certainty about what state the system is in; it expresses the **randomness** of assigning a particular "value" to the parameter "structure".

3.4.2 Random changes and degree of order

The degree of order or, conversely, the degree of randomness is too loose a concept to be of much practical use. Consider a system that can be in any one of n different structures or, otherwise expressed, in any one of n states. Let the probability of it being in state i be denoted by P_i. Then the **entropy** of the system (with respect to structure) is defined by

$$S = -\sum P_i \ln (P_i)$$

If the system is perfectly ordered, P_i is 1 for some values of i and zero for all other i. As $\ln (1) = 0$ and $x \ln x \rightarrow 0$ as $x \rightarrow 0$, S will be zero. On the other hand, complete disorder corresponds to each state being equally likely; that is, $P_i = 1/n$ for all i, which gives $S = \ln (n)$.

3.4.3 Entropy and work

In the types of systems of interest in this text, where the elements are subsystems or equipments, the entropy is somehow a measure of the degree of disorder (or decay) that results from the influence of various, mostly random forces (e.g. temperature, vibration), which manifest themselves as the random failure of individual elements. If no maintenance is carried out, more and more elements will fail; to maintain a steady state it is necessary to do **constructive work** to counteract the **destructive work** done by the random forces.

The ability to do work that decreases the entropy is the hallmark of a living organism; the entropy of a nonliving system will always increase, as is illustrated by the decay following death. No matter what sort of system it is, communications system, a company or a society, its structure will decay unless work is carried out by the people associated with the system. This work is nothing but what is normally called maintenance, and these remarks should serve to emphasize the fact that maintenance must be considered as an integral part of a system—and not only integral but essential; it is the life-giving part of the system. Only through the entropy-decreasing work of its living elements (i.e. humans) can the system remain in a steady state.

3.4.4 Illustration in terms of practical systems

As a first example to illustrate what are undoubtedly some very abstract concepts, consider an organization such as a manufacturing company. On the most detailed level the elements of this system are the individual employees, whose interactions can be both informal and formal. The latter give rise to the structure of the organization, as reflected in an organigram with groups, departments, divisions and so on. The informal interactions, the tendency of individuals to do what suits them best at the moment or that benefits their own private goals, introduce an amount of disorder into the organization. As interactions take place in the corridors or over lunch, the structure is temporarily (and randomly) changed and the entropy is increased. To keep the entropy at some optimum level (which is not zero, as such a rigid organization would be stifling), managers each have to do an amount of work to pull their subordinates into line. Without this constant work, the organization would slide into chaos or anarchy.

A second example is provided by the case of an army in the field. In this system the elements are troop units (perhaps of platoon size) and the interactions are the lines of command, which are supported physically by communications channels. As long as the structure is intact, the army operates as a whole and can carry out such coordinated maneuvers as pincer movements. However, as the temperature of the battle rises, the lines of communication tend to become disrupted, due to both enemy action and random failures of equipment, and the structure starts to break down. The entropy increases, and the army loses its coherent qualities and becomes more and more like a collection of independent units. To prevent the entropy from increasing, work has to be done on the system in the form of repair and replacement.

3.5 Case studies

3.5.1 Electricity supply system

3.5.1.1 Subsystem structures

As was explained in Section 1.4.4.2, the electricity supply system may be considered as consisting of a bulk supply subsystem and one or more distribution subsystems. Each of these can again be subdivided into three subsystems, namely power, control and protection, and each of the resulting six subsystems has a somewhat different structure in order to best fulfill its purpose.

The bulk supply power subsystem has to interconnect the power stations and the bulk supply substations; that is, its main purpose is to overcome distance, but at the same time to ensure a high degree of reliability by providing adequate redundancy. As a result the network structure consists of a meshed core to cover the large distances, and then possibly some radial stubs for short feeders to individual nodes. An example of such a network is shown in Figure 3.13, and the corresponding adjacency matrix is given in Figure 3.14.

As no restriction is placed on power flow, the matrix is symmetric (i.e. $a_{ij} = a_{ji}$, and thus only one-half needs to be shown. Furthermore, the matrix only expresses whether there is a connection or not; it does not say how many parallel lines each connection consists of, and so the elements are restricted to the values 0 and 1. A

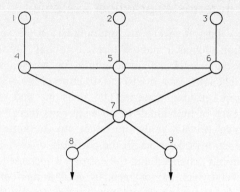

Figure 3.13 Structure of a small bulk-supply system.

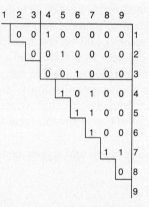

Figure 3.14 Adjacency matrix of the system in Figure 3.13.

division of the matrix into three submatrices has been indicated in order to point out the different types of information carried within each one. The small triangular submatrix reflects the fact that the generating substations do not form a subsystem; the rectangular submatrix reflects the fact that one transmission substation is assigned to each generating station; and it is really only the larger triangular submatrix that reflects the choice of network. In this case that choice is a peculiar mixture of meshed and radial, which leaves each of the sinks dependent on a single connection.

The power network within one distribution area can have several voltage levels and therefore assume a hierarchical superstructure. However, in general the purpose of a distribution network is to cover an area (i.e. to supply a large number of loads distributed over an area). A radial structure does not provide any redundancy, and a fully meshed structure would be much too costly and provide an unnecessary degree of redundancy (due to the large number of loads in close proximity). As a result, distribution networks typically have a multiple ring structure (like a wheel with spokes), as shown in Figure 3.15.

Turning now to the control subsystems, the purpose of the bulk-supply control system is to optimize the supply of power to the distribution subsystems by appropriate utilization of the facilities as required by the demand. This function is by nature a centralized one, as it requires information about the state of the whole network to be brought together in one place in order to be able to make the right decisions, and correspondingly the control network normally has a radial structure (Figure 3.16).

The control of a distribution network is a mixture of local (distributed) and centralized control, with the majority of control functions carried out locally in each substation. The centralized functions are concerned mainly with data logging and maintenance functions. Consequently, the structure of the control system is that of a number of subsystems (which are of course nothing but the substation automation subsystems, the main subject matter of this case study) loosely coupled in a radial fashion to a central control station.

Finally, the protection subsystems are almost completely local, and even within each substation they are subdivided into largely independent parts (e.g. overcurrent protection, overvoltage protection, differential protection). The reason for this disjointed structure is the requirement for extreme reliability; any spatial extension over communications channels or linking to other protection functions makes it difficult to reach that level of reliability. There are some protection functions that require signalling between two substations (e.g. by using pilot wires), but these are in the minority. Thus, in one sense the protection elements do not form a system as they are not connected together, but in another (functional) sense they are connected, namely via the power subsystem. For example, the overcurrent protection on the incoming feeders to a busbar and that on one of the outgoing feeders are **related** by their respective current and / or delay settings, such that, if the outgoing relay fails to trip on a short circuit, the incoming relay will trip and disconnect the whole busbar.

3.5.1.2 Common factors determining structure

The structures of most of the above subsystems are determined by certain common factors. First among these must be geographical situation—that is, the location of the generating stations, the location and distribution of the load, and the features of the landscape in between (e.g. mountains, floodplains, rivers, fjords). A second factor, which influences all subsystems, is reliability (or availability). The electric power system is such an essential part of society's infrastructure that a major failure must be avoided at (almost) all cost. Consequently, the structures of the subsystems will be chosen to provide a suitable degree of redundancy. Finally, a third factor, which influences the choice of structure of every subsystem, is the subsystem's functionality (i.e. its purpose).

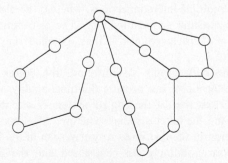

Figure 3.15   Multiple ring structure of distribution networks.

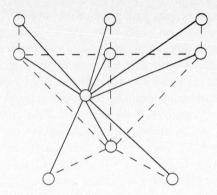

Figure 3.16   Radial structure of a control system that has mainly centralized functions, shown overlaid on the system of Figure 3.13.

For example, the fact that a subsystem's main function is inherently a centralized one will normally lead to a radial structure.

Listing these factors here gives yet another illustration of the systems engineering approach to a problem. Before starting work on any one subsystem, the first step is to consider what factors are common to all the subsystems. It is also necessary to consider what common questions must be answered in the design of all the subsystems (e.g. maintenance policy), whether models can be developed that will apply to all the subsystems (although with different parameter values), and so on. The principle is to do as much work at as high a level as possible; its effect (and thereby its cost-effectiveness) will be multiplied as its results are applied to the lower levels.

3.5.1.3 Essential differences in structure

While the main factors that determine the structures of subsystems are common to all the subsystems, the actual results of applying these factors will differ. Two of the major and more obvious differences are redundancy and centralization.

Although both the transmission network and the distribution network are functionally similar in that they transport the energy between a set of sources and a set of sinks, the relative numbers of elements in these sets as well as their spatial distribution lead to the transmission networks having a high degree of redundancy through meshing and multiple interconnections, whereas the distribution network has a low degree of overall meshing and redundancy. The distribution network structure is usually a collection of "loosely coupled" rings, coupled only at the busbar in the distribution substation.

In the case of the bulk supply system, both the power and control networks interconnect the same locations; but because the functionality of the power system is a decentralized one, whereas that of the control system is a centralized one, the former has a mesh structure and the latter a radial structure. Also, because of the much lower degree of interconnection in the distribution system than in the bulk supply system, the control function is correspondingly less centralized and the resulting control-system structure consists of loosely coupled subnets centered on each substation.

3.5.2 Tactical communications system

3.5.2.1 Problem definition

The performance of any trunk communications network is determined by two major factors: the inherent capability of the equipment involved (e.g. switch, transmission equipment), and the way this equipment is interconnected (i.e. system structure). Consequently, a considerable amount of work goes into determining the best structure, and the main input to this work is the subscriber population; that is, the structure is optimized with respect to a particular subscriber population, and the performance also is therefore determined only relative to this subscriber population.

In the case of the tactical communications system this raises a couple of questions. Seeing that the number of subscribers to the system can vary in a range of 1 to 10 from one deployment to the next, that the mix of different subscribers (e.g. voice, data, high priority) also can vary considerably from deployment to deployment, and that within one deployment the actual locations of the various units can vary on a daily basis, how can a system structure be talked of? And without a given system structure, how is system performance to be defined?

There are at least three different ways to solve this problem:

1. Choose one deployment as being particularly representative, in the sense of occurring frequently, being of special importance with regard to performance or being typical with respect to any other parameter.
2. Choose a number of deployments that together cover the range of possible deployments, and then either form an average for each performance parameter or have a number of values for each parameter (e.g. have three values for each parameter, corresponding to small, medium and large deployments).
3. Seek a **class** of networks whose structure is representative of the structures obtained for deployments within the possible range, and then refer any structure dependency of performance parameters to this class rather than to any actual structure.

The first solution is the simplest, but it is also the most dangerous in the sense that it is very vulnerable to any shift in what is perceived to be the most likely or important deployment. The second solution goes a long way towards alleviating this danger, but it is also more cumbersome. The third solution combines simplicity with the possibility of versatility and low risk, but it depends on finding a suitable class. For the class to provide a simple reference structure, the member networks should be parametrized by a single parameter, namely network **size**. Also, at least one important system characteristic should take on a value that can be considered as a characteristic of this class.

3.5.2.2 Normal networks

The search for a suitable class of networks is aided considerably by the fact that tactical networks of the type considered here must satisfy two **rules**:

1. Each node shall be connected to the network by at least two links.
2. The failure of any one node shall not disconnect any other part of the network.

These rules are by no means trivial; Figure 3.17 shows four networks that violate these rules.

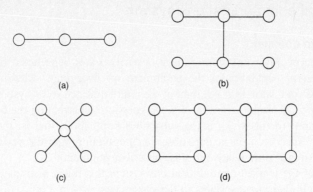

(a)

(b)

(c)

(d)

Figure 3.17 Four networks that do not satisfy the two rules of a tactical network.

Furthermore, the requirement that the networks within the class are to be parametrized by size only implies that the networks must be spatially regular or uniform; that is, a large network must simply be a repetition of a small network. If the network is to be planar, which is reasonable, then it will consist of **rings**; and as shown in Figure 3.18, such rings can have three, four or six sides. Actual tactical networks lie somewhere between three and four sides per ring, but closer to four, and so square rings will be chosen as being most representative.

It then remains to eliminate the shape or aspect ratio (width to length) of the network, and this is done by assuming the networks to be **square**. This requires the total number of nodes in a network to be the square of an integer, and for present purposes the size of the networks will be limited to networks with between 4 and 25 nodes. Of these nodes, the peripheral ones are access nodes, whereas the internal ones are trunk nodes, as shown in Figure 3.19.

The final step in defining the sought-after class of networks is the following: Any parameter of these square networks can be expressed as a function of the number of nodes in the network; such a function will be defined in four points. It is now possible to define a class of networks, such that the size of the network is given by its total number of nodes, and such that the value of any network parameter is the value obtained by interpolating the functions defined for the square networks. This class will be called **normal networks**.

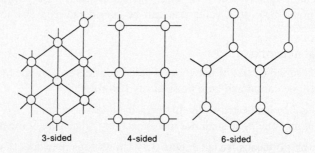

3-sided   4-sided   6-sided

Figure 3.18 Networks consisting of rings with three, four and six sides.

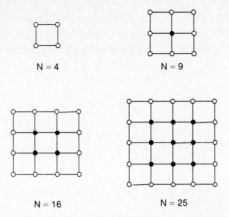

N = 4   N = 9

N = 16   N = 25

Figure 3.19 Square networks. Open circles denote access nodes, and full circles trunk nodes.

As defined here, normal networks are characterized solely by their size, given by the number of nodes, N. This already allows certain system parameters to be determined as functions of the network size, such as:

number of links M $= 2(N - \sqrt{N})$
number of access nodes $N_a = N - (\sqrt{N} - 2)^2$

Furthermore, it is possible to calculate the **average connection length**, defined as the average of the shortest path between any two access nodes, measured in links. It turns out to be:

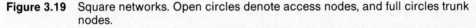

network size N $= 4 \quad 9 \quad 16 \quad 25$
average connection length L $= 1.33 \quad 2.14 \quad 3.03 \quad 3.93$

However, in Chapter 6 it will be shown that, with only a small amount of additional information, a very significant functional property of this class of networks can be calculated.

3.6 Summary

1. **Structure** is the most general characterization of the interactions between the elements. It describes between which elements there is any type of interaction but says nothing about the strength or type of each individual interaction.
2. Structure is an equally useful concept if the interactions **of interest** are limited to a particular type. In this case the same set of elements can have different structures (e.g. signal flow structure and reliability structure).
3. Note that there is a **semantic problem** in talking about systems, which becomes apparent here for the first time (but which comes up again and again). According to the formal definition a system can have only one structure; but it is common practice to use the term "system" more loosely and to mean essentially only the set of elements, in which case a system can have different structures and can change its structure.

4. The concept of structure is important because it can be used to work with systems (i.e. design, optimize, analyze). To do this, structure has to be **represented** in a formal manner. The two most important representations are **digraphs** and **matrices**.

5. On the basis of this representation, **elements** of structure have been developed, such as:

> path
> cycle
> subgraph,

as well as **characteristics**, such as:

> strength
> condensation
> point basis (or contrabasis)
> connectedness

6. The structure of a system is important to systems engineering because there are structure-related system characteristics, such as:

> capacity
> redundancy
> vulnerability.

7. The structure introduces **order** into the set of elements; it is the structure that gives the system characteristics that are different from those of the set of elements.

8. Order tends to be destroyed by random forces; work must be done to maintain a particular order.

3.7 Short questions

1. Are specifications elements or interactions in a system? And in which system?
2. Describe the two principal means of representing structure.
3. How are states related to structure?
4. Can a system have more than one structure?
5. Discuss briefly the difference between the colloquial use of the word "system" and the formal definition.
6. What property of a digraph indicates feedback?
7. What is a cut-set?
8. How is the general definition of a cut-set specialized in the case of a transport network?
9. How are entropy and order related?
10. What type of system can decrease its entropy?
11. Name two factors that can influence or determine a system's structure.
12. Define the vulnerability of a system in terms of its structure.
13. How can hierarchical levels within a system be defined (or characterized)?
14. Can two systems with completely different element sets have the same structure?
15. Are there any system characteristics that depend on structure only?
16. Do projects have a structure? Explain.
17. Explain the relationship between elements and interactions.

3.8 Problems

1. *Structure of a four-node network*: A digital telephone network consists of four interconnected nodes, to each of which are connected N subscribers. Calls between subscribers attached to the same node are called **local calls**; calls between subscribers attached to different nodes are called **remote calls**. Within each group calls are placed at random. The ratio of local calls to the total number of calls placed is the same for every subscriber and is denoted by X. Also, the total number of calls placed by a subscriber per unit time as well as the average duration of a call is the same for every subscriber, and their product is denoted by $d/2$.

   Each of the four nodes consists of an eight-port switch and a number of multiplexers. A multiplexer converts thirty-two subscriber lines to a single time division multiplex (TDM) group, and each switch port handles one TDM group (link). Thus, a switch can establish connections between any of its 256 channels. Within the network the routine algorithm of the switch is such that a path will be established if one exists, and of several it will choose the shortest.

   (a) Consider the network made up of the nodes and their interconnecting cables. Given the above properties of the subscriber population, it should immediately be possible to establish an important property of any sensible network configuration. What is this property?

   (b) How many different configurations satisfy the condition expressed in (a), and how would they be described in a compact form? How many local subscribers n can be connected to node in each configuration?

   (c) Considering the problem of channel utilization and saturation, certain of the configurations in (b) make no sense. How many useful configurations are left to choose from, and which ones are they?

2. *Distribution versus gathering*: Some systems, such as the urban electric grid, can be characterized as **distribution** systems; give another example of such a system. Other systems can be characterized as **gathering** systems; give an example of such a system. How are the structures of distribution and gathering systems related?

3. *Communications network*: Consider a system consisting of $(n + 1)$ elements, of which n, the **peripheral elements**, are (functionally) equal, whereas one, the **central element**, occupies a master role. The nature of the interactions between the elements is that of data transfer; that is, one or more of the functions in one element produce data for, and/or need data from, another element. The interactions between the peripheral elements are weak and local (e.g. nearest neighbor), whereas the interactions between the peripheral elements and the central element are strong. An example of such a system is a SCADA system (e.g. distribution system control).

   (a) Draw graphical representations for at least two different systems of this type.

   (b) What do the interaction (connectivity) matrices for the systems in (a) look like?

   In the realization of such a system the interactions would be taken care of by a communications network consisting of communications **channels** and certain communications **functions**.

   (c) How would you describe the main structure types of such networks?

   (d) Give a few examples of actual (physical) networks of these types.

    (e) How can the interactions be parametrized?

    (f) In terms of the parameters defined in (e), what are the advantages and disadvantages of each structure type? In what circumstances would they be used?

4. *Fast-tracking*: A design project consists of a number of work packages or **tasks**, each of which consists of a number of activities (e.g. modeling, estimation, reviews, producing documentation) and results in a **data package**. The cost of producing this data package is its **value** and is built up linearly during the duration of the task.

    To commence a task, data packages from one or more other tasks may be needed. This results in a structure of the project.

    (a) Describe how this structure could be represented.

    As a consequence of this structure the project will take a certain time to complete. This time will be maximum if all the tasks are in series, minimum if they are all in parallel and in between for any other structure.

    (b) How can the representation in (a) be used to express the duration of the project?

    Instead of waiting for a task to finish it is possible to anticipate the result and use this as the input to the following task(s); this is called **fast-tracking**. However, when the first task finishes, the actual result may be different from the one anticipated, and this will lead to corrective work in the following task(s). Thus it is necessary to weigh the probability and benefit of finishing the project earlier against the probable additional cost.

    (c) For the simplest case of a project consisting of two tasks in series, develop a model that predicts the cost/benefit of the fast-tracking process. The model should be as simple as possible but must be reasonable and reflect all the essential aspects of the process. Any assumptions should be clearly stated.

    (d) In a few paragraphs explain how this model could be applied to a more complex project.

3.9 Reference

Harary, F., Norman, R.Z. and Cartwright, D., *Structure Models: An Introduction to the Theory of Directed Graphs*, New York: Wiley, 1965.

CHAPTER 4

Complexity and generality

4.1 Importance of abstraction

4.1.1 Systems engineering methodology and the use of specificationsas interactions

As discussed in Chapter 2, systems engineering may be viewed as consisting of two parts. Primarily, it is a methodology for designing systems (i.e. for working with complex large-scale issues). In this sense it is a set of rules and procedures for how to organize the work and for how to bring tools and information to bear on a problem, but is not (primarily) concerned with the tools and techniques themselves. A secondary part is then the adaptation and modification of tools and techniques used in other areas (e.g. "conventional engineering", economics, statistics) to suit the particular needs of systems engineering. **Specifications** stand, so to speak, with one foot in each of these parts. On the one hand, they arise as an integral and essential part of the methodology; on the other hand, when properly developed, they become an important tool for carrying out the work in a cost-effective manner.

A central feature of the systems engineering methodology is the stepwise top-down approach, resulting in a set of major **work packages**, related through the work breakdown structure (WBS). Within one branch of the structure the work packages must be carried out sequentially, and the **interaction** between any two of them is the specification produced as the last part of the one work package, which becomes the starting point of the next work package.

The purpose of any specification is to act as a carrier of information from one stage to another within an intelligence-based process. It supports a directed flow of information, from the author to the reader, and its effectiveness is measured by the extent to which the intent of the author is reflected in the results produced by the reader. Consequently, a specification must fulfill at least two conditions:

1. It must be **complete**, in the sense that it treats all questions and areas of concern to the author.

2. It must be **comprehensible to the reader**; it must be expressed in a language and use concepts that will convey the desired meaning to the reader.

The first condition concerns the **content** of the specification, the second its **style**.

For specifications conveying information within the process of developing, designing, and manufacturing devices and equipment, a large body of rules and standards has been developed over the years (e.g. IEEE Std. 830, 1984, MIL-STD-490A, 1985, Supman 1(A) 1978). For example, the contents are usually arranged under the following headings:

Title page and table of contents
1. Scope
2. Applicable documents
3. Requirements
4. Quality assurance provisions
5. Preparation for delivery
6. Notes
10. Appendix

In the past the vast majority of specifications were concerned with conveying information between design and manufacture. Consequently, the main part of the specification, the section "Requirements", was written in terms of parameters and concepts related to the physical properties of the object to be manufactured. These would range from a simple parameter, such as mass, to a complex one, such as corrosion resistance, but they would all be quantified and verified in terms of well-defined measurements. This situation changes significantly when a specification is to convey information between two stages of a design process, particularly systems design. Not only are the hardware-related concepts no longer adequate or applicable, but also, because the design process itself differs from a manufacturing process, the information should be presented differently. In other words, the structure of the specification should reflect the structure of the process it applies to.

The problems associated with this change were most dramatically demonstrated in the area of software engineering in the late 1960s and early 1970s, when the trend in the cost of producing and maintaining software threatened to bring the whole computer-based industry to a grinding halt. Because of this high visibility great effort was put into improving the situation, and as a result software specifications have taken great strides forwards. However, system specifications in general have seen little improvement. Most system specifications are still nothing but great amalgamations of slightly stripped-down equipment specifications, and the results of the system design process are often correspondingly unsatisfactory.

4.1.2 Purpose and function

The **purpose** of an element can only be defined with respect to the system in which it is embedded. For example, a particular resistor may in one circuit be the collector load resistor in a transistor amplifier, whereas in another circuit the same resistor may form part of a spark-suppressing network, or a voltage divider, and so on. In all these applications however, the resistors have something in common, namely the inherent property of being a resistor, of providing a linear relationship between voltage and current. This is called their **function**. The function can in this case be parametrized by a single value: the resistance value.

The process of going from purpose to function is another example of the process of **abstraction**. It is exactly the same as is involved in going from the physical reality of two oranges or two apples to the concept of the number two. Mathematics is the manipulation of numbers without any reference to what they represent; and analogously, resistance values may be manipulated without any reference to their purpose.

The process of abstraction is central to the functioning of the human brain, and the study of it has formed an important part of people's endeavor to understand themselves and their relationship to the world around them. From Plato's doctrine of ideas through Spinoza's concept of substance to the "thing-in-itself" of Kant's transcendental dialectic, the philosophical foundations of abstraction have been expanded and refined. On the other hand, a particular abstraction, that of quantity, has been expanded from its earliest beginnings as simple arithmetic to the present rich and complex subject of mathematics. However, abstraction and reasoning comprise only one-half of the activity of increasing knowledge; the conjugate part is observation and experimentation. Aristotle was an early protagonist of putting observation above contemplation, but it was really the technology-driven advent of sophisticated measuring apparatus from the middle of the last century onwards that gave a decided slant toward experimental science. The operational school of thought, saying in essence that the only reality is whatever can be defined in terms of an experiment, became dominant, particularly after its vindication in such fields as relativity and quantum mechanics.

Engineering has not escaped this trend in the sciences, and the last 50 years have seen an almost explosive development in the performance and sophistication of devices and equipment. As a consequence, engineers today are in a situation where they have more technology than they know how to apply. Landing on the moon, putting 100,000 transistors on a single chip, and creating new living organisms are all being done, but solving urban traffic problems seems an impossible task. The reason is to be found in the fact that the latter problem concerns a very complex system, and the methodology and theoretical foundations for the engineering of systems have, in the heady rush towards new technology with its more immediate and visible results, been allowed to fall far behind. The proper application of the process of abstraction, which was so essential to the development of the foundations of technology, has been forgotten or neglected. Instead of defining and working with the functional elements of a system, engineers try to work with elements defined by their purpose—an almost impossible task due to the proliferation and variability of the latter.

At the end of any one step within the systems engineering process, a system consists of a certain number of interacting functional elements. During the following step each of these elements is treated as a subsystem and is further partitioned into its functional elements. To explain the nature of this partitioning it is easiest to start with an example from the bottom end of the process, the component end, as this part of the process is familiar.

Consider a collection of resistors, capacitors and inductors—three groups of functionally different elements. The elements within a group are characterized by a single parameter (i.e. resistance, capacitance or inductance), the value of which is a number. On the next level up these elements are combined to form a low pass filter. Such a filter may, for given values of input and output impedance, be characterized by one parameter, namely the transfer function. For a given filter the value of this parameter is not a number but a particular function—a particular mapping between two sets of numbers. On the next level up this filter is incorporated into the control unit for a d.c. servomotor; it is used to suppress commutator noise in the motor voltage sensing circuit. Characterizing the control unit as a functional element will probably need a set of functions, higher order entities such as functionals, or a description in terms of operating **modes**. The latter type of description is common for equipment that is too complex to be described by a single set of functions.

The integration into higher and higher levels could continue on, but already at this stage several points have emerged:

1. Each level requires its own concepts in order to characterize the functional elements, and these concepts increase in complexity going up the hierarchy of levels.

2. The design on each level is done in terms of the functional elements, without any reference to the purpose of the elements. For example, the design of the low pass filter uses exactly the same techniques and equations as it would if the filter were part of an audio amplifier in a receiver.

3. At the lowest levels the choice of what constitutes a functional element is either obvious or well defined; at higher levels this partitioning may well depend on the purpose of the parent element on the level above.

The first point can be further illustrated by considering the development of integrated circuits, which has been a process of stepping up to increasingly higher levels within a single device. At SSI level it was necessary to define such elements as AND and OR gates, and to develop Boolean algebra and truth tables to describe them. At MSI level there were such functional elements as counters and shift registers, with such concepts as shift-right (or shift-left) and FIFO. At LSI level arithmetic functions could be realized. Finally, at VLSI level a single device became a sizable part of a computing system, described by high-level concepts such as vectored interrupts, DMA and pipelining.

The systems engineering process does not start at the bottom and work upwards; it starts at the top and works downwards—a characteristic that has earned it the name of **top-down design**. The starting-point is a need, a set of objectives and boundary conditions, or, in terms of the nomenclature introduced earlier, with the **purpose** of the system. By analyzing this purpose a number of **functional** elements (i.e elements of more general significance) can be identified. However, in choosing these elements and, in particular, the concepts that describe them, the systems engineer is guided by, and needs to refer to, the purpose of the system. Then in taking the next step and dividing each of the elements into sets of subelements of lesser complexity, it is necessary to refer to the purpose of the elements in order to make the right choice and find the right concepts or parameters to describe the subelements. Thus, not only does the process of top-down design become a stepwise partitioning, but also at each step it is necessary to develop the proper variables to describe the elements at that stage, and this requires reference to the purpose at each stage. The process becomes a back-and-forth between the **project domain**, in which the purpose can be stated, and the **abstract domain**, where the system exists as a set of functional elements and their interactions, with the **concurrent development** of the system variables at each stage. Only in the abstract domain is it possible to work on and manipulate the system (e.g. build models, calculate system behavior, optimize performance).

This process is illustrated in Figure 4.1. It must be realized that the words "purpose" and "function" refer to successively smaller or less complicated elements as the process progresses downwards. The diagram also shows how at some stage, in this case at the end of stage 3, the complexity has been reduced to a level where the elements can be characterized in a general manner without any reference to the purpose. From here on the design can proceed wholly within the abstract domain, and it is the point at which system design comes to an end and conventional, bottom-up design of the system elements takes over.

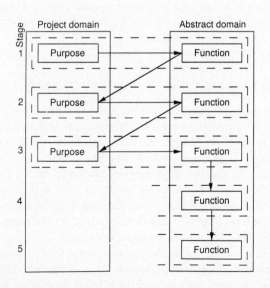

Figure 4.1   System design process: a stepwise process alternating between the project domain and the abstract domain.

4.2 Complexity as a limit to generality

4.2.1 Parameter space and mental capacity

From the discussion in the last section it should be clear that there is an important transition point in the engineering process, namely where it is possible to operate entirely with generally defined concepts and parameters. Where this point occurs depends both on the maturity of the particular field of technology and on the extent to which boundary conditions or assumptions limit the complexity in any given case.

However, there is another factor that influences the partitioning process in a very significant manner: the capacity of the human brain to handle a number of variables at one time. For something to appear to the brain as an entity its characteristics must be described by a limited number of parameters and their relations. This limit is not sharply defined and probably varies considerably from person to person, but most engineers will begin to have problems once the number of parameters reaches something like half a dozen. At this stage they will automatically try to partition the object into two (or more) parts and to consider each part separately.

This limit on mental capacity has at least two consequences. One is that just partitioning a complex problem into less complex parts, even without doing any work in the sense of taking decisions, already **appears** as an overall simplification. It is like trying to eat a watermelon without cutting it up; it is very difficult (and messy) due to the discrepancy between the size of the watermelon and the size of the mouth. Cutting up the watermelon into bite-sized pieces makes it much easier, even though the total volume to be devoured and digested remains the same.

A second consequence is that the concepts themselves cannot be too complex if they are to be useful. For example, if it takes 50 parameters to define a concept such as grade of service, this is not a concept that can be thought about (i.e. handled in one's head). Thinking about it would take at least pencil and paper; subconcepts would have to be defined. This need to structure the set of concepts can be a very significant factor in determining the overall system structure.

4.2.2 Significant parameters and the use of approximations

From the foregoing it may appear that it is necessary to carry out the whole partitioning of a system, down to the level where the elements can be handled as single entities, before a start can be made on doing some work. However, this would negate the main benefit of the systems engineering methodology and is not true. There is of course a second way to reduce the number of parameters associated with a single entity, namely to restrict the questions asked about that entity to such a degree that only a small number of parameters are **significant**. The restrictions fall roughly into two groups: those that restrict the **level of detail**, and those that restrict the **range of applicability** of the results, although this distinction is sometimes blurred.

An example of a restriction in the first group comes from thermodynamics; instead of the six parameters per molecule in a volume of gas, the number of parameters can be reduced to three (i.e. volume, pressure, temperature) if there is no wish to know about the behavior of individual molecules. An example in the second group is the small-signal behavior of a transistor; limiting the applicability to signals that are small compared to the saturation values results in a model described by only a few parameters.

Common to both groups, however, is the fact that the resulting models and concepts are **approximations**, which must be used with the appropriate reservations. The intelligent use of approximations is a significant aspect of the systems engineering methodology; and as it is difficult to formulate generally valid rules for how to find good approximations in any given case, it is an aspect that gives systems engineering the flavor of an art.

4.3 Implications for the engineering process

4.3.1 Need for purpose-related definitions

As was mentioned earlier, general functional concepts are not very useful at system level. As an example, take the concept of system reliability. All would have an intuitive idea of what it means—high reliability is good and low reliability is bad—but aside from such platitudes there is really very little that can be done in an operational sense with this concept. Trying to specify it more precisely, but in such a manner that it is applicable to any system, is a hopeless endeavor. This is obvious in the case of such disparate systems as a society (yes, societies do "fail", as any history book bears witness to) and an electricity supply system; but even for two technical systems, such as a telephone system and the braking system in a car, it becomes very difficult to find a definition of system reliability that applies equally well to both.

The only aproach to this problem is to relate the definition to the purpose of the system (i.e. to the particular application). However, such a definition will again use other concepts (less complex or on a lower level), which also must be defined for the particular case, and so on, until definitions are arrived at that use commonly accepted terms that are valid for any case. Here lies a well-known problem, namely, that the "commonly accepted" or "universally understood" meaning of a concept depends very much on the background of the reader and on the wider context in which it is embedded.

These so-called **pragmatic implications** have always been a source of difficulties in writing specifications. On the one hand, there are those implications that the reader can reasonably be expected to make by virtue of his or her expertise and involvement in the area of engineering being addressed by the specification. While this may not be too serious a problem for equipment specifications and other specifications concerned with a relatively narrow and well-defined field of specialization, it becomes most significant in system specifications that have a strong multidisciplinary character and does require careful consideration of the reader profile. On the other hand, there is the more immediate project-specific context in which a specification is embedded, such as contracts, work in earlier phases and other specifications within the same project. Both of these problems are considerably reduced by making specific detailed reference to the purpose of the particular system.

4.3.2 Structure of system specifications

The structure of a specification should mirror the structure of the system it specifies. In the case of a system specification the object is, on the one hand, a stage of the design process and, on the other hand, the system itself. Consequently, the specification should mirror the duality illustrated in Figure 4.1 as well as the partitioning process (Aslaksen 1987). A suitable structure for a system specification could therefore be as follows:

0. Title page and table of contents
1. Purpose
2. Supporting information
3. System level requirements
4. Functional elements
5. Quality assurance provisions
6. Appendix

The section "Purpose" contains the following information:

- the complete and correct name of the system as well as of the host system into which it fits (i.e. the next higher level);
- a narrative description of how the system relates to the host system and any other entity that influences its performance—this is therefore also a description of the system;
- a narrative description of the system's purpose;
- a list of any assumptions or boundary conditions related to the system's purpose.

The section "Supporting information" consists of such items as:

- applicable documents (i.e. standards, other specifications, any other documents referred to in the text of the specification);
- definitions of terms and concepts that are peculiar to the field of application, or with which the reader is not expected to be familiar, or that are commonly used with different meanings;
- abbreviations—any abbreviation used anywhere in the text of the specification should be written out in full here.

The section "System level requirements" contains the descriptive definition of all input and output parameters of the system. Where a relatively direct relation exists between an output parameter and one or more of the input parameters, this should be included in the definition of that output parameter. Otherwise the definitions will usually be described in terms of the host system; that is, the input parameters are described by showing how or where they are generated, and the output parameters are described by showing how they influence the host system. Although descriptive rather than mathematically rigorous, these definitions should be as complete and exact as possible. Any terms or auxiliary parameters introduced here must also be clearly defined.

The section "Functional elements" constitutes the bulk of the specification. It would normally be structured in an introductory subsection, "4.1 Overview", giving the breakdown of the system into functional elements, their general characteristics and how they are related, and then a subsection devoted to each element. Such a latter subsection would be structured as follows:

4.x.1   Definition of variables
4.x.1.1   External variables
4.x.1.2   Internal variables

4.x.2   Relations
4.x.3   Test requirements

The functional elements exist in the abstract domain and as such must be defined with mathematical precision. Such a definition of a single functional element can be quite extensive and may require several pages. It is exactly at this point that the systems engineering process often breaks down, as providing formal definitions is considered difficult and/or time-consuming. The engineer may be satisfied with descriptive definitions only, which may be ambiguous and preclude any optimization and verification, as was mentioned at the end of Section 4.1.2.

The relations that define the functions of the elements will contain a number of parameters whose values depend on the particular application. These values are determined in the first part of the next step in the design process, when each functional element is restricted, specialized and optimized in the context of the project domain. However, in the specification the parameters involved in the relations between the inputs and outputs of the elements remain abstract, reflecting a validity and applicability that transcend a particular purpose.

When formulating relations for elements in a given stage of the design process, care must be taken to keep the depth or detail of the relations relatively uniform over all the elements. A useful rule is that the depth of the functional definition should be just adequate to distinguish unambiguously the elements from each other. Anything more brings the process out of step and causes confusion.

The interpretation of the terms "variables" and "relations" above must be broad, as the nature of the system under consideration can vary widely, with a correspondingly broad range of element types. Even within a single system there will normally be many different types of elements, such as hardware, software, documentation and activities (i.e. procedures), and the variables will be different for each type. For example, documentation will be characterized by such variables as reading grade, user matrix, availability, format and of course contents, whereas the relations will essentially be the ordering of the contents.

The section "Quality assurance provisions" is concerned with all tests, inspections, reviews and other procedures deemed necessary to ensure that the requirements of the specification, as defined under Section 4 of the specification, have been met once the next stage has been completed. In other words, the quality assurance provisions impose requirements on the way the work in the next stage is carried out—how it is managed, controlled and verified.

The "Appendix" contains any material that does not express a requirement but is deemed to be particularly relevant and helpful in carrying out the work in the next stage.

4.4 Summary

1. Because the systems engineering methodology is a stepwise process, **specifications** are particularly important. They transmit the results of the work done in one step into the next step, forming the basis for the work carried out there.

2. To be effective **interfaces**, specifications must be **complete** and **comprehensible** to the reader.

3. The **function** of an element is its characterization in itself, independently of what it is used for, whereas the **purpose** is related to an application.

4. **Generality** of a concept at the system level is only possible by limiting either the **level of detail** or the **range of applicability**; otherwise the **complexity** would make the concept impractical.

5. System level specifications need a special structure because the **system parameters** need to be **defined** for each particular application.

4.5 Short questions

1. What is the most significant difference between a system level specification and an equipment (or device) level specification?

2. Name two requirements that a specification must fulfill. In the light of these requirements, how could the effectiveness of a specification be defined?

3. How are purpose and function of an object related?

4. How can approximations be said to fall into two major groups?

4.6 Problems

Combustion optimization: Consider the combustion optimization case study, and imagine that you have carried out the work that belongs in the analysis phase up to the point where you are ready to write the system specification. Irrespective of the results of that work, how would you structure the system specification? Write out the headings (i.e. Table of contents) for the first two levels of the contents breakdown.

The system specification must define the system performance. What would be two of the most important system parameters, and how could they be defined for this system?

4.7 References

Aslaksen, E.W., "Systems Engineering and System Specifications", *Journal of Electrical and Electronics Engineering*, Australia, IE Aust. & IREE Aust., vol. 7, no. 3, 1987, pp. 159–65.

MIL-STD-490A, *Specification Practices*, Washington, DC: US Dept. of Defense, 1985.

IEEE Std. 830, IEEE *Guide for Software Requirements Specifications*, New York: IEEE, 1984.

Supman I(A), *Defence Standardisation Manual*, Canberra: Australian Department of Defence, 1978.

CHAPTER 5

Probabilistic nature of systems engineering

5.1 Nature of uncertainty in systems engineering

5.1.1. Sources of uncertainty

The partitioning of a system into smaller and smaller entities during the systems engineering process is not just a dividing up, in the sense of cutting a cake into smaller and smaller pieces. As was emphasized in Chapter 2, decisions are taken at every step in this process, and work is carried out, in the form of modeling and data gathering, to support this decision-making process. However, for several reasons there is inevitably an amount of **uncertainty** associated with these decisions, more so in the early stages and less as the process progresses.

As an aside, it is interesting to note that this uncertainty gives systems engineering a certain management flavor that is not present in, say, equipment engineering. This follows from the fact that it is the element of uncertainty that makes decision-making a management task. For example, no activity could involve decision-making more intensely than mail-sorting, but it is hardly a management activity because it involves very little uncertainty.

The reasons for the uncertainty inherent in the systems engineering process are:

1. Due to the long duration of major projects, systems are designed to perform optimally under conditions that will prevail in many years' time. Estimates of such parameters as interest rate, market volume and the state of technology must be assigned confidence intervals, which increase with distance into the future.

2. Some parameters may simply not be known in detail but known only as averages. They are subject to fluctuations, and suppressing the effects of these fluctuations results in a degree of uncertainty in the results obtained on the basis of the averages alone. Typical examples are ambient temperature and failure statistics.

3. Some of the data on which decisions are based are incomplete at the time the decisions have to be taken. This is especially true in the very early stages of the systems engineering process, and the **risk** involved in not waiting for the data to be completed has to be weighed against the benefit in accelerating the project. This is the classic dilemma involved in **fast-tracking**.

5.1.2 Lack of knowledge versus simplification

The above sources of uncertainty are all due to a **lack of knowledge**, either in the form of an intrinsic lack, as in knowledge of the future, or due to the fact that the data are not available or were not collected in the past and are therefore irretrievably lost. In all cases this is therefore something that cannot be changed. It is a question of making the best out of what is available and being able to realize what the degree of uncertainty is and what its effect is.

A completely different type of uncertainty is that arising from **simplifications**. A typical situation would be when it is necessary to determine the behavior of a system that consists of a very large number of elements. For simplicity assume that the elements are all equal, although this is by no means necessary. The behavior of each element is perfectly well known, say, in the form of differential equations for its various parameters, and the parameters characterizing the behavior of the system have also been defined in terms of the element parameters. In principle the whole set of simultaneous differential equations could be solved, but in practice this would usually be hopelessly costly. Instead a new set of system parameters is defined where, in the definition of any particular parameter, the element parameters occur in the same combination, but instead of now summing over all the individual elements, one sums over all possible parameter value sets for one element, weighted by the probability of a set's occurring. If the number of elements were infinite, the two definitions would be equivalent; but for a finite number of elements, the true value of the system parameter will at any point in time differ from the value predicted by the simplified definition. Conversely, if the simplified definition is used, there will always be some uncertainty about what the true value is.

This is illustrated in Figure 5.1. Part (a) shows the temporal behavior of a particular system parameter $y(t)$ and the value of the same parameter defined in terms of an average $\bar{y}(t)$. At any particular point in time there is a certain probability that the true value will differ from the average value by a certain amount; the true value $y(t')$ is **distributed** around the average value $\bar{y}(t')$, as shown in (b). So, while uncertainty is introduced by using averages, this is not due to any lack of knowledge, and the degree of uncertainty, as given by the distribution in (b), is well-defined. This case is significantly different from the case where the uncertainty arises from a lack of knowledge, and this difference is of major importance in understanding the nature of **risk**, as will be discussed briefly in Section 5.4.1 and more fully in Section 7.4.

Although averaging is a common form of simplification, it is by no means the only one. Another is to ignore higher order effects, as in perturbation theory, and a further one is to ignore certain effects or parameters considered to be irrelevant to the system parameter(s) in question. In all forms of simplification, however, the comments made above about the nature of the uncertainty apply.

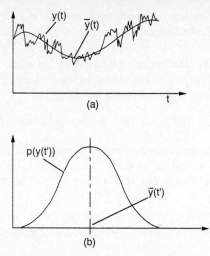

Figure 5.1 Two presentations of a stochastic variable: (a) in real time, (b) as a probability density distribution.

5.1.3 Some simple rules

Whenever there is a lack of knowledge, assumptions have to be made, and this is often done by assigning probabilities to possible parameter values. Assigning probabilities to any model used in decision-making must by definition involve personal judgement. It will be influenced by the engineer's previous experience, interpretation of the present situation and expectations for the future. However, it is sometimes overlooked that probability theory provides a few simple rules and relations that must be satisfied for a set of probabilities to be consistent.

The simplest rule is that the **sum of the probabilities** for all (mutually exclusive) outcomes of an experiment or a given situation must equal 1; that is,

$$\sum_{i=1}^{n} P_i = 1 \quad \text{or} \quad \int_{-\infty}^{\infty} p(x) \, dx = 1$$

Even if only one particular outcome is of interest, it is usually prudent to list all possible outcomes and assign probabilities to them also, just to ensure that the outcome of interest is not given an inconsistent value. Intuition can be deceiving in that a whole class of outcomes can easily be overlooked simply because they are irrelevant to the problem at hand. For example, in talking about two outcomes the expression "it's fifty-fifty" **may** mean that there is a 50% chance of either outcome occurring but it may also just mean that the chance of one outcome occurring is the same as the chance of the other occurring; the actual probabilities may be much less than 0.5 if other outcomes also are possible.

Two further rules arise in connection with **conditional probabilities**. Conditional probabilities play an important role in all applications of statistics and estimating, because the probability of an event is almost always dependent on a number of conditions, such as other events or parameter values. For example, the probability of somebody contracting a particular disease may depend on age, sex, income level and so on; the outcome of a test may depend on environmental conditions, the selection of the test objects and so on.

To review briefly the theoretical background, consider the simplest case of a population whose members are characterized by two parameters u and v. Picking one member at random, the following questions can be asked:

1. What is the probability that u will have a value between u_0 and $(u_0 + du)$? And the same question for v.
2. What is the probability that, having already ascertained that v has the value v_0, u will have a value between u_0 and $(u_0 + du)$?
3. What is the probability that u will have a value between u_0 and $(u_0 + du)$ **and** that v will have a value between v_0 and $(v_0 + dv)$?

The answer to the first question is the simple (or common) **probability** $P[u_0 < u < (u_0 + du)]$, which is expressed in terms of a **probability density function** f(u) as

$$P[u_0 < u < (u_0 + du)] = f(u_0)\, du$$

and similarly for $P(v_0)$ of course. The answer to the second question is the **conditional probability**, $P[u_0 < u < (u_0 + du)|v_0]$, which can be expressed in terms of a **conditional probability density function** f(u|v), as

$$P[u_0 < u < (u_0 + du)|v_0] = f(u_0|v_0)\, du$$

The answer to the third question is the **joint probability** $P[u_0 < u < (u_0 + du)\ v_0 < v < (v_0 + dv)]$, which can be expressed in terms of **joint probability density function** f(uv) as

$$P[u_0 < u < (u_0 + du)\ v_0 < v < (v_0 + dv)] = f(u_0 v_0)du\, dv$$

All of the above holds equally well of course for the case where u and/or v is a discrete variable, but the notation varies a little:

$$f(u_i) \Rightarrow P_i$$
$$f(u_i|v_j) \Rightarrow P_{i;j}$$
$$f(u_i v_j) \Rightarrow P_{ij}$$

To abstract from the type of variable involved it is convenient to introduce the concept of an **event**. An event can be that u lies between u_1 and u_2, that $u = u_i$, or any other condition on the value of a variable. It will be denoted by e_i, with i belonging to an appropriate set. This concept is closely related to one introduced earlier, namely outcome; an outcome may be a single event, but it may also be a combination of events.

Events may be **mutually exclusive**, as is the case with the two events of getting a three and getting a six on a single throw of a die. The conditional probabilities of mutually exclusive events are zero:

$$P(e_i|e_j) = P(e_j|e_i) = 0$$

Events may be **independent**. Two events e_i and e_j are independent if the probability of the one occurring is the same whether the other has occurred or not:

$$P(e_i|e_j) = P(e_i)$$

An example of independence is given by the two events of drawing an ace and drawing a black card from a deck of cards; the conditional probability of drawing an ace given that the card is black is the same as the probability of drawing an ace, namely $1/13$.

Now consider a set of n events, e_1, e_2, . . . , e_n, with the associated probabilities $P(e_i)$. Then there exists a set of $(n^2 - n)$ conditional probabilities of the form $P(e_i|e_j)$, called the **cross-impact matrix**. There are two rules that impose **consistency conditions** on this set.

The first rule is the **rule of the product**:

$$P(e_i e_j) = P(e_i)P(e_j|e_i)$$
$$= P(e_j)P(e_i|e_j)$$

This can be rewritten in the form

$$\frac{P(e_i)}{P(e_j)} = \frac{P(e_j|e_i)}{P(e_i|e_j)}$$

which can be interpreted as asserting that, if $P(e_j|e_i)$ and $P(e_i|e_j)$ are given, the point in the $[P(e_i),P(e_j)]$-plane determined by $P(e_i)$ and $P(e_j)$ must lie on a straight line through the origin with slope $P(e_j|e_i)/P(e_i|e_j)$.

The second rule is the **rule of the triangle**:

$$P(e_i|e_j)P(e_k|e_i)P(e_j|e_k) = P(e_j|e_i)P(e_k|e_j)P(e_i|e_k)$$

This rule states that, for a set of three events represented as the corners of a triangle, the product of the conditional probabilities multiplying around the triangle in one direction is equal to the product of the conditional probabilities going in the other direction.

The rule of the triangle is used to check on the consistency of the cross-impact matrix in a given case and, if the matrix is found to be inconsistent, to adjust it appropriately. Once a consistent matrix has been obtained, the rule of the product is used as a guide in estimating the probabilities. The idea behind this procedure is that it is often easier to estimate conditional probabilities than absolute probabilities, both because a much smaller population of previous occurrences can be concentrated on and because the data available are often in the form of conditional occurrences of events.

5.2 States and transitions

5.2.1 Concept of a state

To ascertain the performance of a system and end up with a practical system specification, it is necessary, at some point in the analysis, to make the transition from a qualitative analysis dealing with concepts, functions and structures to a quantitative analysis in terms of measurable variables. Input and output variables were mentioned earlier. Now it is time to look at how the system transforms the inputs into the outputs, and how this transformation relates to the functions and structure of the system.

Any system (or subsystem) is described by certain sets of variables and certain functions, such as inputs U and outputs Y. A further and very obvious set is the set of time values T. This set can be discrete (i.e. . . . . t_n, t_{n+1}, . . .) or continuous. To express the relations between inputs and outputs a first thought might be to introduce some function h such that

$$h: T \times U \to Y$$

This would not be a very useful representation of the system behavior, however, as it does not distinguish between the effects of the inputs and the effects of the system proper (i.e. the physical system). Also, the inputs are in the form of time sequences of values, whereas what is needed in order to treat the system analytically and gain some insight into its behavior is a description in terms of input functions. These objections are eliminated by the introduction of the concept of a **state**.

Before going any further, note that the use of the word "state" in systems engineering is somewhat different from its use in network design and control theory. In the latter case the state of a system (or network) at a time t_0 is any amount of information that, together with any admissible input function specified for $t_0 < t < t_f$, is adequate to determine uniquely the output function for $t_0 < t < t_f$ for any $t_f > t_0$. In other words, the state variables act as a kind of memory of the past behavior of the system; they include the initial conditions at $t = t_0$. In the case of systems engineering the "state" refers not to a new set of variables but to the system configuration or to the condition of its components. Different states correspond to different operating modes.

Using the concept of the state of a system, the engineering is separated into two parts. One part consists of determining how the system performs in each state (i.e. what are the relations between input and output variables in each state). The second part consists of determining how the system changes state. The latter may take place as a function of the inputs (e.g. putting the defense system in a high alert state if war breaks out in a neighboring country) or because of noise and random influences (e.g. failure of a component).

5.2.2 Element states and observables

The specification of an element (i.e. the characterization of its performance as seen from the outside) may contain parameters that can take on more than one value. Then *a particular set of parameter values identifies a state of the element.* Another way of defining the concept of the state of an element is to say that *the two elements are in the same state if and only if they appear alike to all allowed (i.e. considered) interactions.* Either formulation expresses the fact that the set of states E associated with an element is defined with respect to a set of interactions; and as increasingly detailed interactions are considered, the set E may contain more states. This is analogous to the energy levels of an atom; the number of states resulting from electrostatic interaction is increased only when magnetic coupling and spin are taken into account.

The set E may contain a finite number of states, a countably infinite number of states, a continuum of states, or a combination of these. A simple but important example of a finite set is a set with only two states,

$$E = \{e_0, e_1\}$$

where e_0 corresponds to the failed state and e_1 to the operating state. An example of a continuum is given by the bit error rate (BER) of a communication link, which can take on any value between 0 and 1.

5.2.3 System states

In the top-down process of systems engineering, what is an element at one level becomes a (sub)system at the next lower level. Thus the previous definition can obviously be extended to a system; the state of a system is *its characterization with respect to a subset of the boundary conditions*. This definition reflects the fact, mentioned in Section 1.2, that the boundary conditions fall naturally into subsets and that usually only two of these, the inputs and outputs, correspond to the interactions on the next level up in the top-down decomposition.

The **basic system states** are the points in a set that is the direct product of the sets of element states. Let a system consist of n elements, and let E_i ($i = 1, \ldots, n$) be the sets of element states (for the given interactions). Then the set B of basic system states is given by

$$B = E_1 \times E_2 \times \ldots \times E_n$$

and a basic system state b_j can be represented by an n-dimensional vector,

$$b_j = e_1, e_2, \ldots, e_n$$

where e_i is an element of E_i.

The basic system states reflect the allowable interactions between elements of the system. The **system states** arise out of these as a consequence of interactions between the system as a whole and the outside world. Let U be the set of interactions with respect to which the system states are defined. Then U is a subset of the allowable interactions between elements as well as a subset of the boundary conditions. Under U many of the basic system states b will be equivalent to each other. The set B is decomposed into equivalence classes, each of which represents one system state, denoted by s_i.

The formation of equivalence classes can be looked upon as degeneracy due to the limited interactions contained in U. As more interactions are allowed in U, the degeneracy is gradually lifted; and when U equals the set of all allowable interactions between the elements (i.e. supported by the elements), the system states are identical to the basic system states.

The states introduced so far are all so-called **pure** states; each state can be characterized by definite values of certain interactions. However, in many cases (e.g. when there are frequent transitions between pure states due to a random phenomenon such as component failure) the engineer is interested only in the average state. Such a state, called a **mixed** state P, is a vector in the n-dimensional space spanned by the system states s_i ($i = 1, \ldots, n$),

$$P = P_1, P_2, \ldots, P_n$$

with P_i being the probability of finding the system in the pure state s_i and therefore

$$\sum_{i-1}^{n} P_i = 1$$

To see the difference between pure and mixed states, consider a system consisting of two elements operating in parallel. Under the interaction of interest the elements are either operational or failed, and therefore each is characterized by two states e_1 and e_0. Consequently there are four basic system states:

$$b_0 = (e_0,e_0)$$
$$b_1 = (e_0,e_1)$$
$$b_2 = (e_1,e_0)$$
$$b_3 = (e_1,e_1)$$

However, as the interaction only distinguishes between operating and failed, there are only two system states:

$$s_0 = b_0$$
$$s_1 = \{b_1,b_2,b_3\}$$

The degeneracy of b_1, b_2 and b_3 into s_1 is caused only by the choice of interaction, whereas the introduction of the mixed states,

$$P = (P_0,P_1)$$

or

$$P(x) = (x,(1-x))$$

reflects the fact that the engineer is not interested in the detailed behavior, just the average behavior. The parameter x is obviously nothing but the unavailability.

At this stage it is important to note the following point about the relation between states and systems. From the discussion of the concept of structure in Chapter 3, it is clear that different states can also have different structures, as one of the parametrized characteristics of an element can be its ability to support a particular interaction (e.g. operating and failed state of that interaction). By definition these different states can then be said to be different systems. Therefore there is generally some choice as to what different states of the same system should be called, and what different systems should be called. In other words, what is in one case best treated as states of the same system may in another application be more conveniently treated as different systems. It all depends on the purpose to which the system is being modeled or the aspects the engineer wishes to study. The important thing is to maintain clear and unambiguous definitions.

5.2.4 Transition probabilities

The states of a system may in turn be looked upon as the elements in an abstract system. The relations between them or their interactions (i.e. whether it is possible for the system to move from state i to state j and how large the probability of this transition is) define a structure in a **state space**. The methods used for the physical system (e.g. graph theory) can now be used in state space.

As the most interesting transitions in the present context are those that occur as a result of random forces (i.e. component failures), it will be assumed that the systems can be in a finite number m of states at any given point in time. The probability of the system's being in state i at any fixed time is, as before, denoted by P_i. The **transition probability** p_{ij} is the probability that, if the system is in state i, it will be in state j after the next transition. Obviously,

$$\sum_j p_{ij} = 1$$

If the system is time-invariant, and if the transition probability p_{ij} is a function of the initial state (i.e. state i), but not of the past history of the system, then such a sequence of transitions is called a **Markov chain**.

Within this framework the evolution of the system is characterized by the number of transitions it has undergone, without any reference to the time it has taken. Thus, starting out in some initial state, it makes sense to ask what the probability is of finding the system in state j after n transitions. This probability is denoted by $P_j(n)$, and the probability distribution over all states is represented by a column vector:

$$P(n) = \begin{matrix} P_1(n) \\ \cdot \\ \cdot \\ \cdot \\ P_m(n) \end{matrix}$$

with

$$\sum_{j=1}^{m} P_j(n) = 1$$

The vector $P(n)$ is related to $P(0)$ by

$$P(n)^T = P(0)^T \, \mathbf{P}^n$$

where $\mathbf{P} = p_{ij}$ is the **transition probability matrix** and $\mathbf{P}^n$ is called the **n-step transition probability matrix**.

5.2.5 Equilibrium states

Of particular interest in the analysis of systems is the **steady state** or **stationary** distribution

$$q_j = \lim_{n \to \infty} P_j(n)$$

Most electrical systems exhibit such a steady state distribution by virtue of the fact that they are maintained. Worn or failed parts are replaced, and after an initial settling-in period the system consists of a mixture of old and new parts. Then

$$Q^T = Q^T P$$

or

$$q_j = \sum_{i-1}^{m} q_i p_{ij} \quad j = 1, \dots, m$$

which constitutes m equations with m unknowns, say,

$$
\begin{array}{llll}
q_1(p_{11} - 1) + q_2 p_{21} & + \dots + q_m p_{m1} & = 0 \\
q_1 p_{12} & + q_2(p_{22} - 1) + \dots + q_m p_{m2} & = 0 \\
\quad \cdot & \quad \cdot \\
\quad \cdot & \quad \cdot \\
q_1 p_{1(m-1)} & + q_2 p_{2(m-1)} & + \dots + q_m p_{m(m-1)} = 0 \\
q_1 & + q_2 & + \dots + q_m & = 1
\end{array}
$$

Let **R** be the matrix of coefficients in the above set of equations, and let B be the column vector on the right-hand side of the equality signs. Then

$$Q = \mathbf{R}^{-1}B$$

The Markov chain provides a convenient and easy way to visually represent system behavior, but it ignores time as a variable. Most systems of interest in electrical engineering exist within a continuous time frame; that is, they exist continuously in one system state until a transition occurs to another state, in which they exist continuously until a further transition occurs, and so on. Instead of a Markov chain there is a **Markov process**, and instead of transition probabilities there are **transition rates**. A transition rate is defined as *the number of times a transition occurs from a given state, divided by the time spent in that state.*

This situation can be handled in one of two ways. Firstly, the notion of a Markov chain can be retained by a mental trick; namely, one imagines that the system makes one transition per unit time. Most of these transitions will be from one state to the same state (i.e. a loop), which simply reflects the fact that the system makes a "real" transition only after a number of time intervals. This allows the matrix notation, which is convenient for finding the steady state distribution, to be retained. Secondly, one can operate in real time. In this case the problem is formulated in terms of differential equations, and it is possible not only to obtain the steady state but also explicitly to show transient effects.

5.2.6 Example: a two-state system

As a very simple illustration of the above, consider a system with only two states: s_1 = operating, s_2 = failed. Let the mean time between failure MTBF equal 10,000 h and the mean time to repair MTTR equal 10 h as illustrated in Figure 5.2. Then the **steady state** behavior can be found using

$$\mathbf{P} = \begin{bmatrix} 0.9999 & 0.0001 \\ 0.1 & 0.9 \end{bmatrix}$$

and

$$\mathbf{R} = \begin{bmatrix} -0.0001 & 0.1 \\ 1 & 1 \end{bmatrix}$$

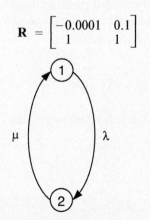

Figure 5.2 State diagram of a two-state Markovian system.

It follows that

$$\mathbf{R} = -0.1001$$

$$\text{adj}(\mathbf{R}) = \begin{bmatrix} 1 & -0.1 \\ -1 & -0.0001 \end{bmatrix}$$

and

$$\mathbf{R}^{-1} = \begin{bmatrix} -10 & 0.999 \\ 10 & 0.001 \end{bmatrix}$$

This gives the result

$$\mathbf{R}^{-1}\mathbf{B} = \begin{bmatrix} 0.999 \\ 0.001 \end{bmatrix}$$

which is exactly the result obtained from the two expressions

$$P_1 = \text{MTBF}/(\text{MTBF} + \text{MTTR})$$
$$P_2 = \text{MTTR}/(\text{MTBF} + \text{MTTR})$$

To carry out the same calculation using a **continuous** time frame, let the two transition rates be designated g and h, with

$$g = 1/\text{MTBF}$$
$$h = 1/\text{MTTR}$$

For time interval dt then

$$P_1(t + dt) = P_1(t)(1 - g\,dt) + P_2(t)h\,dt$$
$$P_2(t + dt) = P_2(t)(1 - h\,dt) + P_1(t)g\,dt$$

In the limit dt $\rightarrow$ 0, these become

$$P_1'(t) = -gP_1(t) + hP_2(t)$$
$$P_2'(t) = -hP_2(t) + gP_1(t)$$

With the boundary condition $P_1(0) = 1$, the solution to this pair of differential equations is

$$P_1(t) = \frac{h + g\exp[-(g + h)t]}{g + h}$$

$$P_2(t) = \frac{g - g\exp[-(g + h)t]}{g + h}$$

The **steady state**, obtained for large values of t, is

$$P_1 = h/(g + h)$$
$$P_2 = g/(g + h)$$

which is exactly the earlier result.

5.3 Statistical models

5.3.1 General

Modeling is a basic activity in the systems engineering process. It will be discussed in some detail in Section 6.3, and models appear in various places throughout this text. In many of these models statistical aspects are important, and it is appropriate to look briefly at these here, in particular with respect to the concepts introduced earlier in this chapter.

Any model can be looked upon as something that transforms a set of inputs into a set of outputs; the two sets are related by the transfer function of the model. If this transfer function is statistical in nature (i.e. formulated in terms of probabilities), then it is a **statistical model**. Two classes of such models are particularly important: performance models and decision models. The fact that they are different arises from the existence of the two types of uncertainty discussed in Section 5.1.2.

5.3.2 Performance models

Returning for a moment to Chapter 4, two matters discussed there become very relevant in the present context. Firstly, the concepts used in the higher levels of the systems engineering process (i.e. the parameters that characterize systems at these levels) need to be defined for each particular system. Secondly, to be convenient to operate with, such concepts must abstract from the myriad of detail present in a complex system, and this process of creating system level concepts out of the properties of the elements most often involves some sort of averaging.

Performance parameters are therefore not simple entities. When the value of such a parameter is presented as a simple function, or even a single numerical value, this is only possible by reference to a particular **performance model** for that parameter.

5.3.3 Decision models

Decision-making involves choosing between two or more **alternatives**. The alternatives are compared, and the basis of this comparison is called the **decision criterion**. Such a decision criterion consists of two principally different parts: a function of certain parameters related to the alternatives, and the sets of values of these parameters for each alternative. The criterion therefore has the character of a model, with the parameter values as inputs and the decision as the output. However, two points merit special mention, and they make decision models somewhat different from other models, such as performance models.

Firstly, as the same function will be applied to all the alternatives, it is necessary to ascertain that the function is valid and makes sense **for all the alternatives**. For example, if the alternatives represent different technologies, it is no good having a function that involves parameters that apply to only one or some of the technologies. As a general rule the function should not contain parameters that are internal to the alternatives; without this rule there is always a danger of having a decision criterion that is inherently biased towards a particular alternative (or set of alternatives).

Secondly, the sets of parameter values used for the alternatives must have been obtained under or must apply to **similar conditions**. For example, reliability values (e.g. MTBF) must be quoted for the same environmental conditions for each alternative. Just as statistics, with a little judicious manipulation, can be used to prove almost any proposition, so a sloppily defined decision criterion will allow the decision to be manipulated by using inconsistent data.

As a little illustration of the importance of the choice of criterion, consider that a stage in the analysis of a system has been arrived at where there exists a choice between three technologies in order to realize a particular element of the system:

A1:   existing technology, features and cost known;
A2:   technology under development, features known, cost uncertain;
A3:   technology to be developed, features able to be specified, cost very uncertain.

The cost of the element **within the system** consists of the total cost of developing, manufacturing and installing it, minus the benefit it brings to the operation of the system. So in this case the alternatives are characterized by one parameter, and for each alternative this parameter has three values according to an optimistic, neutral or pessimistic view, as follows:

	Optimistic	Neutral	Pessimistic
A1	100	100	100
A2	60	80	120
A3	40	70	200

Note that this is a case where it would be necessary to make sure that "optimistic", for example, means the same for each alternative, that the values have been derived in the same way using the same assumptions, and so on.

Several criteria are possible for choosing a "best" solution:

1. *Criterion of optimism (**minimin**):*

Alternative	Lowest cost
A1	120
A2	60
A3	40

Obviously, alternative A3 minimizes the cost for the most optimistic criterion; so this would be the alternative to choose in this case.

2. *Criterion of pessimism (**minimax** or **Wald's criterion**):*

Alternative	Highest cost
A1	100
A2	120
A3	200

This is the "worst case" criterion, and accordingly alternative A1 would be chosen.

3. *Criterion of rationality (**Bayes** or **Laplace criterion**)*: This criterion assumes that each of the future outcomes are equally likely and then minimizes the expected cost.

Alternative	Expected cost
A1	$(100 + 100 + 100)/3 = 100$
A2	$(60 + 80 + 120)/3 = 87$
A3	$(40 + 70 + 200)/3 = 103$

Using this criterion, alternative A2 is the best.

4. ***Hurwicz criterion:*** Let C be a **coefficient of optimism**, with $0 \leqslant C \leqslant 1$. The Hurwicz criterion is to minimize the quantity

$$H = C(\text{min. cost}) + (1 - C)(\text{max. cost})$$

There will of course be one such quantity for each alternative, and the values of these quantities will be functions of C, as shown in Figure 5.3. As a result the choice of alternative will depend on the value of C.

But how is this value to be determined? One method is to pose the following question to a suitable group of people (it may, for example, be included as one question in a Delphi questionnaire; see Section 6.2.1): "If you had two alternatives, A1 and A2, one of which has a cost X independent of your estimation of the future, and the other of which has a cost K_1 for an optimistic estimation and a cost K_2 for a pessimistic evaluation, what value of X would make you indifferent as to choice of alternative?"

As an example, let $K_1 = 60$ and $K_2 = 180$, and let the answer to the question be $X = 110$. Then indifference would mean

$$C(60) + (1 - C)(180) = 110$$
$$C = 0.58$$

Referring back to Figure 5.3, this value of C would mean that alternative A2 would be chosen.

It often happens that decisions are not a one-step process but actually consist of a sequence of interdependent decisions. A most useful approach to this problem is the use of **decision trees**, illustrated in Figure 5.4. Each point at which a decision has to be made is represented by a square, the alternative results of that decision are represented

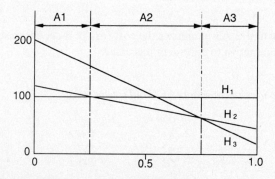

Figure 5.3 Hurwicz functions for the three alternatives.

by circles, and lines emerging from the circles represent the various distinct values that the variable (or variables) characterizing that result can take. Each value is labeled with the probability of its occurrence.

The optimal sequence of decisions in a decision tree is found by starting at the right-hand side, calculating the return at each circle by multiplying the individual return values by their probabilities and adding the results up, and then determining the expected return for the decision points (squares) by taking the highest of the values of the associated circles, until finally the expected return of the left-hand (initial) decision point is obtained.

5.3.4 Evaluation methods

Having constructed a statistical model and obtained the necessary input data, there then remains the task of evaluating the model (i.e. obtaining whatever output the model produces). This is most often a far from trivial task; even seemingly innocuous models soon have the evaluator involved in messy multidimensional integrals and the like. It is not the intention of this text to show how to solve problems once they have reached a mathematical formulation—that is what applied mathematics is all about—but it is probably useful to review quickly two methods that are closely connected with what was discussed earlier in this chapter.

The first of these involves making a transformation from the space of real time to **state space**. Solving statistical problems in real time very often leads to complex differential equations that do not have analytical solutions, but for a single static state of the system the solution may be relatively simple. In such a case the procedure is to solve the problem (i.e. obtain an output from the model) for each system state, and then take care of the dynamics by considering the transition probabilities and the resulting probabilities of finding the system in the various states.

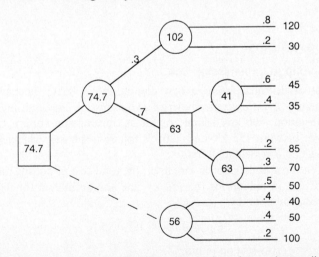

Figure 5.4 A decision tree, with probability values given for each possible outcome of a decision. The numerical values of the individual returns are given on the right. The values inside the squares (decisions) and circles (results of decisions) are the expected returns at those points. The broken paths are not chosen as a result of this analysis.

This procedure should strike a chord with electrical engineers, as it is the exact analog of the Fourier (or Laplace) transform used to determine the dynamic behavior of electrical circuits. Instead of soving complex differential equations in real time, a transform into frequency space is made and the (static) response for any one frequency is found. The response to a more complex waveform is then only the superposition of the solution for each spectral component. The frequency in this case serves exactly the same purpose as the variable labeling the states does in the case of the statistical model, and the spectral components are the counterparts of the states.

The second method is called the **Monte Carlo method**. It is a simulation method and has its theoretical roots in the ergodic theorem, which essentially says that the ensemble average equals the time average. In other words, instead of studying the behavior of a stochastic process in real time, it is possible to study the statistics of a collection of identical (static) systems whose parameters are distributed according to the long-term average of the parameters of the original process.

As a very simple example of this, consider a model that produces a single output y and contains one stochastic parameter x, which is equally likely to lie anywhere in the range from x_1 to x_2. The behavior of y could be studied by running the model in real time, using differential equations and with x making transitions within its range according to given transition rates; but this might not be a simple task, particularly if the model is nonlinear. Using the Monte Carlo method, a set of random x-values in the range x_1 to x_2 is generated, and then a static output value for each x-value in this set is generated. The distribution of y-values obtained in this manner approximates the distribution that would be obtained from solving the model in real time, with the approximation being more accurate the larger the number of x-values in the randomly generated set. The advantage is of course that the Monte Carlo method, relying on repeating the same calculation many times, is ideally suited to be processed by a computer.

5.3.5 Two examples

5.3.5.1 High-voltage transmission line

In the interest of decentralization a new community is being planned in a remote location. To supply it with electrical energy a transmission line is being planned from the nearest substation, 100 km distant. As far as transmission voltage U is concerned there is a choice between 132 kV and 220 kV (phase-to-phase), whereas the choice of copper cross-section is (essentially) free.

Excluding the cost of the copper wire itself, the total cost of putting the transmission line in place, C_1, is $2 million for the 132 kV line and $6 million for the 220 kV line. Copper wire costs are given by

$$C_2 = 0.05 \times 10^6\, A\ [\$]$$

where A is the wire cross-section in mm^2.

The power demand P is known to vary with the time of day according to the curve shown in Figure 5.5, and this curve is the same for every day of the year. The yearly energy demand E is not known, but it is assumed to be equally likely to lie anywhere in the range 100–400 GWh and to be the same every year.

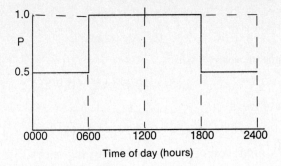

Figure 5.5 Daily demand for power P.

The problem is to decide whether to use 132 kV or 220 kV as the transmission voltage, assuming the following to be true:

1. The transmission line as well as the load are purely resistive.
2. The specific resistance of copper is 0.02 ohm mm^2/m.
3. The design lifetime of the transmission line is 20 years from the day it starts operation. The time it takes to build the line is ignored.
4. The discount rate p is constant over the 20 years, and its value is equally likely to lie anywhere in the range 10–20%.
5. The cost of electricity (to the electricity commission) is constant at 5 cents/kWh (i.e. $0.05/kWh).
6. The two variables E and p are statistically independent.

First of all a decision criterion must be formulated. In this case a very simple one will be used, namely: choose the solution whose total cost, referred to the start of operations (i.e. present value PV), has the greater probability of being the lowest. This is by no means the only criterion that could be chosen, and it ignores a number of factors that could be considered important, such as the voltage drop, the environmental impact of the tower height and so on. However, it does demonstrate that the criterion must be defined in terms of probabilities; it would not be correct simply to specify the lowest cost.

There are three cost factors involved:

- installation cost C_1, which is either $2 million or $6 million;
- copper cost C_2, which is 0.05×10^6 A (in $);
- cost of losses C_3.

To calculate the latter, let I be the maximum current and R the line resistance (per phase). Then the yearly loss L is given by

$$L = 3I^2R \times 15\ 365$$
$$= 16.5I^2R \quad [kWh]$$

But I is related to E by

$$E = 3UI \times 18\ 365$$

or

$$I = 88E/U \quad [A]$$

with E in GWh and U in kV. The resistance R is simply

$$R = 2 \times 10^3/A \quad \text{[ohms]}$$

with A in mm². Thus the yearly loss is

$$L = 2.55 \times 10^8 E^2/AU^2 \quad \text{[kWh]}$$

and the yearly cost of loss is

$$0.05L = 12.8 \times 10^6 E^2/AU^2 \quad \text{[\$]}$$

Finally then, the PV of the cost of losses over 20 years is given by

$$C_3 = 12.8 \times 10^6 E^2/AU^2 \, F(p)$$

where

$$F(p) = \frac{(1 + p)^{20} - 1}{p(1 + p)^{20}}$$

is the PV factor for a series of equal payments (see also Chapter 14).

Now, no matter which of the two transmission voltages is chosen, the value of A to use is the one that minimizes the cost. Thus the requirement is

$$\frac{d}{dA}(C_2 + C_3) = 0$$

which immediately yields

$$A = 16EF^{\frac{1}{2}}/U$$

With this value for A,

$$C_2 = C_3 = 0.8 \times 10^6 EF^{\frac{1}{2}}/U$$

Returning to the decision criterion, the boundary, in the (p,E)-plane, between where one transmission voltage provides the lower cost solution and where the other one does is of course the curve along which they are equal. This curve is therefore determined by setting the PV of the cost of the two solutions equal to each other:

$$2 + 1.6EF^{\frac{1}{2}}/132 = 6 + 1.6EF^{\frac{1}{2}}/220$$
$$E = 825F^{\frac{1}{2}}$$

The corresponding curve is shown in Figure 5.6. As the probability density distribution is a constant, the probability of one solution's having the lower cost is simply proportional to the area in the (p,E)-plane where it is the lower cost one. It is clear that 132 kV is the preferred solution in this case.

5.3.5.2 Software development

A company, prime contractor for a major system, is faced with the decision of either buying or developing a control subsystem. The subsystem is to go into service 6 years from now; any delay beyond this date is to be penalized by \$15,000 per day.

The bought-in subsystem would cost \$7 million, payable when it goes into service, with the subcontractor accepting the \$15,000 penalty clause.

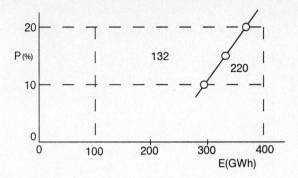

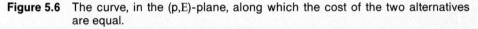

Figure 5.6 The curve, in the (p,E)-plane, along which the cost of the two alternatives are equal.

To develop the subsystem inhouse, an initial cost at the start of the development (i.e. hardware) of $1 million is estimated, followed by a software development effort that is equally likely to lie anywhere in the range 15–30 manyears, at $100,000 per manyear. However, the available manpower is a fixed team of 5 software engineers.

The discount rate p is assumed to be constant at 20% per annum.

As always, the first thing to do is to choose a decision criterion. The one that will be used is as follows: the subsystem will be developed inhouse if the mean development cost is less than the cost of buying the subsystem, with both costs referred to the into-service date. However, again it must be emphasized that this is by no means the only (or even the "best") criterion; it is simple and ignores many aspects of inhouse development, such as proprietary rights, follow-on business and risk associated with personnel (e.g. sickness, leaving the company). It would also be possible to formulate the criterion in terms of the probability that the development cost will be less than $7 million.

Using a discrete time variable with 1-year steps, let n be the number of years the development is commenced before into-service, and let s be the development time. It follows that $s, n = 3, 4, 5, 6$ and that a 4×4 calculation matrix is obtained.

The total development cost, referred to the into-service date, consists of the following three components:

initial cost: $\qquad 1(1 + p)^n$

software cost: $\qquad 0.5 \sum_{i=n-s}^{n-1} (1 + p)^i$

penalty: $\qquad 5.475 \sum_{i=1}^{s-n} (1 + p)^{-i} \quad (s > n)$

Let the total development cost be denoted by $C_n(s)$. Then the mean development cost is given by

$$C_n = \tfrac{1}{6}[C_n(3) + 2C_n(4) + 2C_n(5) + C_n(6)]$$

This function of n is shown in Figure 5.7. It can be seen that there is indeed a range of n-values for which the mean inhouse development cost is less than $7 million. The best time to start the development will be about 5 years and 4 months before the into-service date.

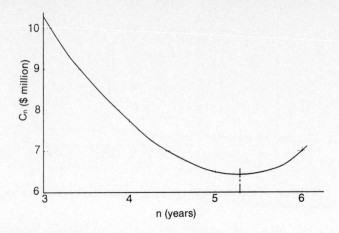

Figure 5.7 Development cost as a function of start date, measured in years prior to into-service date.

5.4 Estimates: consistency and continuity

5.4.1 Sensitivity and risk

From all that has been said about the systems engineering methodology so far, it should be clear that **estimates** play a very significant role. In Section 5.1.3 some rules were introduced that may be used to check on formal aspects of estimates. However, despite satisfying these rules, an estimate may be "wrong"; and as discussed in Section 5.1.2, there are two completely different sources of this "wrongness": the simplification of using averages instead of distributions, and the actual lack of knowledge. Both lead to a chance that the result(s) based on the estimate will be wrong; therefore the result should be expressed in the form of a probability density distribution, often assumed to be **Gaussian**, in which case it is characterized by its **standard deviation**, but any distribution can be characterized in terms of a **confidence interval**. For example, the 95% confidence interval is that interval within which the variable will be found with a 95% probability.

The size of this confidence interval depends on two factors: the **uncertainty** in the estimate of the input variable, and the **sensitivity** of the result to variations in that variable (i.e. the magnitude of the partial derivative). This sensitivity can vary not only from one variable to another within the same model, but also within the range of a variable whenever the result is a nonlinear function of the variable.

However, what is most often of direct interest is not the confidence interval itself but the effect of it. What happens if the real result deviates from the estimated one? The concept of **risk** is an attempt to come to grips with this problem area, but it is by no means a straightforward concept and needs careful definition for each case.

As a simple illustration of the problem, consider a model that predicts the net profit of some venture as a function of a number of variables, some or all of which are uncertain (e.g. future interest rate). Assume that the predicted net profit is given by a particular probability density distribution, as shown in Figure 5.8. This figure also

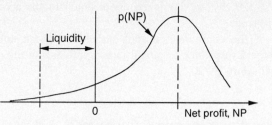

Figure 5.8   Probability density distribution of the net profit, as well as the limit of available funds (i.e. liquidity).

indicates the amount of reserves (i.e. liquidity) available. At least three different measures of risk can now be defined:

- the probability of making a loss;
- the probability of exceeding the liquidity level (i.e. of going broke);
- the probability of making a particular loss, multiplied by that loss, integrated over the whole negative part of the profit axis.

   Defining and calculating the risk in any particular case constitute the activity of **risk assessment** and will be discussed in Section 7.4. However, no matter how risk is defined, it is the sensitivity of the risk to the uncertainty in any one variable that determines how accurate the estimate of that variable ought to be, as will be discussed in the next section.

5.4.2 Consistency

Estimates can be based on anything from a wet finger to lengthy research, but in general it is true that, in serious system design, estimating represents a significant part of the effort. So how much effort should be put into estimating in a particular case? Is there some way of determining the cost-effectiveness of the estimating activity? For the estimating activity as a whole this is largely a question of applying experience and common sense; but when it comes to partitioning the effort among the various variables whose values have to be estimated, there is a requirement for **consistency**. Ideally, each variable should be estimated to such an accuracy that its uncertainty results in the same (incremental) risk.

   A classic example is the estimating of the cost of a system at an early point in the engineering process. This may be done for various purposes, such as tendering, or examining the viability of the project within certain budgetary limits. The total cost is broken down into a number of cost components, such as system design, a number of subsystem design components, a corresponding number of procurement cost components, integration and testing, and commissioning. How far to go in this breakdown is a separate matter; it depends on how similar this system is to previous ones, how well-defined smaller elements are at this stage, to what extent the system contains groups of identical elements and so on. Nevertheless, in any case the costs of a number of components need to be estimated, and to obtain a consistent estimate of total system cost the following rules should be kept in mind:

1. The accuracy (in percent) of the estimate of the cost of a component should be roughly inversely proportional to its value.

2. The accuracy in the costing should be about equal to the accuracy of the system definition at this stage.
3. The accuracy of the component costing should be about equal to the inherent uncertainty of the future value of global factors influencing the overall cost, such as interest rate and salary levels.

5.4.3 Principle of continuous refinement

At the beginning of the top-down design process, the information available is necessarily both very limited and characterized by a high degree of uncertainty. There is therefore a tendency for the people involved, both engineers and managers, to feel that any managed systematic approach is not worthwhile; one just has to muddle along until one has accumulated enough data to do some "real" planning and estimating. Nothing could be further from the truth; it is exactly in the early part of the project that planning and management are most important, because the potential effect of an error is so much larger. However, it is of course true that the parameters used in decision modeling at this early stage are inherently probabilistic in nature, and the techniques used must reflect this.

The fact that the early and later stages of the systems engineering process differ in nature in the above sense must not lead to a discontinuity in the process. One of the most important aspects of the process is **traceability**: the ability to demonstrate at any stage of the process that any decision taken is consistent with all the previous decisions. For example, at some later stage in the system design an engineer may find that for a current isolated problem a certain choice of technology or a certain form of the system architecture would be the most cost-effective. However, these decisions were certainly taken very early in the process, and a traceback would reveal a contradiction or inconsistency. Similarly, a number of assumptions usually have to be made in the early stages, and the later design must remain consistent with these assumptions.

As the systems engineering process progresses, however, new information becomes available, and it should be possible to feed this information into the earlier decision models to improve their accuracy. This is the principle of **successive refinement**; the system concept is refined in the sense that the parameters involved in defining the concept (or, more often, involved in the decision models that led to the definition of the concept) are estimated with higher and higher accuracy (or greater and greater confidence).

A prime example of the principle of successive refinement is the use of **Bayes' equation** to improve on previous assumptions in the light of new evidence. The equation itself is almost trivial; it is nothing but the rule of the product introduced in Section 5.1.3. It is the interpretation that is powerful. Let A and B be two events. Then the rule of the product can be written in the form

$$P(A|B) = P(A)P(B|A)/P(B)$$

This, Bayes' equation, is interpreted as follows:

1. $P(A)$ is called the **a priori** probability; it represents the initial assessment of the plausibility that event A will take place.

2. P(A|B) is called the **a posteriori** probability; it is interpreted as being the revised assessment of the plausibility that event A will take place, after confirmation has been received that event B has taken place.
3. The conditional probability P(B|A) is referred to as the **likelihood**, as it represents the plausibility of the new information (i.e. that B has taken place), given that A is true.
4. It is common to refer to P(B) as the **marginal probability**, as it represents the probability that B will occur whether or not A is true.

The use of Bayes' equation is best explained by a simple example.

5.4.4 Example: missile mission success

Consider the early design stage of the air defense system described in Section 2.2.7. One of the tasks would be to investigate the cost-effectiveness of missiles (SAMs) versus fighter planes in order to decide on the right mix of SAMs and fighters, and one of the parameters needed for such a decision would be the success rate of the SAM, say A. This early in the process nothing may be known about this success rate, except of course that it is a number between 0 and 1, that is, $0 \leq A < 1$. However, this does not mean that the systems engineering process cannot proceed. On the contrary, the fact that nothing is known about A can be expressed exactly by saying that any value of A is equally likely (i.e. $p(A) = 1$), and then this probability density distribution can be used in the decision model.

At some later stage a number of SAMs are test-fired, say m, and of these n are successes and (m − n) are failures. How can this new piece of information, B, be used

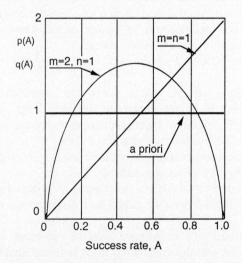

Figure 5.9 One a priori and two a posteriori probability density distributions for the missile success rate A.

to refine the results of the decision model? By revising the estimate for A (i.e. by calculating a revised p(A) using Bayes' equation). The latter is written in the form

$$q(A|n) = p(a)h(n|A)/g(n)$$

with

$$h(n|A) = \binom{m}{n}A^n(1 - A)^{m-n}$$

and

$$g(n) = p(A)h(n|A) \, dA$$

Take the very simplest case, m = n = 1. Then h(A) = A, g = 0.5 and q(A) = 2A. This result is shown in Figure 5.9, and it illustrates how one successful test modifies the estimated success rate by making the higher values of A more probable. The case m = 2, n = 1 also is shown in Figure 5.9. Now h(A) = 2A(1 − A), g(1) = 1/3 and q(A) = 6A(1 − A), and the curve shows how the estimate is refined by peaking q(A) around A = 0.5.

5.5 Summary

1. **Uncertainty** is an inherent feature of systems engineering because systems engineering takes place at the very beginning of a project, at which time the data required for decision-making have to be provided by **estimates** rather than by measurements.
2. Estimates must be **self-consistent**; some simple rules concern the sum of the probabilities of mutually exclusive outcomes, the relationship between absolute and conditional probabilities, and the relations among a set of conditional probabilities.
3. The concept of a **state** is central to the description of system behavior; it introduces a **structure** into the description.
4. A set of states is always defined relative to a set of observables or parameters.
5. The set of **basic system states** is the direct product of the sets of **element states**; the **system states** form a subset with respect to a set of system observables.
6. Uncertainty is taken into account by having **mixed** states; they are probability distributions on the set of **pure** (regular) states.
7. Transitions may be possible between some or all states; this introduces a **structure** in **state space**.
8. Transitions within a set of system states are characterized by a **transition probability matrix**, with the uncertainty arising from randomness in time (transition probability **rates**), or from randomness in branching mechanisms (event-driven probabilities) or from a combination of these.
9. Most systems reach a **steady state** (except for chaotic systems); this mixed state can be found by solving a set of linear equations.
10. The uncertainty inherent in systems is reflected in system models. Two types of **statistical models** are particularly important: **performance** models and **decision** models.
11. The purported accuracy of an estimate must be **consistent** with the data used in forming the estimate, and this consistency must be formulated in terms of **risk**.
12. Estimates must show **continuity** throughout the engineering process; their refinement must be **traceable**.

5.6 Short questions

1. Give two reasons why uncertainty is so important in systems engineering.
2. What is meant by "the principles of continuous refinement"? What is refined, and why is continuity important?
3. How is Bayes' equation derived, and how is it interpreted?
4. What must be defined before a decision can be made?
5. Define "conditional probability", and give an example of the practical application of this concept.
6. Define the two concepts "pure state" and "mixed state".
7. How is a set of element states defined? Can an element have more than one set of states?
8. What is a Markov chain?

5.7 Problems

1. *Three-component system with redundancy.* A system consists of three identical components. Two are connected in parallel and form the operating part of the system, whereas the third is held in store on site as a spare part. The parallel combination will operate as long as one of the two components is operating; so there is redundancy.

   The failure rate of an operating component is $\lambda = 500/10^6$ h, but it is assumed that the failure rate of a component in storage is zero. The mean waiting time to replace a failed component with the spare or, if no spare is available, to remove it is 10 h ($\alpha = 0.1$). Furthermore, there exists a repair facility where a faulty component can be sent for repair and from where it is subsequently returned to site, either to be installed or to be stored as a spare. The fixed turnaround time for this process is 100 h ($\beta = 0.01$).

   (a) Determine the states of this three-component system. (To do this consider first the states in which a single component can be found; then classify the system states by the number of components in each component state.)

   (b) Draw the digraph representing the state diagram.

   (c) Assuming that the system dynamics can be represented by a Markov chain, determine the transition rates in terms of α, β and λ.

   (d) Using the type of equations given in Section 5.2.5, and solving them either with a computer program or by hand using Gaussian elimination, determine the numerical values of the steady state probabilities π_i ($i = 1, \ldots$?).

   (e) Is there a subset of states that can form the basis for an approximate solution? Use this to check the calculations.

   (f) The security of the system can be characterized by three levels, the lowest one being nonoperating (i.e. failed). Which states belong to which security level? What are the probabilities of finding the system in any one of the security levels?

   (g) Show that the system cannot possibly be correctly described by a Markov chain. Try to estimate the effect of the simplifying assumption of Markovian behavior on the results obtained.

2. *Communications system with ten elements*: Consider a communications network consisting of four nodes fully interconnected by six transmission links of equal traffic capacity, each of which may be in one of two states: operating or failed.

(a) What are the basic system states, how can they be identified (i.e. described), and how many are there?

Let the traffic capacity of the network be the only parameter of interest. Then an equivalence relation of the set of basic states can be defined by declaring two basic states to be equal if and only if they have the same traffic capacity. The resulting equivalence classes are then defined to be the states of the system in the present context.

(b) How many system states are there, and how can they be identified?

(c) Draw a state diagram, showing the states and the possible transitions between them.

Now take into account the following additional information:

node: MTBF = 2000 h
        MTTR =   4 h
link: MTBF = 5000 h
        MTTR =   5 h

Using these values, the state diagram in (c) can be simplified by allowing a certain degree of approximation.

(d) Draw the simplified state diagram, and give numerical values for the transition rates.

(e) Determine the steady-state probability distribution vector.

(f) Define the system availability, and determine its value.

3. *Decision tree*: A company is considering undertaking a project, but there are areas of risk and significant liquidated damages penalties if the project is not completed within the specified time. The company must make a decision about the technology it will adopt and the level of resource it will apply to the project (i.e. five or ten engineers). There is also an element of uncertainty in the cost of the new technologies.

The project must be completed within 30 weeks, with penalties of $10,000 per week for every week exceeding the 30 weeks. The contract value is $320,000, and the cost for each engineer on the project is $10,000.

Draw the decision tree for the firm to evaluate the expected return from the optimal decision, and decide whether the firm should undertake the project.

Technology and resource choices

A: Utilize existing technology:

Cost $180,000   Certainty 0.95
Cost $200,000   Certainty 0.05

Resource allocation 5 engineers:

Project time 29 weeks   Probability 0.75
Project time 34 weeks   Probability 0.25

Resource allocation 10 engineers:

Project time 26 weeks   Probability 0.65
Project time 31 weeks   Probability 0.35

B: Utilize technology under development:

Cost $150,000	Certainty 0.35
Cost $180,000	Certainty 0.65

Resource allocation 5 engineers:

Project time 30 weeks	Probability 0.25
Project time 35 weeks	Probability 0.75

Resource allocation 10 engineers:

Project time 27 weeks	Probability 0.85
Project time 32 weeks	Probability 0.15

Would your decision change if the penalties were raised to $20,000 per week?

5.8 References

Fellow, W., *An Introduction to Probability Theory and Its Application*, New York: Wiley, 1968.

Hahn, G.J. and Shapiro, S.S., *Statistical Models in Engineering*, New York: Wiley, 1967.

Schmidt, J.W., *Mathematical Foundations for Management Science and Systems Analysis*, New York: Academic Press, 1974.

Methodology

CHAPTER 6

Analysis

6.1 Overview

6.1.1 Definition

Systems analysis is *the process that produces the system specification, thereby establishing the engineering basis for subsequent system design.* It is structured by the steps required to achieve this outcome, in particular: interpreting the project objectives, gathering further information, recognizing the required system functionality, formulating and refining high-level system models, testing and clarifying critical system options, establishing criteria for high-level decisions about the system, determining the principal system characteristics and documenting the resulting solution.

In essence, the specification achieved through this process is a high level model of the intended system, described by a set of comprehensive and verifiable requirements. It contains a range of key decisions about the system design, such as its configuration and implementation technologies. However, it does not include details that are properly determined by design.

Systems analysis is shaped by previously agreed needs, defined through the objectives of the project. These needs (or end requirements) are the focus of the process—the object of analysis. In other words, definition establishes **what** is needed, systems analysis establishes **how best** to satisfy defined needs. This logical progression creates the **first opportunities for synthesis of the intended system solution.** As such it must be accomplished having due regard for **all the engineering issues implicit in the needs.**

The methodology of systems engineering provides the necessary resources for such an outwardly directed approach, disciplined by the constraints of the project definition. However, the engineering issues themselves have to be recognized, and there are no rules assuring a complete set. Experience and judgement are required to determine the main system functions. Clearly, the repercussions of an inadequate system specification (i.e. one that overlooks or misrepresents engineering requirements) can emerge at any subsequent stage of the system lifecycle, with cost implications that escalate during the development project.

Systems analysis is usually conducted by a small team of systems engineers, sometimes only one, depending on the size and criticality of the project. The essential character of the process is maintained through wide consultation. As discussed in Section 2.4, it takes place during a period of intense project planning. The results of this planning must also be documented before the design phase commences.

6.1.2 Scope

The words "systems analysis" are used in two different contexts. To avoid confusion, these need to be distinguished.

One usage is self-evident. It applies to any analysis of **systems, their characteristics and their performance.** As such, systems analysis becomes part of the design process, using techniques like optimal control theory and numerical analysis. A typical example of this is circuit analysis, where the behavior of a circuit is analyzed in terms of response functions, stability and so on.

The other meaning is the analysis of **objectives and requirements,** or the analysis of the **project definition.** This carries the same meaning as the term "computer systems analysis", where user requirements are analyzed to define software requirements. A description of the actual system, as a well-defined and structured entity, does not emerge until the end of the analysis phase.

This distinction is not always made when systems are being discussed. Undoubtedly, one reason is that some of the relevant techniques are the same or at least closely related, as shown in the next chapter. However, in this text the two meanings are respected, because each requires a completely different point of view and engineering attitude. In this text this distinction accounts for the discussion of analysis and design as separate phases of systems engineering.

6.1.3 Process characteristics

In the analysis phase of a project it is necessary to approach the stated objectives with an open mind and to try to discern all their implications. Typical **questions** to be answered are as follows:

1. What functional elements are involved?
2. What areas of technology could be utilized?
3. What are the social, economical and political consequences?
4. Can the objectives be rephased? Such a rephasing can sometimes lead to a different way of thinking about the objectives.
5. Are there any assumptions implicit in the objectives that must be examined more closely?

Due to the very wide range of these questions, systems analysis needs information from many people in different occupations. A major task is to obtain this information efficiently, to identify its common themes and to seek a consensus of opinion. Furthermore, the system entities used at this stage are defined only in a general way, in terms of their functions or global characteristics, and their interactions are qualitative. However, the whole purpose of systems analysis is to consider widely different system approaches, each satisfying user requirements, before narrowing the choice to the extent required for a system to be sensibly identified and specified.

The decision to choose one possibility over others is usually reached by defining a decision criterion and then constructing a model that generates values of this criterion for the different possibilities. However, such models often require considerable **data**, and the multidisciplinary and probabilistic nature of much of this information makes the data gathering a time-consuming process with which most engineers are not comfortable. Nevertheless, the data-gathering process is obviously of fundamental importance, for on the information obtained will rest most of the decisions taken throughout the project. It therefore pays to go through the data in a conscientious and orderly fashion. Some of the required information may be readily available in handbooks, published reports and previous project documentation, and some may also be available, but more difficult to obtain, in the form of unpublished material. However, a certain part (and it can well be a large proportion) of the necessary information will be present only in the minds of other people.

Two methods, the **Delphi method** and **brainstorming**, are helpful in extracting this kind of information. Each is described below. However, it is usually necessary to spend a great deal of time in relatively unstructured conversation with persons related to the environment in which the system is to operate. In this way the systems engineer is best able to capture all the factors that influence the choice of the most cost-effective system.

6.2 Information gathering

6.2.1 Delphi method

The Delphi method (Linstone and Turoff 1975) is used under one or more of the following conditions:

1. The various persons possessing the desired information are from such varied fields (e.g. legal, medical, political) that they would not communicate very effectively in a face-to-face setting.
2. The persons are scattered over a wide geographical area, and so conferences would be impractical.
3. The formulation of answers requires more time than a conference would allow.
4. The number of people involved is too large for a conference (e.g. more than eight or ten).

The method may be described as using structured communications. Originally it was motivated by the question:

Is it possible, via structured communications, to create any sort of collective human intelligence?

In today's jargon the term "distributed processing" would be used, and the Delphi method would constitute the communications protocol.

The first and very important step is to obtain a suitable group of participants. On one hand, persons likely to have the desired information must be selected; on the other, their willingness to participate must be obtained. The latter may require negotiation, and even some form of remuneration or recognition, such as a copy of the resulting project information or the inclusion of their names on published reports.

In the second step the systems engineer composes an initial questionnaire that outlines the scope of the matter to be treated, be this the solution to a problem, the prediction of future events, attitudes towards a new concept or the like. The engineer then poses one or more general questions, which are primarily intended to motivate the participants, to make it easier for them to get their thought processes moving in the right direction. Questions must be worded in such a manner that they do not preclude unorthodox and novel answers.

Answers are prepared by participants in the third step, which requires a reasonable but definite time limit for its completion. In this step the systems engineer has to ensure discipline within the group. This is not always easy when some of the participants carry a lot more influence than the systems engineer.

In the fourth step the systems engineer evaluates the answers and writes a synopsis showing areas of consensus and appreciable differences.

This synopsis forms the preamble to the next questionnaire, where the questions are reformulated, usually made more specific, and new questions are added. These probe issues indicated by the answers to the first questionnaire as being the most fruitful. This iterative process can be continued a couple of times, as shown in the flow diagram in Figure 6.1, by which time either the desired information has been obtained or it has become obvious that the group has nothing more to offer.

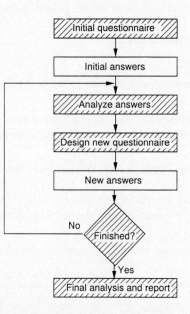

Figure 6.1 Block diagram of the Delphi method. The steps carried out by the systems engineer are indicated by shading.

The final step consists of a thorough analysis of all the material and the formulation of conclusions. This report should of course be distributed to all participants, with a suitable acknowledgement of their efforts.

Summarizing, the Delphi method may be said to consist of four phases:

- exploring the subject under discussion, with each participant contributing what he or she feels is pertinent;
- formulating a group view and identifying significant disagreements;
- resolving the disagreements as far as possible;
- evaluating all the information and determining its consequences.

6.2.2 Brainstorming

Brainstorming also is a collective process, based on the principle that brains working together perform better than the same number of brains working separately (Dehlbecq and Van DeVen 1970). It is a process for generating new ideas or new points of view, as distinct from gathering information. It relies on direct, informal but intensive interaction among the participants.

Again, the selection of a suitable group of participants is most important, but this is now constrained by the fact that the participants have to be able to get together at the same time for the brainstorming session. The group should not be too large, usually less than ten persons.

Brainstorming is less formal than the Delphi method, but it is by no means an unstructured process. It requires considerable skill on the part of the leader (i.e. the systems engineer) to prevent it from becoming an unproductive free-for-all. If the group has had little or no previous experience with brainstorming, the first step is for the leader to explain the purpose of the process and its rules. The objective is to stimulate creative thinking, to let new ideas emerge by association with previous ones. The associative operations are typically to substitute, rearrange, reverse the order of functions, specialize, generalize and so on. It must be impressed on the participants that this is not an analytic process; consequently, no judgement can be passed on new ideas. The leader must be strict about this and cut off any discussion of feasibility or merit.

However, to ensure that the associative process really progresses and does not become circular, a degree of formality must be introduced. This consists of writing down and displaying each new idea as it emerges. (This is best done on sheets of newsprint or something similar. A blackboard is not so suitable because of the danger of erasure.) Displaying the ideas in writing in front of the whole group not only curtails the re-emerging idea but also stimulates the associative process. By rearranging the order or grouping of the ideas, a certain trend or train of thought may become visible, which may then be actively pursued.

A brainstorming session progresses through three phases:

In the first phase the leader gets the process started by introducing the participants to each other and explaining the subject matter to be brainstormed. During this first phase it is important not to be formal, but to introduce a friendly light-hearted atmosphere, so that the participants become relaxed and spontaneous.

In the second phase, the free-wheeling phase, the participants contribute any idea that comes to mind. To ensure participation opportunities for all, it is usually advantageous to give each person a chance to contribute in a sequential order (e.g. by going round the table). If someone has no idea to contribute, he or she just lets the opportunity pass by; another opportunity will always come again on the next round. On the other hand, if someone suddenly has an inspiration, it is perfectly all right for him or her to interrupt the sequence. The leader (or possibly a secretary) writes down each idea without any comment or judgement; the purpose is simply to get as many ideas as possible. A typical number would be 100 ideas in half an hour.

When the flow of ideas starts to taper, the leader unobtrusively begins to order or group the ideas, discarding any duplications and clarifying any unclear formulations. This is the beginning of the third phase, in which the most promising ideas are crystallized and then described in more quantitative terms.

The final result of a brainstorming session is a set of ideas, maybe half a dozen, which are produced at a total cost of, say, 30 manhours. However, it is senseless (and demoralizing) to hold a brainstorming session without also being prepared to follow up the resultant ideas. This requires a much larger effort, at least tenfold. The implications of each idea must be studied seriously, which usually entails a literature search and considerable study. A formal written opinion should be produced and issued to the group members (if at all possible).

6.3 High-level modeling

6.3.1 Importance of models

Models form an essential part of systems analysis, for two reasons. Firstly, in the analysis phase the information available on the exact detailed behavior of the system is very limited. Indeed, the analysis phase is the process of defining the system requirements; so a model is used to express the macroscopic behavior of the system, disregarding underlying details. By starting with a very simple model, representing only the most global features of the system, the results of the analysis are incorporated successively in a process known as **refining** the model. Secondly, systems engineering is concerned with the future — in particular, with the **future conditions** under which the system must perform. Thus, the system to be designed is only part of the model; an equally important part is the modeling of the interactions with the external world (i.e. the boundary conditions).

To a certain extent systems engineering is an iterative process, alternating between synthesis and analysis. At each stage of the top-down process, an attempt is made to verify that the system developed so far does indeed meet the requirements of the project definition. This is a major area for the application of models. Models are also useful for examining or predicting the effects of various alternatives. If a certain part of a system can be realized in more than one way, or with different parameter values, the engineer would want to determine how the different choices will affect the system performance. To succeed in this endeavor, it is necessary to know how that part interacts with the rest of the system and with the external world. A model is then constructed that contains just enough detail to show the effects of the variations under consideration.

The use of models in the analysis phase is analogous to thermodynamics. If it is wished to examine how temperature affects the relation between volume and pressure of a gas, the equation $PV = RT$, which models the behavior of the gas under certain assumptions, can be used. It is not necessary to worry about the underlying dynamics of the individual gas molecules.

Very many concepts used in everyday life are based on simplifications (i.e. on models). This fact is often forgotten and can lead to some surprising results when the assumptions of the model are no longer valid.

6.3.2 Input and output identification

The purpose of any system is to produce something, to satisfy certain requirements laid down in the objectives of the project. This must necessarily result in some interaction with the surrounding world, and those interactions that fulfill the objectives are called the system **outputs.**

In choosing the set of output variables the following rules should be observed:

1. Each variable must be precisely and unambiguously defined.
2. The set of variables must be as small as possible; that is, the number of variables must be reduced to a minimum, consistent with providing a complete characterization of the objectives.
3. The variables must be as independent (i.e. orthogonal) as possible.
4. The variables must be directly relevant to the objectives.

With regard to rule 1, the quality of the definition must not be confused with its level of detail. In the analysis phase, details about the physical form of the variable (e.g. voltage levels, pulse risetimes, protocols) normally are not available. The definition is concerned only with defining the information content of the variable. For example, a variable may be defined as "the rate of mass throughput in the main feed line, denoted by G and measured in kg/s, and with $0 \leqslant G \leqslant G_0$, where G_0 is the nominal rating of the feed pump". This definition describes what the variable represents, specifies where and in what units it is measured, uses all available information in order to narrow down its range, and assigns a label to it. However, it leaves open for the later design stages any decision about how the measurement is carried out and in what form the result is presented.

Rules 2 and 3 are related, in the sense that an orthogonal set is also the smallest possible set. However, in dealing with high-level concepts it is generally impossible to assign an unambiguous dimensionality to the objectives. It is therefore useful to keep both points in mind.

The word "directly" in rule 4 is important. If any output must be transformed by part of the external world before it corresponds to the objective, then that part of the external world actually belongs to the system. In other words, the outputs define a boundary between the system and the external world, and the choice of variables should make it as easy as possible to verify that the system meets the objectives.

An example that illustrates, to some extent, all three of the last points is given by a system that produces some product and in which variables are needed to describe the dependability of that process. Three variables present themselves: mean time between failure (MTBF), mean time to repair (MTTR) and availability. However, as they are related, it would be incorrect to use all three; any two determine the third. The choice depends on how they relate to the objective of the system, and it may well be that one variable is adequate. If the overriding aspect is the productivity of the process,

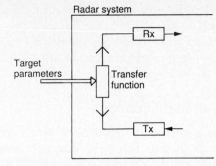

Figure 6.2   Target parameters as input variables for a radar system.

availability will be the prime variable. If the process supplies a vital facility (e.g. power to a life support system), then MTBF will be the prime variable. (Once a patient is dead due to a system failure, it does not matter whether it takes a little longer to repair the power supply!) Finally, if the process supplies a reservoir, MTTR will be the prime variable, as short interruptions may have no influence at all, as long as the availability does not fall below a certain value.

To generate the outputs the system needs certain **inputs**, such as power and various items of information. (It is possible to conceive of systems that need no inputs, e.g. ones that operate with an internal clock and battery, but these are of no practical interest in the present context.) In general, the same rules and comments apply to the choice of **input variables** as to the choice of output variables.

Again, an example illustrates how the choice of variables and the definition of the system boundary are inseparable. Consider a radar: the electromagnetic signal picked up by the antenna and receiver is not the input signal. This signal is generated by the system itself and is internal to the system, as shown in Figure 6.2. The appropriate input variables are, among others, the size and reflectivity of the radar targets, their speed and elevation, their number and so on, and these enter as coefficients in an equation that describes the relation between the output and input signals at the radar antenna.

6.3.3 Types of models

At the highest level a system model can be considered as a single element, with defined inputs and outputs. The purpose of this single element system is to **transform** the inputs into the outputs. The function that carries out this transformation is the **transfer function**, and the nature of this function provides a classification of the model. Aside from the aspects of system behavior treated by a model (e.g. reliability, grade of service, cost-effectiveness), models can be divided into four large categories: linear/nonlinear and static/dynamic transfer functions, as shown in Figure 6.3.

The concept of **linearity** is one of the most important in the classification of relations between variables. There are two underlying reasons for this importance. Firstly, linear models can be treated relatively easily with mathematical rigor. Secondly, **any** system can be approximated by a linear model if the ranges of the variables are suitably restricted. In linear systems the relations between the variables are given by constants,

	Linear	Nonlinear
Static	Easy	Difficult
Dynamic	More difficult	Very difficult

Figure 6.3 Subdivision of models into four categories.

which are the **parameters** of the system. The relations between the inputs u_i and the outputs y_i are the linear equations

$$y_i = b_i + a_{i2}u_2 + \ldots + a_{in}u_n$$

or

$$Y = B + AU$$

where A is the parameter matrix.

The theory of linear equations (or linear transformations) is very well known, and there exists a vast literature on how to solve practical problems. This theory can still be applied when the variables are not simply numbers, but statistical variables characterized by distributions. The design and analysis of linear statistical models using such concepts as regression analysis, covariance, confidence intervals, and significance levels, are treated in many books, and suitable computer programs are available.

This comfortable state of affairs changes drastically when the model becomes **nonlinear**. The solutions are usually much harder to find, and it may be difficult to show that a particular model has physically realizable solutions or even any solutions at all. Moreover, nonlinear behavior can be difficult to grasp intuitively and can be quite startling. The relations between the variables are no longer constants, but functions, and the interpretation of the parameters can be quite involved. The modern computer allows very complex nonlinear models to be studied; but the price paid is the lack of insight afforded by numerical solutions relative to analytic solutions. However, nonlinearity is also very fundamental, because all physical systems are finite and therefore **saturate** at some point, and because many interesting effects arise only as a result of nonlinearity. Examples are modulation and frequency multiplication.

So far it has been assumed implicitly that the model, or, more precisely, the situation being modeled, is **static**. Time has not entered into the description of either the variables or the system parameters. This assumption can be valid even though elements of the model are dynamic in nature. Take the case of failure, surely an event inherently connected with the passage of time. A model giving the failure rate as a function of certain variables, such as temperature and voltage, can still be a static model. However, if one or more of the variables are functions of time, two cases must be distinguished.

If the response of the system to the **rate of change** of its variables is zero or negligible compared to its response to the variables themselves, a static model may be used to relate the variables to each other at any instant in time. This applies in the case of a model giving the voltage resulting from a current passing through a pure resistance. However, if the response of the system to the rate of change of its variables is appreciable, a static description is not directly applicable. In this case the model is dynamic and the relations between the variables are described generally by **differential equations**. Two special dynamic cases are well known: periodic variables, in which case static relations again hold for the amplitudes and phases of the variables, but with frequency as an added variable; and transitions between equilibrium states (as discussed in Section 5.2), in which case the relations between the variables are again time-independent, but with the state label as an added variable. In other words, the parameters of the model become functions of the state label.

The general case, that of the nonlinear dynamic model, can be very difficult, and only exceptional models have analytic solutions. The availability of cheap computing power has revolutionized modeling of this case and placed it within the means of small firms and even individuals. By introducing time derivatives of the variables as "dummy" variables, the differential equations are transformed into algebraic equations valid at one instant in time. By dividing time into discrete intervals, the dummy variables can then be expressed as functions of the variables at two or more adjacent moments in time. Finally a (possibly very complex) set of recursive relations for the values of the variables at different times is arrived at. The set of recursive relations is readily handled with a computer; even a desktop computer can step through such models at the rate of several hundred thousand floating-point operations per second.

The above classification of system models arose from a discussion of the system as a single functional element, without any internal structure. Once the main functional elements of a system have been identified, the same considerations apply to each element. However, the **structure** of the system, as discussed in Chapter 3, now leads to a further division of system models into two classes: those with a cyclic and those with an acyclic structure.

In the case of a model with an **acyclic** structure, the overall system behavior follows from the behavior of the elements in an almost trivial manner, simply by "chaining" or "linking" them together. The effort involved in obtaining the system behavior is just the sum of the efforts involved in obtaining the behavior of each element. However, if the structure involves **cycles** (i.e. contains feed-back), the system behavior may be much more complex than that of any of its elements, and completely new aspects may be introduced, such as instability. Furthermore, the numerical evaluation of the model also becomes more complex, and special consideration must be given to data consistency and to avoiding locked states or eternal loops.

6.3.4 Example: buffering system

At this stage it may be helpful to illustrate the above ideas on modeling by returning to the small system already introduced in Section 1.3.3, namely the one consisting of a charger and a reservoir. Considering the system as a single functional element, its purpose is to supply a sink with a commodity from a source; that is, it falls into the

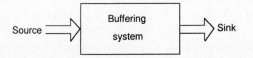

Figure 6.4   Buffering system considered as a single element with a transport function.

large class of transport systems, and the output is therefore of the same nature as the input, as shown in Figure 6.4.

If the system performance is defined as the ability of the system to satisfy the demand, the aspect of interest is the relation between output Y and demand D. As there are three different cases of this relationship, it defines three states:

state 0: Y = 0
state 1: Y = S < D
state 2: Y = D.

The two main elements of the system are a charger and a reservoir, as previously shown in Figure 1.3. The charger is either operational or not, and the reservoir is either empty or not; so the two **state variables** U and V can be introduced:

U = 1 if charger operational, 0 otherwise
V = 0 if reservoir empty, 1 otherwise.

Since the performance is defined relative to the demand, the state of the demand also plays a role; so one more state variable W is needed:

W = 1 if D > S, 0 otherwise.

The state of the system can now be related to the states of the elements as follows:

System state	U	V	W
0	0	0	0
2	1	0	0
2	0	1	0
2	1	1	0
0	0	0	1
1	1	0	1
2	0	1	1
2	1	1	1

The three system states can be represented in a **state diagram**, as shown in Figure 6.5. Transitions are possible in both directions between any two states. However, the resulting chain of states under the influence of time is not a Markov chain. Due to the reservoir the transition probabilities depend very much on the past history of the system.

So far, the aspect of system behavior being modeled, namely system performance, has been defined only in a very general way. It must now be defined in terms of the

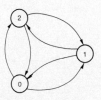

Figure 6.5 State diagram of the buffering system.

observable system states. However, to do this it is necessary to know more about the particular application of the system, as already indicated in Section 1.3.2. Of the many possibilities available, assume that the system is supplying an essential load, so that whenever the demand cannot be met in full a catastrophic failure occurs. In this case the essential feature of system performance is its **reliability**, which shall be defined as the mean time of operation for the system to leave state 2, assuming a full reservoir at time $t = 0$. Thus, with this more limited view of system performance, the state space degenerates into two states only: the operating state, say $H = 1$, and the failed state, $H = 0$.

The system model, consisting of only two elements and having easily visualized relations between the variables, is deceptively simple. It turns out that even the formulation of the equations governing system behavior in an analytic form is very difficult, and that solving them is even more so. The system is a classic candidate for simulation, and the development of a simulation model can be tackled as follows.

At any instant in time the system state (i.e. the value of H) is determined by the values of U, V and W, in the manner shown in Figure 6.6. The problem is therefore reduced to the three subproblems of determining U, V and W at any instant in time.

Starting with V, let time be a discrete variable, and introduce, as an auxiliary variable, the level of the reservoir L, with L bounded by 0 and C inclusive. Then $V = 0$ if $L = 0$, and $V = 1$ for $L > 0$. If L is the level of the reservoir at any one instant in time, let LA be the value of the level 1 unit of time earlier. Then

$$L = LA - (D - US)$$

but

$$\text{if } L > C, \text{ then } L = C$$

and

$$\text{if } L < 0, \text{ then } L = 0$$

To determine the value of U it is necessary to generate a random number between 0 and 1, RND, and to choose any interval in that same range of length f (where f is the failure rate of the charger, measured in probability per unit time). Also, to keep track of the time remaining until a failed charger comes back into service, the dummy variable n

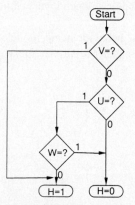

Figure 6.6 Relations between state variables.

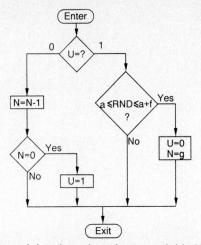

Figure 6.7 Model for determining the value of state variable U.

is introduced; it can take on integer values in the range 0 to g (where g is the repair time in units of time). The value of U then follows as shown in Figure 6.7.

The value of the demand D in the unit of time preceding the determination of the state is again found by generating a random number in the range D_1 to D_2, and the value of W follows immediately by comparison with S.

The complete model for determining the MTTF is shown in Figure 6.8, and the result of running this model repeatedly for different values of S and C is shown graphically in Figure 6.9.

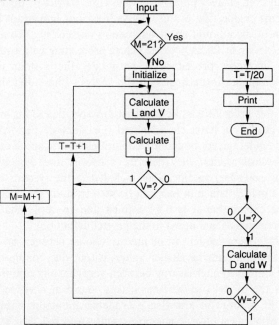

Figure 6.8 Model for determining the value of MTTF (here T) as an average over 20 runs. (M is the run counter.)

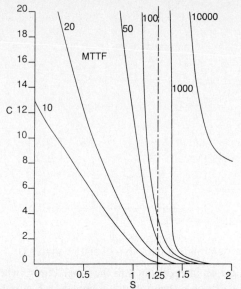

Figure 6.9 An example of results obtained using the model of the buffering system. MTTF is determined as a function of the two system parameters C and S.

6.3.5 Specifications as models

The systems engineering effort up to and including detailed specifications for all the system elements may account for only 10% of the total engineering effort involved in the project and an even smaller part of the total cost. However, the consequences of an error in the systems engineering are usually very serious and result in additional costs out of proportion to the amount spent on systems engineering. Thus, it is necessary to verify the systems design as far as possible before proceeding with equipment design and manufacture. A common procedure is to model the system using the element specifications. This applies in particular to the analysis phase and the resultant system specification.

The basis for this procedure is the fact that any specification is automatically also a description of a model. It states the idealized (i.e. desired) performance (e.g. input/output relations) under given boundary conditions. Both specifications and models abstract from physical reality; they describe relations and dependencies. Thus, the element (including software) specifications included in the system specification can be used to build up a model, which in turn can be used to study the overall performance of the system to the extent that it is now defined and to verify that this performance satisfies the requirements as laid down in the project definition.

Conversely, the system model can be used to choose between several configurations or to determine the optimum parameter values within one configuration, and usually the model will be employed alternatively between verifying and optimizing.

The extent of the model (i.e. whether it is a coarse or a refined model) does not depend only on the amount of knowledge available about the system at this stage; it also depends on what questions the engineer wants answered. For example, if only the static performance is of interest, a model can be constructed that will reflect the static performance **exactly**. It will generally be very much simpler than a full dynamic model, and the limitations to its validity (i.e. the timescale on which it is valid) can be unambiguously stated.

The latter point presupposes that the system is **stable** over the whole range of input values of interest. In that case the result of any change in these values will consist of two parts: a transient (i.e. dynamic) part and a static part. The transient part will die out (often exponentially) and become negligible after a **characteristic time interval**, leaving the new static value of the outputs. The transient phenomena are considered to be **nonessential** to the questions the model is being used to answer. A typical example of such models is afforded by classic thermodynamics; it considers only equilibrium states, and its variables (e.g. temperature and pressure) are strictly defined only in the static case. This does not prevent it from being called thermo**dynamics** and being used to describe dynamic situations; it only means that the nonequilibrium phenomena are nonessential on the timescale of interest.

In addition to modeling the system under consideration in its narrower sense (i.e. as defined by the functional part of the system specification), the model must often take into account the environment in which it is to operate. The various parameters describing the environment may in principle be independent variables, but may exhibit correlations that must be described by sub-models of their own. For example, temperature and supply voltage may be correlated, with high ambient temperature resulting in low supply voltage due to a large air-conditioning load.

Furthermore, the system may be only a subsystem of a larger (possibly existing) system, and this larger system may impose relations and limitations on inputs, outputs and boundary conditions. Thus, a complete model can consist of several parts, as shown in Figure 6.10. The **environment model** reflects the environment within which the system operates. The **driving model** provides the inputs (or what in some cases is called the **scenario**) to which the system under consideration reacts and produces corresponding outputs. These outputs are evaluated in the **performance evaluation** program, which may entail checking accuracy, stability, speed of response, mission success and so on. The **host system model** reflects the interactions between the system under consideration and the system in which it is embedded (i.e. the host system). In particular, the outputs of the system model will often modify the state of the host system which in turn influences some or all of the inputs, thereby providing feedback paths that can have a major influence on the dynamic performance of the system.

The performance evaluation program can represent a significant part of the systems engineering effort. In some cases it may be almost trivial to set criteria for system

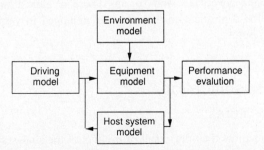

Figure 6.10   Several auxiliary models may be needed to simulate the actual operation of any given system (here represented by the equipment model).

performance, such as minimum error; but in other cases it can be very difficult to extract from the project definition quantitative criteria to allow it to be said that the objectives have been reached or that one version is better than another. This is especially the case when the system cannot be tested in actual operation, as is the case with some weapons systems or safety systems for nuclear reactors. In such cases it falls to the systems engineer to define what is a realistic simulation and what will be deemed an acceptable outcome. The evaluation program would normally be a part of the system specification.

The driving model entails one aspect that merits special mention, namely the **completeness** of the performance evaluation. How can one be certain that the inputs provided by the driving model represent all possible inputs? If some values of the input parameter space are not tested, these could be exactly the ones for which the system does not operate satisfactorily. For example, this problem was encountered in the testing of VLSI chips, such as RAM with 64 kbit or more per chip, where it became practically impossible to test every bit pattern. Designing the driving model becomes an optimization problem in itself: minimizing the overall cost of possible inadequate system performance and the simulated performance testing.

In addition to the problems of specifying the driving model and of reducing the model to eliminate non-essential phenomena, there remains the problem of the statistical or uncertain nature of some of the input data, which leads to the question of the **significance** of the results. There is no sense in designing a model of a system that describes its performance with an accuracy of a fraction of a percent, when the input data will result in an uncertainty in the output of 5% or 10%. As with any task that consists of several parts or steps, it is necessary to balance the efforts that go into each part so as to end up with the most valuable result (in terms of applicability, relevance and significance) for a given total effort.

Finally, the system modeling serves the very important function of verifying the system **operating procedures**. These procedures, laid down in system operating manuals, must be developed phase by phase, and they constitute just as important an element of the system as does the hardware or the software. If they are left out of the performance verification, it is too easy to overlook such important problem areas as: Does the equipment present adequate information for the operator to take the required action? Is the information presented in a form that is directly relevant to the operator's action? Is it presented where the operator needs it? Does the operator have adequate intervention possibility? Do the procedures minimize the chance of operator error? Are the communications links adequate? And so on. There is almost no limit (except an economic one) to the sophistication that may be employed to verify or study system performance at this point.

6.4 System specification

6.4.1 Special features

The major, and sometimes the only, result or product of the analysis phase is the **system specification**. All that was said about systems-related specifications in Chapter 4 applies to the system specification; but as it occupies a very special position in the hierarchy of specifications, it has certain additional requirements attached. In the words of MIL-STD-490:

> This type of specification states the technical and mission requirements for a system as an entity, allocates requirements to functional areas, and defines the interfaces between or among the functional areas. Normally, the initial version of a system specification is based on parameters developed during the concept formulation period or an exploratory preliminary design period of feasibility studies and analyses. This specification (initial version) is used to establish the general nature of the system that is to be further defined during a contract definition, development, or contract design period. The system specification is maintained current during the contract period, culminating in a revision that forms the future performance base for the development and production of the prime items and subsystems (configuration items), the performance of such items being allocated from the system performance requirements.

The above description clearly reflects the major aspect of the system specification, namely, that it is the **base or point of departure for all the subsequent engineering effort**. Therefore, in addition to being **correct**, in the sense that its requirements accurately reflect the objectives in the project definition, the system specification must be **complete**. Once the system specification has been approved, and as long as it is not revised (see Section 6.4.2), all the methods and tools used to control the engineering process are aimed at achieving conformance with the system specification; the completeness and correctness of the specification are not questioned. Anything left out of the system specification is likely to be missing when the system is put into service.

The contents of a system specification will of course depend on the particular system, as regards both the areas to be addressed as well as the level of detail appropriate to each of these areas. However, as a general guideline, the following issues should be considered when writing a system specification:

1. A clear, concise and complete description of the **purpose** of the system as a whole. If such a description cannot be provided, the reason is most often that the engineer has concentrated on certain aspects of the system only, instead of first developing the highest possible level view of the system to its fullest extent.
2. Development and definition of all **concepts** necessary to specify the system performance. This issue is particularly important in a system specification because there is often a choice of how to define performance parameters at this level and because, once that choice has been made, it can have a significant influence on the design.
3. Specification of the system performance **parameter values**. This issue is separate from defining the concepts and is much more directly related to the objectives of the project definition.
4. Specification of the **environment** in which the system is to perform. In the case of equipment-based systems this will normally entail specifying the climatic conditions, mechanical requirements, EMC requirements, and human resources available to operate the system. For a software system it may include specifying the hardware environment and the operating system, and for a management information system it may include specifying the organizational environment.

5. Specification of any **boundary conditions**, such as a particular technology or mandatory use of a particular programming language.
6. Specification of all **support requirements** (i.e. everything needed to operate and maintain the system over its lifetime). This includes the repair subsystem, training, spare parts stocking and supply, all documentation, and the management subsystems needed for the operation of the whole support effort.
7. **Test requirements**. This part must specify how the finished system should be tested to verify that all the above requirements have been met.

6.4.2 Configuration control

Any project document, as soon as it is approved, becomes a source of information for the future. However, as a project is a "living" thing, with changes taking place continually as a result of changing requirements, new technology, changing environment and so on, it is necessary to ensure that every document is revised to reflect the current status of the project. This activity is called **configuration control** and constitutes a significant part of the project management effort. Systems engineering is heavily involved in providing the input to this process, but in addition to the formal control of the whole set of documents within a project, configuration control takes on a special significance as a tool within the systems engineering methodology.

Ideally, systems engineering is a top-down process, with each stage building on the previous one and providing the definite basis for the next one. However, in practice there is always a certain amount of **feedback** from one stage to previous stages, due to new information and to a better understanding of the problems involved as the design progresses. As long as the changes are confined to levels below the system specification, this is a relatively straightforward matter and one that is well known from ordinary equipment engineering. However, if the changes require changes to the system specification, special care is needed, for the following two reasons:

1. As the basis for the whole design effort, the system specification ties together a number of different areas and disciplines that work more or less autonomously. Any change arising in one of these areas that requires a change to the system specification therefore needs to be studied with respect to its influence on **all** the system. The consequences of such a proposed change will often not be perceived by the originator of the proposal, and while it may be beneficial to one area of the system, its overall effect may be negative, leading to its rejection at the system level.
2. Any proposed change to the system specification needs to be considered from the point of view of how it affects the degree to which the system satisfies the objectives of the project definition. Usually it will be necessary to prove to the client (or end user) that the changed specification will meet the objectives at least as well as the original specification.

It is evident that changes to the system specification are very costly and need to be carried out in a strictly controlled fashion. The corresponding procedures form part of normal project management tools; but the work is usually carried out by systems engineers, and good systems engineering practice includes taking the requirements of configuration control into account when developing models and other evaluation and design tools.

6.5 Case studies

6.5.1 Combustion optimization: economic model

6.5.1.1 Background

The motivation for introducing the combustion optimization system is clear: to save on fuel costs by reducing flue losses, and to do this by minimizing the excess oxygen in the flue by using feedback control. However, is this also an economically viable proposition? To answer that question it is first necessary to construct a model that will relate a measure of economic viability to the system parameters.

In general the cost-effectiveness of the system would be looked at; but for small cost-saving acquisitions of this type a simplified measure is often used: the so-called **payback period** N, defined as the ratio of acquisition cost A to the effective yearly savings E and measured in years. In judging an investment of this sort a rule of thumb is:

N	Judgement
1	excellent
2	good
3	doubtful
>4	forget it

As a result it is possible to define a value of N, say N = 3, that represents the border between a viable and a nonviable system.

The effective yearly savings are the yearly decrease in fuel costs C minus the yearly maintenance costs. The latter are essentially the sensor replacement costs (including installation, test and recalibration), divided by the lifetime of the sensor (in years). The yearly maintenance costs can be written as sA, where $0 < s < 1$.

The decrease in fuel costs will depend on the yearly fuel costs Q and the extent of the flue losses L. The latter will depend on the oxygen level in the flue x and the flue temperature. Based on market information, it will be assumed that $L = 0.05Q$ for $x = 0$. It then follows, using the fact that the oxygen content of air is 21%, that

$$L(x) = 0.05Q \frac{21}{21 - x}$$

with x in percent.

Ideally, a burner should operate with $x = x_0$, where x_0 is the minimum oxygen value consistent with an acceptable pollution level. In practice, however, to compensate for uncontrolled variations in the combustion parameters, the burner will operate at an average value x_1. Again based on market information, it will be assumed that $x_0 = 1.0\%$ and $x_1 = 2.5\%$.

The combustion optimization system reduces the difference between x and x_0 by a factor $(1 - q)$; that is,

$$x = x_0 + q(x_1 - x_0)$$
$$= (1 + 1.5q) \quad (0 < q < 1)$$

The factor $(1 - q)$ is a measure of the **accuracy** of the optimization system; q is a measure of the errors in the system, with $q = 0$ corresponding to the highest accuracy (i.e. no error).

6.5.1.2 Model

With the above background information it is now straightforward to construct a model that determines the minimum value of Q, Q_{min}, that will give a 3 year payback period, as a function of the three system parameters A, s and q. Let the decrease in fuel cost be denoted by C. Then

$$C = L(x_1) - L(x)$$

$$= QF(q)$$

with

$$F(q) = 1.05 \left(\frac{1}{18.5} - \frac{1}{20 - 1.5\,q} \right)$$

However,

$$N = A/(C - sA)$$

and setting $N = 3$ the desired relation

$$Q_{min} = A(s + 0.333)/F(q)$$

is immediately obtained.

6.5.1.3 Discussion

The value of Q_{min} determines what proportion of the total burner market can be addressed; the higher the value, the smaller the effective market. Conversely, to address the market down to a particular Q value, the above relation becomes a relation between the three system parameters. This is most conveniently expressed as curves of constant A/Q_{min} values in the (s,q)-plane, as shown in Figure 6.11. This figure shows, for example, that the acquisition cost A must be less than 1% of the yearly fuel bill Q for the optimization system to have a chance of being viable, and this is clearly a very useful piece of information to have before proceeding any further into the design. The figure also shows the trade-offs possible between the three system parameters.

This modeling exercise demonstrates a point made earlier, namely, the great importance of nontechnical information in the early stages of the systems engineering process. The assumptions made for $L(0)$, x_0, and x_1 were based on knowledge of the market, of actual operating burner installations, not on any theoretical considerations. Obtaining such information may be tedious, especially for a keen design engineer, but it is nonetheless vital to a properly structured engineering process.

6.5.2 Tactical communications system

6.5.2.1 Switch capacity

In Section 3.5.2.2 the groundwork for modeling tactical networks was done by defining the structure of such models; they will be called **normal networks**. To proceed further with the modeling and to obtain some quantitative results, it is necessary to narrow the applicability somewhat and to introduce an actual switch. The switch used in this particular case has eight **ports**, with each port supporting one time division multiplex (TDM) group. At each port this group can be either a trunk group or a loop group. In the former case there will be 29 channels available to traffic; in the latter case the group

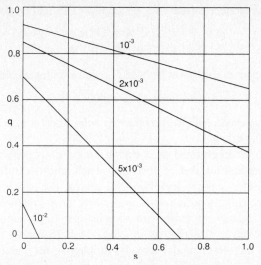

Figure 6.11 Curves of constant A/Q_{min} values in the (s,q)-plane.

will carry 30 individual loops (i.e. subscriber channels). Also, the switch is **nonblocking**, which means that any channel can be connected to any other channel.

When the switch is placed in a node and connected into the network, each link to that node will occupy one port. The remaining ports are available for subscribers, and throughout this modeling exercise the channels in these ports will be called subscribers, irrespective of whether subscribers are actually attached or not. In other words, it is what the network offers that is of interest, not whether this capacity is also used.

A similar approach is taken with respect to the traffic generated by the subscribers. The main assumption that will be made in developing the present model of the trunk network is that the traffic generated by the subscribers is such that all the trunks are equally and constantly loaded; that is, neither temporal fluctuations nor spatial inhomogeneities are allowed, and as a consequence the model will reflect the upper limit on network performance. In a real case there will of course be both temporal fluctuations in the level of generated traffic and unequal average loading of the trunks, but these will be treated as "second-order" effects in Section 9.3.1.

6.5.2.2 System characteristics

Under the conditions defined in the last section, the total **traffic capacity** C of the network can be defined by

$$C = 29M/L$$

where M is the number of links and L is the average connection length; that is, C is the maximum number of connections that can be established simultaneously in the network. The traffic capacity will of course increase with increasing network size, but so will the number of subscribers. It is easy to see that the number of subscribers U is given by

$$U = 720 + (N - 2)600$$

and therefore the **subscriber capacity** C_s, defined as C/U, can be determined:

N	4	9	16	25
C_s	0.121	0.123	0.120	0.117

The remarkable aspect of this result is that C_s is almost independent of N; that is,

$$C_s = 0.12 \pm 0.003$$

In other words, this **class** of networks offers its subscribers a maximum or ideal traffic capacity of about 0.1 erlang.

This result can be interpreted as a design relationship between the two network parameters structure and number of ports per switch, and the traffic generated by each subscriber. If the latter is much less than 0.01 erlang, the network will never be fully utilized. In this case either a switch with more ports, or a structure with a lower degree of meshing (e.g. a regular structure with six sides per ring), should be chosen. If the traffic generated by each subscriber is in excess of 0.01 erlang, the converse applies. Thus, this simple high-level model already supplies a very useful guideline for the system design.

6.6 Summary

1. In the **analysis phase** a system is identified that will produce the service or product required by the objectives contained in the **project definition**.
2. The result of the analysis is the **system specification**.
3. Although analysis and design use many similar techniques and tools, analysis takes a much broader view. It is **multidisciplinary** in the widest sense and takes social, economical and political factors into account.
4. An essential part of analysis is **information gathering**, and two useful procedures are the Delphi method and brainstorming.
5. In identifying a particular system and determining the system requirements, many **decisions** have to be taken. **Models** are particularly important in guiding the decision-making at this stage.
6. The analysis phase can also be seen as the **transformation** of user requirements into system requirements; this emphasizes the importance of the system specification as the basis for all further work in creating the system.
7. The definition of a system consists of identifying the inputs and outputs, and determining the main functional elements and their interactions.

6.7 Short questions

1. What is analyzed in the analysis phase?
2. What signifies the end of the analysis phase?
3. What are the major activities involved in systems analysis?
4. Give a simple high-level classification of models.
5. What types of models are required in order to model system performance?

6. In what sense can specifications be considered to be models?
7. What are the main issues that need to be considered when writing a system specification? (Keywords only.)
8. Explain the special problems involved in controlling the system specification.
9. What is the Delphi method, and under what circumstances would it be used?

6.8 Problems

1. *Anticollision radar*: To prevent the collision of ships at sea, an anticollision radar is to be developed. The microwave part is identical to that of a normal radar and consists of a continuously rotating narrow-beam antenna, a transmitter that emits a continuous train of narrow (typically 1 μS) pulses of microwave energy, and a receiver. However, the signal processing is much more sophisticated than that of a normal radar, in that it identifies and tracks all other vessels (and/or objects) within the range of the radar, calculates their courses relative to its own vessel, and classifies their predicted paths according to whether a collision may occur or not. Furthermore, such a system can also assist the duty officer in determining an appropriate course change by simulating the results of a range of changes.

    (a) Formulate the main requirements of such a system, as they might be formulated by a user in the definition phase of a project to develop the system. (Not more than 80 words.)

    (b) Identify the main functional elements involved in satisfying these requirements.

    (c) What are some of the most important inputs (i.e. information, parameters) to the system design process? (It may be helpful to think about other radar collision-avoidance systems, e.g. for automobiles, and to recognize their inherent similarities and differences.)

    (d) Draw a block diagram of the system (showing the main physical components) and identify the main characteristics of each block and/or parameters transferred between blocks.

    (e) Mention any potential problem areas that will have to be addressed during the system design.

2. *Personal computer system*: Large computer installations are obviously systems, in the sense of large-scale combinations of complex interacting elements. However, as the demands placed on personal computers increase, these home installations become more and more like systems also. They consist of many elements, both hardware (e.g. CPU, monitor, mass storage, laser printer, scanner) and software (operating system, graphic package, database management system); their interactions are complex; and the functionality is structured (i.e. forms a system in itself).

    (a) Structure the "Requirements" section of a **system specification**. On the one hand, there is the structure as seen by the user (i.e. the various functions available to the user and their relations). On the other hand there is the structure as seen by the system designer (i.e. the hardware and software modules). The latter is obviously important as a basis for doing the actual design, but the former is equally important in being able to verify that the design meets requirements. The structuring of the system specification should cater for both needs.

(b) Produce the headings of the complete system specification for the first three levels. Emphasis should be placed on a choice of headings that carries maximum information and on a consistent subdivision into levels (i.e. roughly equal importance of the elements on each level).

3. *Power supply system for a mine*: A new mine and ore-processing plant are to be constructed in the Australian outback, many hundreds of kilometers from the nearest town. Consequently, all power needed for the mine, plant and associated facilities will have to be generated locally. A project is defined for the purpose of designing the **power supply system**. The objectives are as follows:

- Determine the requirements placed on the supply of power and the boundary conditions under which the power supply must operate.
- Design the power supply system, including its logistic support, that will meet the requirements in the most cost-efficient manner.
- Provide full procurement specifications for all equipment and installations.

With this very minimal amount of information:
(a) Determine the contents of the first few steps of the analysis phase leading to a high-level requirement specification. The description of each step should consist of a description of the work to be undertaken in that step and an exact definition of what information is to result from that work.
(b) Determine the main elements of the system, with a clear delineation between what belongs to the system and what does not.

6.9 References

Atwood, J.W., *The Systems Analyst*, Hayden Book Company, 1977.

Buffa, E.S. and Dyer, I.S.,, *Management Science/Operations Research*New York: Wiley, 1977.

Bunn, D., *Analysis for Optimal Decisions*, Wiley, 1982.

Dehlbecq, A.C. and Van DeVen, A., *A Group Process Model for Problem Identification and Program Planning*, University of Wisconsin, 1970.

Linstone, H.A. and Turoff, M., *The Delphi Method*, Adison-Wesley, 1975.

MIL-STD-490 *Specification Practices*, Washington DC: US Department of Defense, 1985.

Optner, S.L., *Systems Analysis for Business and Industrial Problem Solving*, Englewood Cliffs, NJ: Prentice Hall, 1965.

Rawlinson, J.G., *Creative Thinking and Brainstorming*, Gower Publishing, 1981.

Shamblin, J.E. and Stevens, Jr G.T., *Operations Research*, New York: McGraw-Hill, 1974.

Wilkes, F.M., *Elements of Operational Research*, New York: McGraw-Hill, 1980.

CHAPTER 7

Top-down design

7.1 Overview

7.1.1 Definition

Top-down design is *the stepwise process of partitioning a system specification into a self-consistent set of specifications for all the elements of the intended solution.* The elements include the physical elements of the system directly involved in providing the required system functionality and performance, usually called **prime equipments,** together with elements of the repair subsystem or logistic support elements, such as training and documentation.

In each step of this process, concepts affording a more detailed definition of the system must be formulated, refined, endorsed and documented. The last of these actions produces a new set of specifications, which in turn supports the next iteration. The process develops the design in a disciplined way, while ensuring due recognition of the additional requirements and constraints created at each level. Modeling and simulation are used widely to ensure an optimum solution.

The design phase terminates when the partitioning has achieved elements of a **generally recognized type** (e.g. transmitter, antenna, data storage unit or terminal) that can be **specified fully by parameters common to that element type**. Here it must be emphasized that the criterion—not the name given to a particular element—should always be used to determine whether further partitioning is required. The boundary between system and equipment must be judged finally on a case by case basis. Moreover, the criterion recognizes that today's system can become tomorrow's equipment. Equipment design (i.e. designing and developing hardware or writing software) is part of the implementation phase, as discussed in Section 2.2.5.

The process described in this chapter implies that the system design occurs within one organization. However, the same principles are applicable where contractual interfaces are involved and design responsibilities are shared across a number of enterprises. In large projects, several layers of contractual interfaces are common (e.g. government agency/prime contractor, prime contractor/subcontractor, subcontractor/ supplier). Issues specific to these contractual interfaces are not discussed in this chapter.

7.1.2 Applicability

A separate design phase is warranted whenever the natural description of the system functionality involves a hierarchical structure—that is, where there is at least one level of functional subsystems between the system, seen as a single entity, and elements that can be produced in a normal implementation program.

In general, formal top-down design becomes more important as:

- the size of the project increases;
- the diversity of issues increases;
- the number of "specialist" design personnel increases (i.e. the project becomes increasingly multidisciplinary);
- the number of inexperienced project staff increases;
- the available timescales decrease (i.e. more personnel are required).

A word of caution is appropriate in relation to the final condition of applicability in this list. Top-down design is no panacea if project timescales are inadequate. While legitimate pressures to reduce timescales are part of most projects, the cost, inefficiency and management overhead of attempting excessive parallel activities should always be recognized. A task that one person can complete in 12 months is not necessarily achievable by 12 (or even 24!) people in just one month. Engineering projects require a certain "natural" time for problem mastery.

7.1.3 Purpose

From an **engineering standpoint**, top-down design allows decisions that:

- improve access to essential design information on a "need-to-know" basis;
- retain a common focus for design, based on system requirements;
- locate interfaces in a logical manner, exploiting functional independence where possible;
- facilitate the testing, integration, operation and maintenance of the intended system.

In relation to the first of these, a properly informed design team is clearly essential. Where large numbers are involved, this inevitably means thorough documentation, as other forms of communication (although crucial) are notoriously unreliable with the passage of time.

From a **management standpoint**, top-down design secures **accountability** through the work packages created by partitioning. These offer the logical structure for monitoring progress within a project. Consequently, top-down design has strong links with project management (see Section 2.5.4), affording:

- visibility and traceability in relation to expenditure;
- an early warning of problems;
- isolation of elements to be subcontracted;
- a structure for delegating design responsibility.

The process can also be used to build confidence in the design itself, particularly in the future capacity of the elements to come together in a successful system solution. Risk assessment trends are important in this regard.

7.2 Partitioning process

7.2.1 Partitioning requirements

The term "top-down design" suggests an essential dichotomy. On one hand, reference to "design" pinpoints the need for creativity, technical knowledge and sound engineering judgement—qualities always required for work involving analysis or synthesis. The methodology is a means of utilizing these capabilities to good effect. On the other hand, the "top-down" descriptor underscores the importance of an ordered approach to a process that would otherwise be unwieldy and inefficient at best, and unsuccessful or even impossible at worst. Above all it emphasizes a **"way of thinking" about design**.

As already stated, the objective of top-down design is to partition the system into a logical and complete set of elements in a step-by-step process. In the first subdivision, a set of subsystems must be identified that offers an optimum solution, judged in terms of the system requirements. (The notion of an "optimum" solution is used here to emphasize the need for choice among a set of recognized and realizable options.) In some cases an appropriate partitioning may be almost self-evident; in others several genuine options may be available, with the final choice requiring the prior development of special decision criteria. Whatever the circumstances, the function of each resulting subsystem should be distinctive and capable of individual definition within a set of subsystem specifications. The subsystems must also be realizable. Collectively, their specifications should form a complete set and describe the emerging design in greater detail than the initial system specification.

After review, the design proceeds to the next level of definition, with the further partitioning of the several subsystems as required.

As this process of subdivision continues through several stages, functional dependencies between elements should be limited to the branches identified by a family tree structure. Figure 7.1 illustrates this idea, with the principal interactions shown by the **hierarchial organization of subsystems and units**. This in turn limits the impact of design decisions, including any changes, to immediately related elements of the system (i.e. elements of the family tree that are horizontally connected or vertically dependent). For most systems, total compliance with this simple structure is rarely achievable, due to both inherent and practical constraints on the design. For example, performance requirements subject to budget allocations can affect many elements.

The changing character of the specifications supporting the top-down process is also illustrated in Figure 7.1. The system specification establishes all the essential characteristics of the system and expresses them as quantitative functional requirements. At the subsystem and unit level each specification contains additional and distinctive requirements, reflecting the progressive detailing of the system design. Typically, these specifications establish functional requirements in specialist fields or particular technologies. At the element level the specifications deal selectively with detailed design

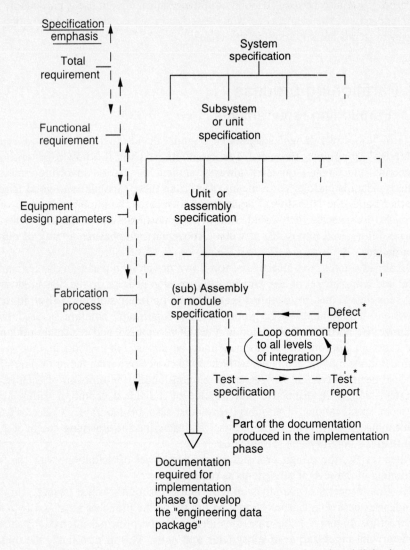

Figure 7.1 Family tree structure for specifications, showing functional links between different hierarchical levels.

requirements, including the definition and specification of design parameters for hardware and software items, and with the processes applicable to implementation. These specifications must be appropriate for equipment design and manufacture.

A vital part of the design phase is the **preparation of specifications** governing the testing, integration, transportation, installation, operation, maintenance and retirement of the system. Full documentation covering these aspects of the design must be prepared prior to the implementation phase, to avoid unnecessary compromises in system effectiveness. Some major problems arising with VLSI technology can be traced to the historical tendency among engineers to bypass this requirement.

7.2.2 Partitioning criteria

The partitioning process itself is shaped by a number of standard factors or **partitioning criteria**. In many cases, their application greatly simplifies the partitioning process. They include (but are not limited to):

- the physical separation of system elements;
- organizational division and responsibility for implementation or supply;
- the expertise required for design and manufacture;
- power or information flow across interfaces;
- constraints inherent in "post-design" requirements (e.g. testing, operation, maintenance);
- established practice and existing technology base.

Some divisions become self-evident due to the overriding importance of one or more of the above, while others still involve a critical choice. Judgement (or compromise) tied to **decision criteria** can be necessary to resolve conflict in difficult cases.

Examples showing the application of these criteria are given throughout this text, particularly in Chapters 1 and 2.

For large projects the engineering expertise and technology base are usually advanced incrementally, thereby containing the otherwise daunting prospect associated with this process. (Where this qualification has not always applied, e.g. in some aerospace programs, costs and timescales have proved extremely difficult to predict or contain.)

7.2.3 Process characteristics

At each stage of the partitioning process, the design should be examined for:

- its completeness;
- its superiority (e.g. in terms of cost and effectiveness);
- the intended commonality of elements, as a cost factor;
- the control of key parameters;
- the definition of interfaces (e.g. their adequacy);
- technological risk.

These aspects will be considered separately in subsequent sections of this chapter. All feature in **design review**, discussed in Section 7.3. At this point it suffices to note that this systematic review of the design places an increasing burden on the quality of the project definition as a faithful, complete and unambiguous interpretation of customer need. A disciplined top-down process cannot lead to a system that corrects deficiencies (e.g. omissions or misjudgements) in the initial phase of a project. It may of course lead to revision of the system specification, as the implications of particular parameters become clearer. Such revision requires careful control and will be discussed in Section 7.2.6.

Apart from experience, the engineering "tools" of modeling and simulation afford many of the insights needed for good design decisions at all levels. These were reviewed in Chapter 6. Typically, important questions relating to parameter limits and performance sensitivities cannot be resolved without their extended use, often with system-specific software.

The design process is characterized by a transition from serial to parallel activity, and it requires a corresponding expansion of the engineering team engaged on the project. While the efficiency and effectiveness of this team are properly the concerns of the project manager, the influence of the systems engineering process can be a significant factor. From a design viewpoint the need for effective **communication** is essential. It promotes a shared understanding of options and constraints within the team, and it limits the risk of uncoordinated decisions by design engineers working in isolation and relying on prior experience or preconceived ideas. Face-to-face discussion should occur when difficult decisions are required, and decisions should be made in ways that encourage maximum cooperation at an engineering level. Given the different cultural perspectives of specialist design engineers, the systems engineer has an important role at this stage. The design must be directed by the system requirements, not by the technological possibilities.

The preparation, approval and controlled distribution of all specifications and other relevant documents must be the first priority during the design phase. No other strategy is effective as a basis for engineering communication and for coordination of the emerging design. Formal methods for managing the design documentation will be discussed in Section 7.2.6. At this point the need to formulate, refine and finalize design solutions in a systematic top-down way must be emphasized, notwithstanding the difficulties that may surround decisions at any level. Under many contracts this involves customer approval by scheduled dates. Thereafter, proposed changes must be subject to the closest examination, in accordance with the procedures for configuration control. This is neccessary because of their potential cost to the project at the implementation phase.

Regrettably, the development of complex systems is seldom a progressive step-by-step sequence free from all backtracking. At any level of partitioning, unanticipated problems can be discovered for which redesign is the only solution. Effective containment of the resulting changes, especially those affecting interfaces, is a project management task—one facilitated in part by the thoroughness and insight of the initial design.

Design reviews should lead to the early discovery of engineering oversights, thereby containing their cost and time impact. However, many critical judgements cannot be fully tested until the verification phase of a project (or even thereafter in some instances). On an engineering level the resulting uncertainties about the effectiveness of the design can be monitored formally by **technical performance assessments**. These provide data for **risk analysis**, a process described in Section 7.4.

Effective design always requires decisions based on incomplete data. In terms of **risk management**, this means that the risk of an inappropriate technical decision (e.g. due to inadequate analysis) has to be balanced by the risk to timescales and costs (due to indecision).

7.2.4 Interface control

Effective interface control in an era of multidisciplinary projects requiring teams of design specialists has become an important challenge in systems engineering. While communication is often a problem between engineers with similar training and expertise, the process is much more demanding in a multidisciplinary setting.

Strategies for improving interface control include the following:

1. The **range of required expertise** should be established as objectively as possible, to contain design problems of two types. The first occurs when a few of the system requirements receive inadequate attention due to an imbalance in design expertise (e.g. decisions dominated by mainstream specialists). The second arises from excessive dependence on "generalists". Notwithstanding their value to many projects, generalists cannot solve all complex problems with acceptable insight. The design expertise required by multidisciplinary projects cannot be provided by one person, as the limits of technology continually expand.

2. Regardless of the individual expertise of team members, familiarity with the broad class of work is desirable within engineering teams and their host organizations. Projects attempted without this infrastructure of knowledge and practice will be loaded by substantial cost and timescale penalties, by comparison with projects drawing on established **multidisciplinary capabilities**. (For any organization this observation has significant management implications, involving personnel and technology development policies.) For example, many companies employ professional mechanical engineers to undertake the "packaging" of electronic and electrical systems. Their professional expertise becomes a cornerstone of product development programs—but this is not achieved overnight.

3. A natural yet vital extension of this argument is the encouragement of **overlapping expertise** in development projects. Technical interfaces that ignore the risk of "no man's land" are seldom satisfactory. Consequently, an antenna engineer may well require a basic knowledge of mechanical and structural design, while a mechanical engineer develops a similar capacity for electromagnetic principles. In many organizations this process allows the identification of future systems engineers, whose background includes projects in several specialist fields. The often salutary experience of working professionally under an engineer with a different background is now more commonplace in electrical engineering, given the opportunities created by its more pervasive technologies (i.e. electronics, communications and software).

4. Ultimately, the use of a **common language** must be achieved for successful interface control. For engineers, basic physics should offer the means of achieving an unambiguous technical understanding. Specifications prepared by specialists should be examined carefully for clarity, as misunderstandings arise where jargon is used or esoteric knowledge assumed.

   For example, the realization of a microwave lens requires several exchanges of effective engineering responsibility. In the first instance an electromagnetic solution is synthesized by an antenna engineer. This design must first be translated into a physical solution, described by a set of mechanical drawings that detail its method of fabrication from metal sheet and bars. Any testing during manufacture is limited to metrology and related procedures. Finally, the finished lens assembly has to be tested for function; this requires the antenna engineer to conduct microwave measurements. Given the timescales and costs in this process, the specification of the lens at the mechanical/antenna engineering interface is crucial. It should allow for a proper exercise of mechanical design trade-offs (particularly tolerances), yet seek to guarantee the required electromagnetic performance. A physical requirements specification can meet this need in an even-handed manner, although its preparation requires a clear definition of the fundamental design issues.

5. The **time** required for face-to-face discussions across multidisciplinary interfaces can be a major component of top-down design. Comparison with the time required to resolve single discipline interfaces is a poor guide. The main risk is underestimation of the time required for the former.

7.2.5 Documentation

The systems engineer must be responsible for releasing a range of design documentation that defines the system solution. Foremost among these documents would be:

1. *Block diagrams:* These name the functional elements used for a particular description of a system (or subsystem), show the principal interactions (typically but not necessarily information or power flow), and define relevant inputs and outputs. They provide a high level diagrammatic description masking all unnecessary complexity and are widely used by systems engineers. System block diagrams may be included in system specifications to illustrate required design features.

2. *Family trees:* These provide a hierarchical description of a system in terms of all its intended subsystems, units, assemblies and subassemblies. They are developed to the lowest level item to be tested during subsequent integration. Family trees not only summarize the top-down design but also provide the highest level access to the documentation package required for manufacture, assembly and testing. Consequently, family trees should show the name, part number and model number of each item as well as the quantity required to effect the next level of integration. Each item of the family tree is then described by its own parts list under normal drawing office practice.

3. *Budgets:* Budgeting becomes necessary when the attainment of system requirements depends on distributed design decisions. The range and character of budgets are clearly multiple, depending on the type of system. Some important examples of performance budgets will be given in the next section, but reliability, maintainability, space or power consumption budgets could be equally appropriate among a long list of possibilities.

   Budgets should distribute the design challenges equitably and be reviewed for validity as difficulties arise. The methodology for combining budget terms should be properly developed, and the anticipated outcomes (including safety margins) should remain consistent with overall system requirements. Individual terms should be established as acceptance criteria for subsystems or assemblies where appropriate.

4. *Flow diagrams and schematics:* The complexities of signal paths within a system can necessitate these additional system documents. Flow diagrams, for both hardware and software, identify the logical interconnections and give signal details. Schematics offer a more physical interpretation of the design and can be essential for signal routing (i.e. as a lead-in to wiring diagrams). Physical compatibility between hardware interfaces can be facilitated by the identification of connector types on schematic drawings.

As with budgets, the variety and purpose of flow diagrams and schematics cannot be captured in a few sentences. However, the need within project teams for documents that summarize the design should be addressed regularly during the design phase. In many cases of course, these same documents form a useful part of required customer documentation. However, it is their value to the design process that is stressed here.

5. *System tests:* System integration and testing requirements usually impinge on design. The stages of integration and the methods of performance verification must be known for the development of suitable test facilities, such as emulation capabilities. The design of special facilities can become another project requiring extended timescales and considerable resources.

6. *Design standards:* Project design standards must be established in accordance with the system specifications, having regard for existing practices within the design organization. A range of decisions is usually required at project level to ensure an integrated approach to implementation and a sensible rationalization of preferred hardware. Where performance implications appear quite limited, cost trade-off studies should be used as a basis for choice.

In addition to the documents already listed, day-to-day design and development activities should be ordered by a mix of plans and directives providing the framework for coordination and control. Plans form part of the activities of **systems engineering management**, and were discussed in Section 2.4. Typically, directives may cover design documentation practice (e.g. use of logbooks, preparation of design reports), changes in configuration management procedures, tests and trials on electromagnetic compatibility, and so on. Many issues relevant to the design can be addressed. In each instance the directive should express a logical response to the system requirements and detail the methods by which compliance is to be demonstrated.

7.2.6 Example: real-time budget

A challenge in the design of a substation automation system lies in satisfying the real-time operation constraints. As already explained, the functions to be carried out may be divided into two groups: the time-critical functions (e.g. auto reclose and breaker backup protection) and background functions (e.g. tapchanger control and high machine interaction). Correspondingly, the system configuration consists of a background subsystem and the so-called "peripheral" system. However, all the data going to and coming from the substation equipment (i.e. up to 960 digital outputs, 1776 digital inputs and 256 analog inputs) have to pass through the peripheral system; it is the peripheral system that updates the database.

The various processes that take place in a substation and their typical operating times (e.g. breaker operating times and arc deionization times) mean that a maximum acceptable cycle time for the fast functions has to be input and output, as well as processed, every 40 ms, and still leave enough time within each 40 ms cycle for the rest of the data to be handled within a reasonable time, say, 1 s. As a result of this real-time constraint it becomes necessary to allocate the cycle time to the various functions—that is, to establish a real-time budget.

The budget arising out of the top-down design is as follows:

Function	Allocated time [ms]
Reading all digital inputs and detecting any change of state	12.0
Updating database on change of state	0.5
Fast circuit-breaker routines:	
• command routine	4.0
• error routine	4.5
• back-up protection	2.5
• auto reclose	1.0
Output handler	6.5
Commands from background system	1.0
Copying database to background system	2.0
Reading 8 analog values into database	3.0
Scheduler (overhead)	1.0
	38.0

7.2.7 Configuration management

7.2.7.1 Introduction

The **system configuration** denotes the totality of all documentation (or data items) that defines the system at any one time in its lifecycle. Changes in the configuration are greatest during system development (i.e. the analysis and design phases) but continue to some extent throughout the operational lifetime. However, the **level of detail** of the definition at any one time, and therefore the level of detail in the configuration, increase throughout the design phase. The level of detail is defined by the **configuration items (CIs)**, which are essentially (but not necessarily) identical with the system elements.

Configuration management (CM) denotes the totality of activities that are necessary to control the documentation defining all the CIs. These activities fall into four groups:

configuration identification (or definition)
configuration control
configuration status accounting
configuration audits.

A project-specific **Configuration Management Plan (CMP)** defines how these activities are to be carried out. In particular, it defines the activities pertinent to the particular project, describes the organizational elements responsible for the various activities, contains or references standard procedures for their execution, and details the appropriate requirements for the whole process.

7.2.7.2 Configuration identification

Configuration identification is the listing of all the documents that define the system—a deceptively simple activity, but one that forms the basis of CM. Clearly, if an item of documentation is not recognized as being part of the configuration, it will escape all the subsequent control procedures.

The first CI is normally the system specification. The analysis phase ends, and CM commences, when the system specification is placed under CM. As the system is partitioned into finer elements during the design phase, additional documentation is produced, mainly in the form of specifications. This documentation also is placed under CM by listing it in the configuration identification. However, while higher level CIs disappear as the partitioning proceeds, so that in the end the set of CIs corresponds to the set of system elements, some higher level documents, such as the system specification, can never be deleted from the configuration identification list.

It might reasonably be asked why the project definition is not placed under CM. The answer lies in an understanding of the contractual aspects of a typical project. A contractor can bid for the execution of a project only when a project definition is included as part of the **request for tender (RFT)**. The development of the project definition is usually a separate contract, often on a time and materials basis, or a task carried out by the client's organization if the necessary expertise is available therein. The project definition, possibly modified during contract negotiations, becomes part of the contract, and any subsequent change to the project definition requires a **contract amendment** to be negotiated between the client and the contractor. Such negotiations lie outside the scope of CM, and therefore the contract (and within it the project definition) is not subject to CM. However, it is subject to the special process of contract management, which forms part of project management activities.

7.2.7.3 Configuration control

Configuration control consists of the activities and procedures needed to document all changes to the configuration and ensure that such changes take place only after appropriate approvals have been obtained. It thereby provides **traceability** and **accountability** for the evolution of the configuration throughout the system lifecycle. It is often supported and formalized by software packages.

The central entity in configuration control is the **Configuration Control Board (CCB)**. In recognition of its importance the CCB is chaired by the project manager. Its other members include the (systems) engineering manager, the integrated logistic support (ILS) manager, the test and trials manager, and, as its secretary, the assistant project manager in charge of CM. The latter prepares the paperwork for and coordinates the agenda of the CCB meetings and produces the minutes of the meetings. The CCB minutes constitute the most authoritative record of the system evolution.

The matters treated by the CCB are all related to changes and are presented in the form of various standard documents:

    engineering change proposal (ECP)
    document change notice (DCN)
    request for deviation.

Any change originates as a change request, produced by any organizational entity within either the client's or the contractor's organization. This request is discussed and processed by the appropriate parts of the contractor's organization to determine all the implications of the requested change, resulting in a formal document, the engineering change proposal (ECP).

This is presented to the CCB, which first of all decides whether to accept the ECP, based on formal grounds (e.g. has the request been processed correctly, has it been signed off by the right positions). The CCB then decides how to **classify** the changes, according to the following criteria:

- *class I*: changes that affect cost and/or schedule and therefore require a contract amendment;
- *class II*: changes that do not affect cost or schedule but do affect the form, fit or function of at least one system element;
- *class III*: changes that affect none of the above, such as an editorial change in a specification or the substitution of a more modern, but equivalent part for an existing part.

Class I changes are referred to the contract manager for action. If a contract amendment is negotiated, the ECP is returned, together with a copy of the contract amendment, to the CCB, where it is put on the agenda for a subsequent meeting and treated in the same manner as a class II change. If such a change (i.e. class I or II) is approved at a CCB meeting, the secretary arranges for all affected documentation to be modified. In a final action the CCB approves each of these changes, usually in the form of one or more **document change notices (DCNs)**. Class III changes are sent directly for processing; no further CCB approval is needed.

A **request for deviation** is used to request and document a temporary departure from a requirement at any particular point in the systems engineering process, when a permanent change is not acceptable. Such a request must specify when or under what conditions the deviation is to be rectified; it must not be open-ended. The phrase "until further notice" should be avoided, as this makes it difficult to carry out an audit.

7.2.7.4 Configuration status accounting

Configuration status accounting is a management information system that keeps track of the current status (version) of every document under CM. It also contains a record of all previous changes to the status. A good system for configuration status accounting makes unrecorded document changes all but impossible and requires careful design. It may at first sight seem paradoxical, but it is less serious to have an incorrect, unauthorized but recorded change than to have a correct, properly authorized but unrecorded change. The latter is like an activated timebomb. At any time it can be converted into a major problem when an old copy of the unchanged document resurfaces and is accepted as the current version.

For documents stored electronically there are commercially available software packages that provide a secure and detailed record of any changes to the documents under their control.

7.2.7.5 Configuration audits

Within the systems engineering process, **configuration audits** are primarily carried out to ensure the following:

1. Documentation for a CI that is under CM is complete. For example, the specifications for a CI may be placed correctly under CM, but certain drawings pertaining to the CI (e.g. installation drawings) may have escaped the system and may be simply kept in a "miscellaneous" drawer in the drawing office.

2. Changes recorded in the CCB minutes are reflected correctly in the current version of the documentation, as issued to users. The process being audited here is a multistep one, and there are numerous possibilities for errors or omissions to occur.
3. The configuration status reflects the current situation.

7.3 Design reviews

7.3.1 Purpose and format

Systematic structured **audits** of design work hold a prominent place in large projects. In some organizations, reviews are used to assess individual engineering performance; in others they are regarded as a screening procedure benefiting both engineer and employer.

Design review meetings can be conducted by an external agency (typically a government customer) or by an internal group of (preferably project independent) personnel and others qualified by appropriate knowledge. They are held to review the evolving design in the context of system requirements at strategic stages of a project. In most of the literature these reviews are termed:

- preliminary or **system requirements review (SRR)** (when work packages have been defined, and program perspectives and engineering preparation can be reviewed, e.g. adequacy of requirement specification);
- intermediate or **system design review (SDR)** (when substantial progress has been made and problems can be highlighted);
- critical (when design is essentially complete);
- final (when the system design has been completely documented and release for implementation is proposed).

Separate agenda (including checklists reflecting the different emphases of each review) should be used to structure the process and ensure an informative exchange. As follow-up is an essential component of review, all meetings should be minuted by action lists of corrective or investigative tasks and should identify responsibilities and completion dates.

The character of design reviews should be determined carefully in terms of expected preparation, participation, time allocation, intensity and formality. In many organizations these aspects, together with detailed checklists, are covered by standard practice manuals. Decisions should be shaped by the cost and importance of the review to the project. Procedures should be based on the assumption (requirement?) of good design practice, such as laboratory logbooks detailing progress.

For design reviews to maintain an appropriate objectivity, a value system for judging the quality of design is required. As individual engineers tend to weight design attributes according to their own perspectives, and design always involves trade-offs, a **design philosophy** can be a valuable aid in resolving conflicting assessments. Principles that need to be ordered by priority (notwithstanding their interdependences) could include:

> minimization of unit cost
> timescales for development
> performance characteristics
> development costs
> producibility
> flexibility
> technology development
> availability
> product growth
> documentation.

In all these initiatives, however, one basic principle should be observed at all times: **design reviews are not occasions for "second-guessing" the design itself.**

7.3.2 Requirements

The SRR reviews the work done prior to the start of the design phase. It may take place towards the end of the analysis phase or, more commonly, in the very early part of the design phase. The work reviewed falls into three groups.

The first group consists of the results of the following **analysis** tasks:

- requirements analysis: the work that led to the system requirements;
- functional flow analysis: the work that led to the first partitioning and associated interface definitions;
- preliminary requirements allocation: as it is applied to the first level of system partitioning;
- system cost-effectiveness analysis: in particular the economic feasibility analysis that led to the project proceeding.

The second group includes any **studies** that have been carried out. They may typically be related to such topics as risk, engineering specialties, testing (or test facilities) and design trade-off.

The third group is perhaps the most important one, yet probably the one most often neglected. It consists of all the engineering and engineering management **plans**, namely:

> System Engineering Management Plan
> Configuration Management Plan
> Reliability, Maintainability and Availability (RMA) Plan(s)
> Human Engineering Plan
> Data Management Plan
> Inegrated Logistic Support Plan
> Test Plan.

No work should be authorized before the plan under which it is to proceed has been reviewed and approved.

The SDR is normally carried out halfway through the design phase. It considers the design work done up to that point with regard to such characteristics as:

- optimization (i.e. consideration of alternative designs);
- traceability (from the system specification);
- completeness (i.e. conformance to the plans);
- risk.

The SDR also provides a major opportunity for coordination with related projects or with project activities outside system design, as well as a convenient interface with any independent verification and validation process.

An important aspect of any review is that it should result in a complete documentation of the **changes** that have taken place in the system during the period under review. The system design process is never a straightforward top-down process; there is always a certain amount of corrective feedback, and the requirements may also change during the time it takes to complete the design. It is a "living" process, and the design reviews should act as checks that ensure that this aspect of the process remains under control.

The critical design review is the last opportunity to question the system design and to consider introducing changes; at the conclusion of this review the system design is **frozen**. Otherwise the review is conducted along similar lines to the SDR.

The final design review has a completely different character from the previous design reviews. At this stage there is no longer any question of reviewing the quality of the design work, to question its adequacy or to suggest improvements. The purpose of this review is to check that:

- all work resulting from previous reviews has been carried out;
- the documentation (or data packages) is complete;
- all required approvals have been given.

As far as the engineering of a system is concerned, the **final design review marks the end of the main part of the systems engineering process.** The design is released for implementation, and equipment design and development commence.

7.3.3 Checklist

Preliminary review questions:

1. Has the purpose of the system been defined?
2. Has the planned operational deployment (i.e. usage pattern, operating mode) been defined?
3. Has the system lifecycle been defined?
4. Have all basic system performance parameters been defined?
5. Has the operational environment been defined in terms of temperature extremes, humidity, shock and vibration, storage, transportation and handling?
6. Has the maintenance concept been adequately documented?
7. Have all applicable standards been identified?
8. Have all boundary conditions (e.g. presented equipment, mandatory work practices and prohibited materials/substances) been properly identified and documented?
9. Are all necessary engineering and support plans complete and available?

Intermediate review questions:

1. Have quantitative reliability and maintainability factors been specified?
2. Has the level of maintenance been defined for each repairable item?
3. Have criteria been established for test and support equipment at each level of maintenance?
4. Have system operational and maintenance functions been defined?
5. Have operating and maintenance procedure requirements been defined? Have the necessary procedures been prepared?

7.4 Risk analysis

7.4.1 Definition

While a clear distinction has been drawn between system analysis and top-down design in this book, the topic of risk assessment is common to both. In Chapter 6 some of the methods for gathering informed opinion and formulating decisions were presented. These do not eliminate risks of course; they are merely tools that can be used to quantify and then contain **recognized** risks. This section examines the functional characteristics of control techniques applicable to design and development projects, and illustrates the use of these tools in a systems engineering context.

Risk analysis is an *iterative process that attempts to identify what could go wrong and to do something about it*. Risk analysis **identifies** potential problem areas, **quantifies** risks associated with these problems, **assesses** the effects of these risks and generates alternative actions to **reduce** risks.

Risk assessment *is an integrated judgement involving the main concerns in any project: technical performance, development costs and timescales*. It is a component of risk analysis, and can be expressed either qualitatively or quantitatively at a probability or confidence level. It requires the technical evaluation of an engineering capability for achieving specified requirements or objectives with a particular system design. Subsequently, the impact of this evaluation on both costs and schedules should be estimated to provide a single measure of the "system development risk".

Superimposed on the analysis activity is **risk management**, which includes *all management aspects of risk detection, analysis and reduction*, including the *determination* of alternative actions to reduce risks and the *selection* of criteria to implement these actions.

7.4.2 Purpose and importance

Risk assessment establishes a measure of the system development risk at the different stages of a project. It is used to monitor any engineering concerns impinging on the system design, performance, costs or development timescales, and the acceptability of expected outcomes. Drawing on the advice of management and engineering personnel, it is most frequently used during the design phase, to detect any loss of confidence.

With the growing complexity of systems there is an understandable desire for accurate risk assessment. Historically, the development of major systems has revealed a near universal weakness in the ability of systems engineers to identify risks accurately.

Many contributing factors can be cited, starting with the optimism usually demanded to win competitive contracts.

On a more positive note, however, one of the cherished but unfulfilled goals of systems engineering should now be noted. If risk assessment could provide the much required bridge between management and engineering, the prospects of marrying sound business management and technological innovation would be enhanced immensely.

Greater competence in the area of risk assessment is bound to be required of future systems engineers, in both the private and public sectors.

7.4.3 Prerequisites

System performance requirements must be clearly defined before an initial assessments of risk can be attempted. Thereafter, trends in the assessed risk can be measured and interpreted meaningfully only if these remain unchanged. Clearly, a new baseline for trend measurements is required if the performance requirements are changed during the design phase.

For credibility, risk assessment requires a negative viewpoint when analyzing and estimating probabilities. It is therefore concerned with judging the probability that a proposed course of action will **not** succeed within a particular combination of technical, cost and timescale requirements.

In many cases risk assessments are presented through sensitivity analyses, to expose the relationship between development risk and the requirements on system performance, project costs and development timescales. Such analyses require a small number of carefully determined risk parameters, specific to the project and the system. Some of the factors to be recognized in the definition of these risk parameters are given in Figure 7.2, which is reproduced from Chase (Chase 1974).

The concept of **technical performance measurement** is introduced by Figure 7.2. This term is concerned with the factual status of the design and is a measure applied during design reviews, following the collection of data from project design personnel. It provides information for design and project control and will be discussed further in Section 9.2.3.

7.4.4 Methodology

7.4.4.1 Definition of risk factor

If credible risk assessments are required during the system design phase, the methodological problem can be quite overwhelming.

In principle, every design parameter becomes a probability function in a risk assessment. In turn, every assembly, subsystem and overall system-performance characteristic is then a probability function resulting from an interaction of probability functions. Systematically identifying the "weak links" is a meticulous and time-consuming task. Clearly, all uncertainties can never be identified without an extended and repeated test and evaluation of a system under operational conditions. Fortunately, the "weak links" are usually more readily obtained; these then become the focus for more detailed review (e.g. by means of sensitivity analyses).

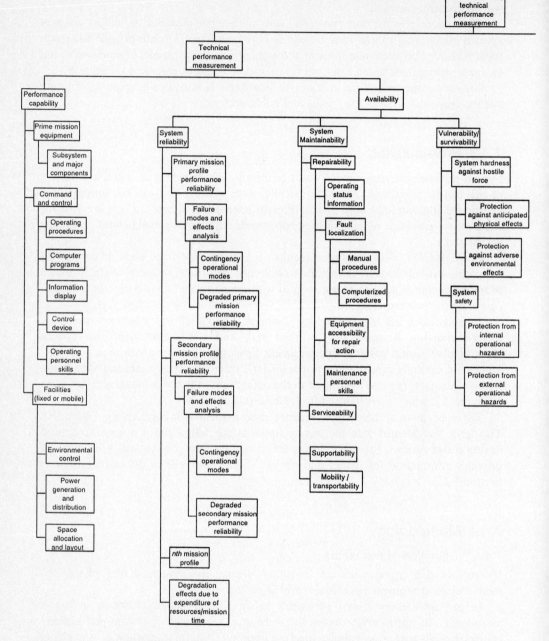

Figure 7.2 Elements involved in the definition and analysis of risk. (Reproduced from Chase 1974.)

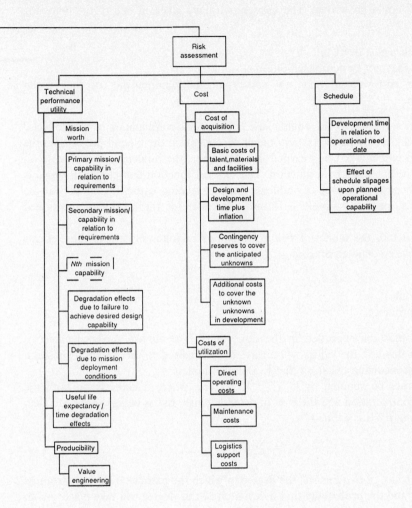

Approaching the analysis activity in a top-down fashion, the first question to be asked is: What could go wrong? The answer is usually given in one of the following ways:

1. There may be a cost overrun.
2. The project may not be completed on time.
3. The system performance may not satisfy all the requirements of the system specification.

The next question then is: What would cause any one or a combination of these effects? The answer is a list of parameters, very much dependent on the type of project and the environment in which it is being carried out. However, the influence of any cause on any effect can be described by a function. The simplest function that can be assumed is linear. On this basis a matrix of influence factors relating causes to effects can be developed. This matrix is simply a formal expression of the sensitivity mentioned earlier.

In greater detail, the model of risk is developed as follows. Let the three effects introduced above be denoted by E_i ($i = 1, 2, 3$), with

E_1 = cost
E_2 = schedule
E_3 = performance.

Each of these must be expressed in the same units. The almost universal choice is dollars, and in this text the values will always be normalized to the system acquisition cost (or the non-recurring costs), so that E_i is dimensionless.

Let the causes be denoted by C_i ($i = 1, \ldots, n$), where n depends on both the system under consideration and the level of detail at which risk is being examined. Then the sensitivity matrix R is defined by

$$E_i = \sum_j R_{ij} C_j$$

A cause is the result of two factors: the **degree** to which the parameter deviates from its planned value, and the **probability** that a deviation of this degree will take place. As an example, consider a task (i.e. work package) that needs a certain number of specialists working on it for it to be completed as planned. When the time comes to execute the task, this number of specialists may not be available. There is a certain probability of being one short, another probability of being two short, and so on. The result is a probability distribution $p(x)$, where x is the shortage. This is illustrated in Figure 7.3. The cause will in this case be defined by

$$C = \sum_x p(x)x$$

which is just the average value of x. It is equally possible and quite common to express C as the product of a single value of x, say X (usually the most probable value), and a probability P of this value occurring. Then the total effect of a given cause

$$E'_j = \sum_{i=1}^{3} E_{i,j}$$

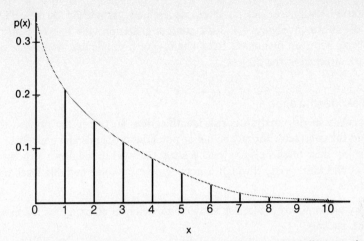

Figure 7.3 Probability distribution p(x) of being short x specialists at the time of execution of a task.

can be expressed as

$$E'_j = C_j \sum_{i=1}^{3} R_{ij}$$

$$= P_j X_j \sum_{i=1}^{3} R_{ij}$$

At this stage it is important to recognize that, conceptually, two probabilities are involved in this measure of effect. One is explicitly accounted for by the probability P of the cause; the other is contained within the sensitivity matrix R. It is the probability Q that a given effect will indeed **occur** as a result of the cause. It is expressed by the equations

$$Q_j = Q'_j / \max Q'_j$$

$$Q'_j = X_j \sum_{i-1}^{3} R_{ij}$$

where "max" indicates "worst case" and ensures that $0 \leq Q_j \leq 1$. This normalization is needed because R as defined above is unconstrained in terms of the input-output parameters which it relates. A practical method of achieving this normalization will be discussed in the next section.

With these definitions of the probabilities P and Q it is now possible to define the **risk factor** (RF) by

$$RF_j = P_j + Q_j - P_j Q_j$$

where $P_j Q_j$ is the joint probability of P_j and Q_j. This definition arises from the desire to express risk as a probability of anything going wrong, or

$$RF = 1 - \text{probability (nothing going wrong)}$$
$$= 1 - (1 - P)(1 - Q)$$

If P_j and Q_j are mutually exclusive, their joint probability is zero. This occurs only when there is no effect (i.e. $Q_j = 0$), in which case $RF_j = P_j$.

The above relation for RF is often represented graphically by so-called **isorisk contours**, as shown in Figure 7.4. Risk contour maps provide one of many ways to display all risk items on one chart. Such charts give a simple and immediate picture of the total risk attached to the project.

7.4.4.2 Risk identification

The first activity in risk analysis is **risk identification**: surveying the project, the client, the user and the contractor for any actual or potential problems and concerns.

The project may often be surveyed by using the work breakdown structure (WBS) as a type of checklist, going through it item by item at some suitable level and asking such questions as:

1. Were any assumptions made, explicitly or implicity, in planning for the execution of this work package?
2. How well founded were such assumptions?
3. Where data were used as input to the planning process, how reliable were they?
4. How applicable were the planning methods used for this particular project?

The client, the user and the contractor's own organization may prove to be much more difficult when it comes to identifying risks. Potential problem areas are connected with organizational instability (e.g. reorganization, takeovers, high personnel turnover rate). Another group of potential risks arises from the degree of experience and expertise available within these organizations for this particular type of project, and from their capacity to respond quickly and adequately to the changes in requirements that will inevitably occur.

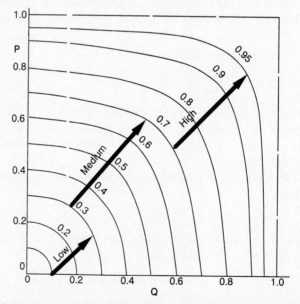

Figure 7.4 Isorisk contours in the (P,Q)-plane, where P is the probability that a cause (i.e. deviation to a certain degree) will appear, and Q is the probability that a certain effect will result from that cause.

7.4.4.3 Risk assessment

Risk assessment is concerned with the assignment of actual values to the causes and effects. In practice, this is done by assigning values to the probabilities P_j and Q_j. The following procedure is typical, but by no means unique:

1. For each identified risk (i.e. cause) define its magnitude X in a quantitative and measurable manner. Examples are:
   - an initial shortfall of three software engineers;
   - a change of technical director within the first year;
   - a change of government in the next election;
   - a 2-month delay in the supply of GaAs devices.
2. Determine the three coefficients R_i (i = 1, 2, 3); that is, ask: If the cause takes place, what will the effect be? The effect will of course depend on the circumstances prevailing at the time, on the scenario assumed. First try to find the worst case scenario, and then try to determine (i.e. estimate or guess) the effect under the scenario most likely to prevail, given the form of the project as it stands. As a result a value of Q between 0 and 1 is obtained.
3. Finally, assess how likely it is that the case under consideration really will occur; that is, estimate a value for P. Depending on the type of risk, there exist various models for obtaining the value of P.

   For example, if the risk of a piece of software being defective is under consideration, a simple model is

$$P = \tfrac{1}{3}(P_m + P_c + P_d)$$

where P_m = probability of failure due to lack of maturity
P_c = probability of failure due to complexity
P_d = probability of failure due to dependency on other items

and where the value of these probabilities are chosen according to Table 7.1.

Table 7.1 Probability of failure in software

Value	P_m	P_c	P_d
0.1	Existing	Simple	Independent
0.3	Minor redesign	Slightly complex	Schedule, existing factors
0.5	Major change	Moderate complexity	Performance, existing factors
0.7	New, but similar to existing	Complex	Schedule, new factors
0.9	Totally new	Very complex	Performance, new factors

The above procedure can be straightforward. However, in addition to determining a model for P, a major difficulty lies in determining the sensitivity (i.e. the factors R_{ij}), because R_{ij} generally expresses a relation between two completely different considerations. One, E, is always expressed in dollars, whereas the dimension of X can be anything, as can be seen from the examples in Step 1 above. Take the case when X is the fact that there is an initial shortfall of three software engineers. What is the likely effect of this on the project cost? Clearly, the answer depends on how the situation would be handled—on assuming a certain scenario.

7.4.4.4 Risk abatement

Having identified the risks and determined their importance or criticality, the next step is to try to find ways of reducing both the probability that the cause will occur and the severity of its effect. These **risk abatement** measures again depend on the specific case and on the flexibility of the project plans, but three types of measures are of general applicability:

1. Pull the risky activities forwards in the program. This reduces the effect of an unfavorable outcome on the project schedule.
2. Develop a fallback solution in parallel with the preferred one. This will of course mean increased cost, but may on the average be advantageous. (It is exactly the same argument as is used to justify insurance.)
3. Shift resources from noncritical activities to critical ones (e.g. by shifting the emphasis on reviews and independent validations to the critical activities). (This is analogous to CPM analysis to shorten the project duration.)

7.5 Summary

1. **Top-down design** is the stepwise process of **partitioning** a system specification into a self-consistent set of specifications for all the elements of the intended solution.
2. The **benefits** of top-down design are:
   • improved control of the design process;
   • reduced time period needed for design by doing design tasks in parallel;
   • reduced design effort by eliminating suboptimal solutions early in the process.
3. The partitioning on one particular level should conform to a single **partitioning criterion**; however, different criteria may be appropriate on different levels.
4. The design process must be adequately **documented**. Standard types of documents include:

   > system specifications
   > equipment specifications
   > interface specifications
   > test specifications
   > block diagrams
   > family trees
   > budgets
   > flow diagrams
   > manual of standards.

5. An important tool in controlling the process is **configuration management**, which is the management of all the documentation defining the system elements.
6. Another tool consists of formal **design reviews**.
7. The uncertainty inherent in systems engineering is reflected in **risk**—that is, the probability that things will not turn out as planned. **Risk analysis** should therefore be an inseparable part of any system design effort.

7.6 Short questions

1. What is the starting-point of the design process?
2. Why is the converse of top-down, bottom-up, not suitable for system design? Under what circumstances would bottom-up be an appropriate approach? Give an example.
3. Name two important tools or procedures used to control the design process.
4. Name three types of documentation produced within the design process.
5. List the main benefits arising from the application of top-down design procedures in an engineering project. Consider both the technical and managerial aspects.
6. Compare the design phase with the analysis phase, noting major similarities and differences.
7. What aspects should typically be checked at each level of partitioning during design?

7.7 Problems

Automobile electrical-system design: The electrical system within a modern car provides for the generation, distribution, control and utilization of electrical power. Typically, power is needed for internal and external lighting, indicators and instrumentation; for occupant comfort and entertainment; for power-assisted throttle actuation, fuel injection, braking, engine cooling, steering and suspension; for vision improvement in rain or mist; for the programmable (or "intelligent") control of vehicle performance parameters; for security and sensing purposes; and for other distributed loads enhancing vehicle appeal (e.g. power-assisted window or sunroof operation). In general, power is consumed only following the activation of individual subsystems by the occupant (e.g. by switching).

Overall, the system design reflects the evolutionary development of electrical and electronic equipment and its increased use in vehicles.

(a) Draw a block diagram that shows the functional **structure** and principal interactions of the electrical system found in a modern car.
(b) Identify twelve technical or performance issues relevant to a balanced high-level set of design requirements for the electrical system.
(c) Identify three distinctive elements of the electrical system found in all cars that have been re-engineered over the last 10–20 years, to take advantage of technology developments. (Give a brief explanation of the resulting benefits, especially in relation to your answer to (b).)
(d) Suggest three widely "standardized" **design philosophies** accepted by carmakers that could be changed with potential economic and technical justification in an unconstrained top-down design of an entirely new electrical system exploiting current technology. (Comment on the possible obstacles to the implementation of your proposals as part of your answer.)

7.8 References

Acquilar, R.J., *Systems Analysis and Design*, Englewood Cliffs, N.J.: Prentice Hall, 1977.

Chase, W.P., *Management of Systems Engineering*, New York: Wiley, 1974.

Deutsch, R., *System Analysis Techniques*, Englewood Cliffs, N.J.: Prentice Hall, 1969.

Director, S.W. and Rohrer, R.A., *Introduction to System Theory*, New York: McGraw-Hill, 1972.

IEE Proceedings, vol. 130, part A, no. 4, June 1983 (special issue on design).

Shinners, S.M., *Techniques of System Engineering*, New York: McGraw-Hill, 1967.

Wymore, A.W., *Systems Engineering Methodology for Interdisciplinary Teams*, New York: Wiley, 1976.

Zanger, H., *Electronic Systems, Theory and Applications*, Englewood Cliffs, N.J.: Prentice Hall, 1977.

CHAPTER 8

Verification of system performance

8.1 Overview

8.1.1 Definition

The verification phase of an engineering project is *a planned process of testing, evaluating, rectifying and documenting system performance at successive levels of system integration.* It is a structured process, which is planned and executed systematically with due regard for potential design modifications at any level. It occurs after the system design has been implemented. Its purpose is to show, as efficiently and effectively as possible, that the requirements of the system specification, including those relating to the system lifecycle, have been or will be met by the system as implemented. Verification is planned top-down but occurs bottom-up, with an emphasis on testing at the lowest level of system integration afforded by each requirement of the specification and by the system design. The process should facilitate system integration without the duplication of tests at successive stages of assembly.

8.1.2 Engineering requirements

The need for engineering judgement is a critical aspect of the verification phase of projects, where its consequences assume new implications. The challenge is essentially twofold.

The first area of judgement concerns the **validity** of the test and evaluation procedures themselves, and their capacity to establish or predict the "true" performance of the system over its operational lifetime. Typical issues to be considered in the context of validity include:

- the use of emulated interfaces or simulated conditions during testing, often with "worst case" values of specified parameters;

- the simplification of interdependencies, with potential sensitivities being masked by test procedures;
- the accuracy of instrumentation subsystems, their perturbation of the system under test, and the effects of any post-processing in the assessment of performance;
- any legal obligations (e.g. those involving the safety of electrical systems).

Where issues of legal obligation or human safety are involved, the actual test procedures normally have to be approved in advance by an external authority.

The second area of judgement (sometimes in conflict with the demands of validity) concerns the **economy** of the verification process. This involves more than cost containment. In major projects, economic considerations can take many forms and cover:

- risk assessment, weighing the costs of extended testing against the benefits of improved design confidence;
- cost minimization, linking projected verification costs to system design and implementation options;
- process vulnerability, particularly at higher levels of system integration, where verification can be affected by poor planning;
- alternative verification strategies and the search for cost-effectiveness.

The first of these considerations would be a concern in the testing of preproduction prototype systems. Typically, the risks would be recognized from the earliest stages of the development project and spread accordingly, with due allowance made in the establishment of the project definition and system requirements. The systems engineering basis of this spread risk will be considered in the next section.

A cost-effective verification strategy (i.e. one that is technically and economically sound) is as challenging to achieve as a successful project definition, system analysis or top-down design. It requires engineering creativity, insight and discipline, any of which can be overlooked if the systems engineering focus is limited to the early phases. In competitive tendering, for example, innovative design proposals are offered regularly in response to specified performance requirements, to secure a desired advantage. The associated effort involved in developing the proposed design is considered (quite properly) acceptable and even necessary. Unfortunately, the same approach is not always evident in the corresponding response to quality assurance requirements, despite the expected cost of compliance during an extended verification program. Such downgrading of verification as a systems engineering activity is unwarranted.

8.1.3 Relationship to earlier phases of systems engineering

In previous chapters it has been stated that a typical engineering project (i.e. one leading to the development of a new system) supports five sequential phases of systems engineering activity. Verification has been introduced logically as the last of these distinctive high-level activities.

A distinction must again be made here (and throughout this chapter) between the "verification phase" of an engineering project and "verification procedures" of the kind encountered at various stages of projects (e.g. as a part of configuration management or design reviews). The same ambiguity arises with the terms "analysis phase" and "design phase", which each carry a distinctive meaning in this text, not to be confused with the general processes of analysis and synthesis, which are applied in varying degrees at all stages of engineering projects. It should also be noted that the verification phase, as defined in this text, is sometimes called "integration and test" phase; in certain specialized fields, such as software engineering, this may even be the accepted usage. The term "verification" was chosen in this text as best suited in the context of a universally applicable methodology.

The character and (ultimately) the effectiveness of the verification phase are very much predetermined by the systems engineering decisions taken in earlier phases of the project. In particular:

1. The **philosophy** shaping performance verification as a project activity is established during the definition phase (and included in the project definition statement).
2. The **quality assurance requirements** applying to the system are established during the analysis phase (and included in the system specification).
3. The **plan** ordering the resulting processes of verification is determined also as part of the analysis phase (and issued as the System Test Plan).
4. The **preparation** needed to execute the plan is undertaken during the design and implementation phases (and often becomes a significant project in its own right).
5. The **test specifications** applying to every level of system integration are developed during the design phase (and form part of the overall documentation required).

Every requirement established through the system specification should be verifiable under the System Test Plan.

8.1.4 Scope and focus

The nature and purpose of the system itself strongly influence the verification process and its systems engineering responsibilities. Despite this diversity, some general features can be established about the process and its structure.

For example, consider the verification of:

- the functional performance of a critical subsystem;
- the environmental performance of the "prime equipment";
- the operational performance of the complete system, including its logistic support elements.

Each of the required set of tests occurs at a different stage of system integration, as part of the stepwise process of verification. While the steps themselves tend to be dictated by the system design (i.e. by its structure), typical **tests at each stage of integration can be distinguished** by:

- the type of requirements to be measured;
- the nature of the required test facilities;
- the choice of test venue(s);
- the extent of any logistic support;
- the elapsed time required to establish and conduct tests and assess their results;
- the cost to the project;
- the effectiveness of the tests as a verification of complete system performance.

These characteristics offer a natural basis for partitioning the overall process. For major systems the number of steps may be three to five, with a variety of tests scheduled at each step. The resulting **sequence of tests** may be classified into:

- tests on design models, mockups and so on—typically **laboratory tests**;
- tests on prototypes at subsystem and prime equipment level—typically **factory tests**;
- tests on complete systems under simulated operating conditions, utilizing logistic support—typically **field tests**;
- tests on complete systems under actual operating conditions, for design upgrades or improved effectiveness—typically **inservice tests**.

A potentially different partitioning of tests and engineering responsibilities can be established for systems being developed for production. Successive versions of these systems may be described by the terms "engineering model", "prototype", "preproduction model" and "production unit". Here the partitioning focuses on a **functional classification of tests** and would recognize, in the design field:

- **research and development tests**, frequently influencing decisions during the analysis and design phases;
- **design qualification tests** for the first completed system of a new type, demonstrating its performance in terms of functional, environmental, reliability, maintainability and compatibility requirements under prescribed conditions;
- **type tests** to establish or confirm type characteristics for a system (e.g. environmental performance under a quality assurance program);
- **design modification tests**, for an established system requiring improvements to its performance, reliability or maintainability, technology upgrades for manufacturing reproducibility or cost savings, and so on.

In practice, this classification of tests fits comfortably within the multi-step sequence introduced above. Subsequent material in this chapter dealing with planning, coordination and documentation applies generally to all tests, regardless of partitioning or classification detail.

Beyond the design realm, a variety of further tests is often needed to:

- establish the process conditions and limits demanded by production yield requirements;
- ensure the maintenance of these process conditions as part of quality control;
- confirm market projections for system sales;
- certify compliance with legal requirements (e.g. safety standards).

Tests of this kind are not considered explicity in this chapter.

8.1.5 Verification as an embedded project

Verification is often affected by increased pressures on project personnel. At a time when individual designers are looking ahead to their next project, there may be a tendency towards overconfidence and inadequate evaluation on their part. If this outlook is matched by management concerns about project timescales or costs, the psychological ingredients are in place for an unsatisfactory project outcome. This is particularly true if the steps outlined in Section 8.1.2 have not been undertaken competently and thoroughly. Sadly, many engineers only accept the discipline of systems engineering after personal experience of a project in difficulty—specifically, of system requirements unable to be demonstrated without time or cost overruns. In this respect it must be noted that the verification phase tends to expose poor decisions taken in the earlier stages of a project, including previously unrecognized or unacknowledged ones.

To counter these risks many project managers designate a single engineer (or group) to be responsible for the verification phase as an **embedded project** (within the host project). The role and principal interest of the specialist designer can then be balanced during design against the planning required for the phased integration, testing and evaluation of the system following its implementation.

Under this arrangement considerable attention must be paid to defining the interfaces between the "embedded" project and its "host", which it eventually dominates during system test and integration. At first the appointment of staff to such functions in a design team may be challenged as unnecessary, but experience with systems of varying complexity shows that the role is a vital one, requiring a substantial and focused effort from project inception. Above all it emphasizes that a **systematic approach bearing upon the system design at all levels is essential** to permit the complete and effective verification of system requirements. The widespread appreciation of this principle in fields such as VLSI and software engineering may be seen as an encouraging development, albeit one triggered by lessons learnt the hard way.

8.1.6 Trends

All phases of modern systems engineering reflect the increasing complexity of systems and the resulting criticality of their implementation technologies. Both of these influences can be recognized in the verification phase of engineering projects, through:

- the extended range of system performance requirements to be demonstrated;
- the increased amount of test data to be obtained;
- the greater accuracy/validity to be achieved;
- the increasing reference to "in-service" performance requirements.

Under this combination of pressures, verification is becoming more technology-dependent, as evidenced by the widespread use of computer-based tools. Concurrently, its links with design and implementation are being strengthened through the acceptance of integrated system databases. With this CADMAT (i.e. computer-aided design, manufacturing and test) technology the design engineer is able to develop automated test procedures with the very tools and models required initially for simulation studies. In many projects it is now accepted that the exploitation of this technology is all but essential, even for "one-off" systems (e.g. software systems). However, its full potential lies with systems intended for production, where its effects on quality assurance and system costs are then very significant (Carter 1987).

8.2 Evaluation of systems

8.2.1 Planning issues

As indicated in Section 8.1.2, the planning required as a lead-in to the verification phase commences in the early stages of an engineering program. It is guided by:

- the project definition statement, in terms of philosophy; and
- the system specification, particularly the quality assurance requirements;

and results in a document called the **System Test Plan**.

The System Test Plan establishes the set of activities needed to demonstrate the compliance of the system with the performance requirements of the system specification. It is written *before* the system is designed, for application *after* the system elements have been implemented. It is concerned with objectives, responsibilities and techniques, and offers a framework for a systematic and integrated approach to testing and evaluation in a stepwise process. It recognizes that every system performance requirement must be verified, at least during design qualification, always at the lowest level of integration consistent with technical feasibility. It also establishes the organizational policies and arrangements for testing and evaluation to occur as an embedded project.

The importance of this **structured (top-down) approach to planning** cannot be overemphasized. As all designers contribute to the testability of a system, and most participate in the verification program, the plan is an essential reference during design. It must therefore be produced concurrently with the system specification—that is, during the analysis phase of the project.

While the details of any plan must be project-specific, the following issues should be considered in its preparation:

- specific objectives of the system integration, test and evaluation program, viewed as an embedded project;
- any constraints and assumptions applying only to it;
- any applicable standards and definitions;
- the definition and purpose of tests at each level of system integration;
- the intended sequence of tests, with due regard for their possible duration;
- acceptable methods of testing, in relation to individual and possibly combined performance requirements;
- any special factors crucial to the validity of required tests;
- the definition of interfaces needed to support emulation;
- the organizational structure operating before and during the test program, with particular regard to responsibilities and delegations;
- venues for all tests, together with the required levels of logistic support in each case;
- standard test procedures and conditions;
- documentation requirements specific to the test program;
- reporting and follow-up arrangements, including the scheduling of required modifications; and
- the coordination and approval of procedures.

It is important to recognize here what can and cannot be included in the System Test Plan. As the highest-level project document dealing with system integration, testing and evaluation, it must recognize—primarily through its specified requirements—all issues relevant to the verification phase. **In a structured way, it must establish what is to be achieved, what is to be tested, who must do it, when and where it should be done, and how it should be done, in a process that delves into the issues and thereby builds a comprehensive definition of the verification phase**.

The means of checking for this implicit "completeness" are limited in much the same way as they are in other phases of systems engineering. Understanding and judgement are again crucial. Clearly, a well-formulated plan is more than a management strategy for maintaining project discipline!

By virtue of its release with the system specification, the System Test Plan is normally a mix of prescriptive and indicative requirements. The former are externally imposed; the latter relate to the system design and can therefore be finalized during the design phase. The schedule of subsystem factory tests is a typical example of the latter; the number and definition of required emulators may be another. Notwithstanding any subsequent revision, the System Test Plan remains limited in its scope to issues of the type listed above. It is not a test specification, and it does not set test paramaters.

8.2.2 Preparation

In accordance with the concept of an embedded project (Section 8.1.4), appropriate time and resources need to be allocated to prepare for system testing and performance verification. From the test plan a range of typical engineering tasks can be identified, covering:

- the determination of all required test specimens, with due regard for their expected availability by type, model number, change order status and so on;
- the preparation of all test specifications, including tests under simulated or actual operating conditions;
- the development of evaluation rules and procedures, for high-level tests permitting subjective interpretation or customer participation;
- the design and acquisition of special test facilities, appropriate to each stage of the test plan;
- the specification and development of emulation capabilities;
- the development of an information management system, covering the collection, validation, processing, analysis, reporting and maintenance of test results;
- the specification and establishment of the associated logistic support for field and operational tests, including the selection and training of the required personnel.

With the exception of the test specifications, responsibility for this preparation is typically assigned to the system test engineer(s). Regardless of the delegation, however, the interfaces between such activities and the (host) project require detailed agreement.

The development of test specifications, for each level of system partitioning (or stage of integration), is unarguably a responsibility of the design engineers. The progressive issue of test specifications as approved configuration items is evidence of the following:

- The intended design can be tested at the corresponding level of system integration.
- The proposed tests are consistent with the requirements of the System Test Plan.
- The specifications have been endorsed by the system test engineer(s).

Engineering preparation carried out under the System Test Plan (as part of the embedded project described in Section 8.1.5) should be further guided by the need for:

- validity, as described in Section 8.1.2;
- economy and cost-effectiveness, again as described in Section 8.1.2;
- the early identification of system deficiencies in either design or implementation;
- the elimination of unnecessarily duplicated tests;
- scheduling flexibility, to accommodate modifications and later retesting;
- the protection of critical items subject to deterioration or wear through repeated adjustment (e.g. connectors);
- an evolutionary transition from prototype test procedures to subsequent production testing.

8.2.3 Methods of verification

The main methods used by engineers for verifying properly defined system requirements are:

experiment
analysis
inspection,

with physical or theoretical modeling as a special subcategory of experiment and analysis.

8.2.3.1 Experiment

Experimental work involves direct measurement using equipment and facilities appropriate to the test being conducted. It requires suitably calibrated acquisition, processing and recording systems, and an understanding of error mechanisms and measurement accuracy, parameter and data rate limits, the effects of the test environment, and so on.

Through a judicious use of technology, measurement practice continues to match the demand for more complete and accurate information about system performance. Many test engineers now depend on computer-controlled measurements to meet expectations of greater quality assurance and improved personal or organizational productivity. With the resulting push-button convenience, fundamental insights probably receive less attention. Current generation engineers and technicians tend to assume the accuracy of their sophisticated instruments rather more readily than their predecessors were inclined (or entitled) to do with often simpler equipment. The corollary, however, is seen in the demand for more complex systems.

Despite the availability of better tools, the need for professional competence and the challenge of achieving it are clearly undiminished. In much the same way that future engineers are more likely to be involved with systems than with equipment, their interest in testing will center on complete instrumentation systems, as distinct from the individual instruments that form these systems (e.g. oscilloscopes, meters, bridges). (Detailed knowledge of the latter may reside with specialist technicians.) Examples of this perspective already abound, for example, in the testing of communication networks, navigation systems, power supply systems and satellite links.

The potentially sharp distinction between measurement and **analysis** is reduced by the need to process the raw experimental data in many test programs. Excluding tests that involve sampling, this may call for the elimination of known systematic errors or require the assumption of a further transfer function that (in practical terms) cannot be measured.

For example, in the flight testing of aircraft landing systems employing radio wave transmissions, the instantaneous (reference) bearing of a suitably instrumented aircraft is determined optically from the ground, relayed to the aircraft, and compared there with the measured bearing obtained through an onboard calibrated receiver of the system under test. Analysis of this measurement technique (developed in response to a very difficult test requirement) reveals many potential sources of error. Some of these are systematic and can be quantified. For example, the known delay time associated with the measurement, transmission, reception and recording of the optical reference data is larger than the corresponding delay in the test data obtained from the onboard receiver of the system under test. If the bearing of the aircraft used in such tests changes rapidly (e.g. during a test orbit at close range), any uncompensated comparison of the test and reference data "as recorded" involves a measurement error substantially larger than the prescribed accuracy limits for the ground system under test. The cost of the additional instrumentation required for accurate real-time comparisons is unjustified. In fact analysis shows that such action would simply substitute one set of errors with another. Compromises of this kind are commonplace in data acquisition systems, where the expected errors caused by additional instrumentation can exceed those in post-processed data.

8.2.3.2 Analysis

Analysis can be appropriate where direct measurement would involve the destruction of a system, a safety hazard or unacceptable cost. In these circumstances analysis alone may suffice, or it may be part of a broader verification strategy.

For example, an antenna to be installed adjacent to an aircraft runway is required to exhibit a prescribed frangibility, which can be defined as the turning moment causing the structure to collapse under impact. The cost of the antenna system precludes destructive testing, even on a one-off basis (i.e. by type testing). Typically, an analysis of the frangibility afforded by the design, backed by relevant test data for shear pins and structural models of the antenna, is acceptable for verification purposes.

A requirement for self-diagnostic features in various electronic systems offers another example. The effectiveness of the system design may be verified by an analysis of failure modes and their effects, supported by statistical data collected under the Maintainability Demonstration Plan.

With large software systems the sheer variety of input conditions can make experimental verification at the highest levels impracticable. Analyses illustrated by representative simulations (e.g. benchmark tests) must then provide the principal means of verification.

8.2.3.3 Inspection

Inspection is the obvious method of verifying many requirements included in system specifications. It normally involves a systematic verification of checklist features consistent with the intent of each requirement. Restrictions on the choice of parts, requirements covering physical characteristics or implementation details, or compliance with human engineering considerations in operation and maintenance, are typically checked by inspection. Documentation requirements are also checked in this way. Inspection remains a vital element in the controlled inflow of parts and materials into organizations.

8.2.3.4 Modeling

Modeling has already been considered as a tool for analysis and design. However where a model is known to be sufficiently complete and representative, it can also be used to establish system characteristics. While this perspective is a little unusual, the opportunities afforded by geometrical or electrical scaling illustrate this point.

For example, the scaling possibilities afforded by Maxwell's equations can be useful for precisely representing a linear electromagnetic system. If the geometrical scaling factor p is constant under a transformation from full-scale to model coordinate system—that is,

$$\begin{bmatrix} x \\ y \\ z \end{bmatrix} = \begin{bmatrix} p & 0 & 0 \\ 0 & p & 0 \\ 0 & 0 & p \end{bmatrix} \begin{bmatrix} x' \\ y' \\ z' \end{bmatrix} \tag{1}$$

—and if the other independent variables (t, μ, ϵ) within the equations are also subject to scaling, then a variety of different systems can be defined (constrained by the equations themselves) that are electromagnetically equivalent. In the most common, where μ and ϵ are regarded as invariant (due to the difficulty of simulating the atmosphere with any other medium),

$$t = pt' \tag{2}$$

and

$$\sigma = \frac{1}{p} \sigma' \tag{3}$$

Equations 2 and 3 point to the need to scale time (or frequency) and conductivity in a direct and inverse manner respectively, matching the change in physical dimensions of the system. This relationship between size and frequency is of course a familiar one.

Software modeling of a system whose performance can then be determined by simulation is another possibility for verification purposes, although it also is more commonly identified as a design tool when the deliverable is hardware.

8.2.4 Standards

In many countries standards defining acceptable (or prescribed) test methods have been prepared as a means of disseminating hard-won information and establishing common practice. Foremost among these are various USA military standards (MIL-STDs) concerned with test methods—part of the extensive documentation available to the systems engineer.

Standards can be particularly useful for tests involving:

environmental performance measurements
electromagnetic compatibility measurements
reliability and maintainability demonstrations
network performance demonstrations (under OSI),

to illustrate but a few classes relevant to electrical systems engineers. Standards should of course be used selectively.

8.2.5 Documentation

Two categories of documents required under all System Test Plans are:

test specifications
test reports.

While these vary widely according to the system and the associated level of testing, the following issues should be considered in their respective preparation.

8.2.5.1 Test specifications

A test specifiction will normally have the following sections:

1. *Introduction*:

   • states the purpose of the test, cites applicable documents (e.g. operating manuals, standards), identifies (sub)system by type or model number, describes the intended function and operation of the (sub)system, and gives the number of required test specimens.

2. *Requirements*:

   • cites the applicable specification of the system under test and lists the performance, environmental or general requirements to be confirmed.

3. *Test procedure*:

   • lists and describes in detail how the tests are to be conducted, specifying instruments, emulation interfaces, fixtures, circuit diagrams, measurement methods, recording procedures and so on;
   • designates when measurements are to be made, with respect to environmental cycling;
   • describes the method and sequence of individual environmental tests, test setups and facilities;
   • describes specific methods for evaluating the effects of environmental exposure;
   • details supporting procedures to be followed;
   • describes and formats records to be developed.

4. *Evaluation*:

   • describes the criteria for a successful test, which may be either on a pass/fail attribute basis or demonstrating a confidence level from a statistical analysis of the data.

5. *Appendices*:

- may be provided to illustrate sample data, details of experimental arrangements or other relevant material.

8.2.5.2 Test reports

Test reports should be prepared for each test or group of related tests. They should contain:

- the results of analyses on data obtained during testing;
- processing of results in a prescribed or standard manner in respect of each test objective (e.g. worst or average of several measurements);
- a statement about the compliance or otherwise of the system under test with the specified requirements.

All pertinent data, tapes, graphs, charts, photographs, computer printouts and references (e.g. inputs from failure and statistical analysis support activities) should be included.

In some instances reports should include recommendations for system changes to achieve performance, reliability, cost-effectiveness or other requirements.

8.3 Interdisciplinary linkages

8.3.1 Context

Verification links engineering design with a range of specialized disciplines. Two of the more important here are **quality assurance** and **quality control**. While the preceding material in this chapter applies equally to one-off and production systems, the linkages described below mainly concern the latter.

8.3.2 Type testing

Exhaustive testing of a system or item of equipment is a very expensive operation. In some instances the cost may exceed the value of the item by a large factor. Furthermore, a discriminating test may damage or weaken equipment (e.g. repeated testing for dielectric strength).

As a result of these and similar considerations, the concept of a **type** has been developed, and the proper use of **type testing** is a major factor in the economics of testing. By definition, a system or a piece of equipment (hardware and software) belongs to a type with respect to a set of **parameters** if *all the factors that determine these parameters are common to all members of the type*.

The simplest example of this is of course items that are exactly alike, such as mass-produced electronic equipment (e.g. radios, taperecorders); in this case all system parameters are characteristic of the type.

A more complex example is afforded by a family of modules, say, in the form of printed circuitboards, which can be combined to form equipment with different functions. In this case only some parameters are characteristic of the type (e.g. ambient temperature performance, shock resistance), whereas other parameters (e.g. accuracy, processing speed) vary according to the particular combination of modules.

The determination of type characteristics becomes even more complex when the system or equipment involves software. To what extent can the software be changed by the user, and which parameters must remain unchanged as type requirements? The difficulties are readily illustrated by examples (e.g. a programmable controller with embedded data acquisition and process control software).

The final complication arises with systems that include people (e.g. the system consisting of pilot and fighter plane). However, even here certain performance parameters are characteristic of a system type—the combination of an aeroplane type (e.g. long-range bomber, high-performance interceptor, ground-support attack plane, or all-purpose fighter) with pilot type (e.g. militia, reservist or active professional), for example.

Using the concept of types, testing falls into two categories:

- **type tests**, which are carried out on the first product/model(s);
- **performance tests**, which relate to a particular application.

The correct partitioning of testing requirements in this way is of great importance to the economics of production systems. Clearly, overall testing costs across production runs will diminish as more of the testing is done as type testing, but the certainty and effectiveness are also diminished. Decisions on this partitioning of tests require an understanding of quality assurance.

8.3.3 Quality assurance

In engineering enterprises a **quality assurance system** ensures that products supplied to clients perform **according to specification**. In this context quality is measured in terms of product acceptability, by the extent to which specified requirements are satisfied.

Quality assurance systems influence product quality in a number of distinctive ways:

1. They define the approved **methods and procedures** of an enterprise in all phases of its operation, at every level.

   For the design engineer these procedures are evident from the beginning of the system development cycle. They range across all support processes, from the acquisition of materials and parts to the shipment of completed products, and impose a mandatory discipline on engineering practices.

2. They provide for the management of approved methods and procedures through a **quality organization**, extending ultimately to all employees of the enterprise.

   The management of approved methods and procedures is achieved in part through **quality control**. This process is no longer limited in its meaning or influence to production areas and to the containment of product variability due to such factors as human operations. Sometimes termed "total quality control", it requires the continual monitoring, investigation, assessment and improvement of all activities (including design) that contribute to product quality.

3. They use **testing** as a tool, with an emphasis in many situations on sampling methods.

   One of the main issues here is increased reliance on product type testing and the resulting need for conformance through effectively controlled engineering and manufacturing processes.

The best quality-assurance systems build a "quality culture" in which all personnel within an enterprise contribute effectively to the maintenance and improvement of processes, products and services. Typically, they require the chief executive to accept direct responsibility for quality.

8.3.4 Legal aspects

Before accepting a system (or a piece of equipment) from a supplier, a customer will perform certain **acceptance tests**, which were agreed to at the time the contract was awarded. Should the system fail any of these tests, the supplier is required to take appropriate corrective measures, and the tests are repeated. Should the system turn out to be unusable, the customer may turn to another supplier, and the original supplier must withdraw the rejected system without any compensation. Only in rare cases will the questions of **consequential damages** arise; they are most often excluded anyway.

The situation is somewhat different in the case of type tests. If a piece of equipment fails, and it can be shown that this failure is not due to random causes but to the fact that the equipment does not conform to its type (e.g. through the intentional use of a lower grade of material, or through skipping certain procedures), then the manufacturer may very well be liable for consequential damages. In other words, saying that a product conforms to a type places a much heavier responsibility on the manufacturer.

8.4 Sampling methods

8.4.1 Rationale for sampling

The need for decisions amid uncertainty is a common theme within systems engineering. For quantitative models it leads to a dependence on statistical methods and probability theory. In this chapter it is seen in the use of acceptance sampling procedures as a method of quality assurance. The procedures introduce **criteria for acceptance or rejection with statistically known risks**, on the basis of sample test data.

Sampling provides an attractive strategy for several reasons:

1. Sampling limits the amount of testing required and so has a definite economic appeal. In production, only a small number of randomly selected items need to be tested from a larger population (i.e. lot or batch) submitted for acceptance. This is not to suggest that 100% testing is inappropriate in production; clearly, circumstances can be cited where such a procedure offers the only satisfactory standard. The verification of functional performance in complex printed circuitboard assemblies is such an example.
2. Sampling is the only feasible method where testing damages (or destroys) the item itself. Examples of this problem arise with the testing of various weapon, safety and switching systems.
3. Sampling recognizes that certainty is not achieved through testing unless the test process is itself perfect. Experimental studies of the accuracy achieved under 100% repetitive **manual** inspection suggest that errors at the rate of 10% can occur. In such circumstances, sampling techniques can be more reliable as a basis for decisions and for the protection of quality standards.

4. Properly designed and executed sampling schemes can contribute on a psychological level to the maintenance of quality, especially in situations requiring repetitive activity.

The extensive literature on statistics offers a range of viewpoints on the design and utility of sample test plans. For example, in quality control, testing is focused on product variability and on the resulting need to estimate (with a prescribed level of confidence) the achieved mean values of parameters or to identify temporal shifts in their distribution. Apart from their comparative ease of understanding, the generality of these issues presents a natural "entry point" for the following review of sampling techniques, which concentrates on the **formation of test strategies and the validity of inferences drawn from test results**.

Unfortunately, the discipline-specific terminology of classic quality control tends to mask the wider potential of sample testing within systems engineering. The latter extends to the demonstration of lifecycle performance parameters (e.g. system reliability) under assumptions of stationary point-of-time processes, and to the improvement of such parameters through "growth plans" (by system design and/or manufacturing modifications) (O'Connor 1985). Sampling is also required in Bayesian decision modeling, introduced in Chapter 5. For sophisticated applications not amenable to standard test procedures (e.g. those given in MIL-STD-105), specialist expertise should be sought in the development and interpretation of sampling schemes.

8.4.2 Attributes and variables

In the development of a sample test plan, a decision must be taken at the outset about the type of information to be used in the assessment of samples. If items are classified simply as effective or defective in respect of a desired property, but no other information is recorded, the description **"testing by attribute"** is applied to the process. When the measured value of the desired property is obtained as additional information, the term **"testing by variable"** is used to describe the sampling scheme. While intermediate classifications are possible (e.g. when items are graded), only the major division just defined is considered further in this section.

Generally, the distinction between attribute and variable stems from the chosen method of testing, as distinct from some intrinsic difficulty in performing measurements. For example, a semiconductor device capable of working at, say, 120°C may be regarded as satisfactory (or effective) for a particular requirement but as unsatisfactory (or defective) otherwise. (This does not mean that the rejected device is of no value for all other purposes! It is important to recognize the strict technical sense in which the word "defective" is generally used in the literature.) If the maximum temperature of satisfactory operation T_{max} can be measured, however, the further information points to "how good" or "how poor" the device happens to be. This measured quantitative characteristic (T_{max}) is the variable. The attribute in this case is defined by a lower specification limit of the variable, while the area under the lot probability density function for T_{max} below 120°C is a measure of the percentage defective (see Figure 8.1).

Similar interpretations are used to define attributes associated with upper, or combined upper and lower, specification limits. These relationships allow attribute

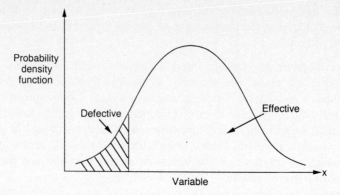

Figure 8.1 Lower limit specification of percentage defective, for a continuous variable.

testing to be regarded as a special case of variable testing, in which the only allowed measurements on a continuous scale of, say, 0 to 1 are 1 (for effective) or 0 (for defective). Of course, more information is obtained from a test by variable, and the required number of items per sample for equivalent protection is smaller.

Any measured variable can be converted to an attribute (e.g. by reference to specification limits that are "met" or "not met").

8.4.3 Risks in sampling

If items in a given population were all identical, testing just one would be enough to characterize the lot. In practice, however, intended copies of most items differ to an extent that precludes such a simple sampling strategy. The number of items to be tested (i.e. sample size) is therefore a vital aspect of statistically based methods, since it provides one measure of the information available for making decisions or inferences about the population.

Two particular sensitivities drive the design of sampling schemes where batch acceptance or rejection is envisaged. Historically, these have been identified as the **producer's risk (α or alpha factor)** and the **consumer's risk (β or beta factor)**, in recognition of their origins and character. The former is concerned with the assurance of batch acceptance if the percentage actually defective is less than, say θ_0, and the latter with the assurance of rejection if the percentage defective is greater than, say θ_1, where $\theta_1 > \theta_0$. Use of the terms "producer" and "consumer" is still widespread, although the terminology itself is somewhat unfortunate, suggesting very narrow interests.

The significance of these two risk factors, α and β, under a given sampling plan is revealed in the **operating characteristic (OC)**, which plots the probability of batch acceptance against the actual percentage of batch defective. The basis for deriving such curves will be considered in subsequent sections for specific test plans; meanwhile, Figure 8.2 shows that the probability of acceptance P_{acc} is a continuous function of the percentage defective.

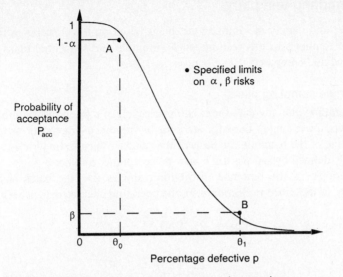

Figure 8.2 Typical operating characteristic for a sample test plan.

Assuming that α, β, θ_0 and θ_1 are given, the resultant curve is satisfactory in terms of both producer and consumer risks if:

for $\theta < \theta_0$, $\quad P_{acc} > 1 - \alpha$
for $\theta > \theta_1$, $\quad P_{acc} < \beta$

where P_{acc} is the probability of acceptance.

Operating characteristics can be produced for sampling schemes involving attribute or variable testing. In the latter case, appropriate extension of the concept defined in Figure 8.1 is required.

Schemes utilizing operating characteristics in the context of **hypothesis testing** form an important part of sampling theory. They are widely used to establish confidence levels in the value of a distributed parameter. Their limitations lie in the practical difficulties of **justifying the choices of** α **and** β risk points and in their **disregard of batch size**. Various alternative schemes are available, such as those utilizing **decision theory** and **average outgoing quality level (AOQL)** principles, both of which are described in the references (e.g. Bowker 1977).

8.4.4 Mathematical modeling

Given the variety of probability distributions for statistically defined characteristics, a correct mathematical model is needed to ensure that inferences drawn from sampling tests are valid. The binomial, hypergeometric and Poisson distributions are well suited to most attribute tests, while the normal and (to a lesser extent) the logarithmic distributions are appropriate for variable tests. This selection does not preclude the use of other distributions. The conditions applying to the use of these distributions, together with their statistical properties, are summarized in Section 8.9.

8.4.5 Sampling test plans

The number and variety of sampling test plans based on the concepts in the preceding sections are multiple. In this section, single, multiple and sequential plans are defined, characterized and compared.

8.4.5.1 Single sampling plans

A **single sampling** plan involves the random selection of a prescribed number of items n from a batch, the testing of these for attribute or variable characteristic as required, and the sentencing of the batch on the basis of the results. When testing for an attribute, the test is totally defined by sample size n and the acceptance number c.

For example, if the binomial distribution applies (i.e. the batch size N is large relative to n, as described in Section 8.9), the operating characteristic is seen to be

$$P_{acc} = \sum_{x=0}^{c} P(x)$$

$$= \sum_{x=0}^{c} \binom{n}{x} p^x (1 - p)^{n-x} \tag{4}$$

where p = actual (but unknown) fraction of population that is defective
x = number of sample defectives.

Where the batch size N is not large relative to the sample size n, the probability of acceptance has to be established under the hypergeometric distribution. It then takes the form (see Section 8.9)

$$P_{acc} = \sum_{x=0}^{c} P(x)$$

$$= \sum_{x=0}^{c} \left[\frac{\binom{k}{x} \binom{N-k}{n-x}}{\binom{N}{n}} \right] \tag{5}$$

where k = actual (but unknown) number of defectives in N.

The ratio k/N serves in this case to define the percentage defective for the operating characteristic.

Figure 8.3 shows plots of P_{acc} (as the accumulative binomial of p) for a series of similar test plans, where the sample size is taken as 10, 20, 30 or 40, and the number of allowable defectives for lot acceptance is kept to maxima of 1, 2, 3 and 4 respectively. The several plots show that the batch is always accepted if the actual percentage defective is zero and, correspondingly, always rejected if all items are defective. Between these extremes, however, the transition from high (say > 0.9) to low (say < 0.1) probability of acceptance becomes sharper as the sample size increases.

The last point illustrates one of the intuitively predictable features of this and other sampling schemes, namely that **discrimination** improves with sample size. In the context of assumed α and β risk factors, this can be seen in Figure 8.3 (through the points A and B) by the diminishing ratio of $\theta_1 : \theta_0$ as n increases. It is important to realize,

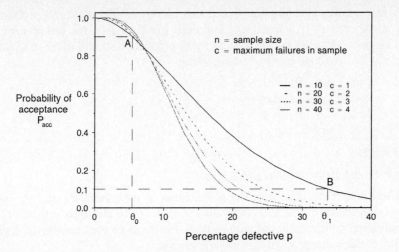

Figure 8.3   Operating characteristic for a single-sample test plan:

$$P_{acc} = \sum_{x=0}^{c} \binom{n}{x}(p)^x(1-p)^{n-x}$$

however, that this improved discrimination (with increasing sample size) does not imply reduced risks of "wrong decisions". The potentially ideal operating characteristics suggested by Figure 8.4 are only achievable (in theory) with 100% testing—that is, under strategies that defeat the reason for sampling.

Notwithstanding the insight afforded by sensitivity studies (through the variables n and c for single-sample attribute tests), the **synthesis** of test plans, as distinct from their **analysis**, is of greater concern to the system test engineer. With α, β, θ_0 and θ_1 specified, the task of designing a single-sample test procedure amounts to defining n and c in an optimum way. This is usually taken to mean the smallest sample.

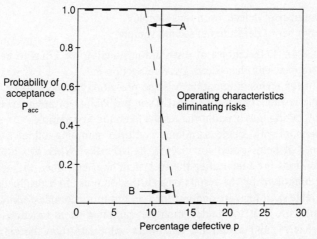

Figure 8.4   "Ideal" operating characteristics.

Extensive tables have been developed that allow the solution to this problem on a direct reference basis. An alternative approach is afforded by the use of Figure 8.5, which assumes a Poisson approximation to the binomial distribution (i.e. θ small). The results can also be found using χ^2 tables, where

$$\chi^2 = 2n\theta_0 \quad \text{and} \quad P_A = 1 - \alpha$$
$$= \text{level of significance of } \chi^2$$
$$\chi^2 = 2n\theta_1 \quad \text{and} \quad P_B = \beta$$
$$= \text{level of significance of } \chi^2$$

$$\vartheta = 2c + 2$$

where ϑ = degrees of freedom and is even.

These relations mean that, with the probabilities of acceptance P_A and P_B taken as levels of significance in χ^2 tables, two values of χ^2 are to be identified at these two levels of significance, which are in the ratio of $\theta_0:\theta_1$ for the **same** even number of degrees of freedom. Either value of χ^2 gives n, while ϑ gives c.

8.4.5.2 Multiple sampling plans

In **multiple sampling**, one or more samples are tested before an accept or reject decision is taken. However, the number of samples (or tests) is known.

The most common example of multiple sampling is double sampling. Under such a plan a first sample n_1 is tested. If c_1 or fewer failures occur, the lot is immediately accepted; correspondingly, if the number of failures exceeds r_1, where $r_1 > c_1$, the lot is rejected. When the number of failures from the first sample exceeds c_1 but not r_1, a second sample n_2 is tested. The lot is then accepted only if the total failures from both samples do not exceed c_2.

The probability of acceptance can be shown to be

$$P_{acc} = P \{x_1 \leq c_1 \quad \text{or} \quad (x_1 + x_2) \leq c_2\}$$

$$= \sum_{x_1=0}^{c_1} P(x_1) + \sum_{x=c_1+1}^{r_1} P(x_1) . \sum_{x_2=0}^{c_2-x_1} P(x_2) \tag{6}$$

where x_1 = number of defectives in first sample
x_2 = number of defectives in second sample.

As in Section 8.4.5.1, the choice of distribution function for x has to be appropriately made, in accordance with the criteria given in Section 8.7.

The first of the terms in equation 6 is the probability of acceptance based only on the first sample results. The second term is the probability of acceptance based on the second sample, taking into account the results from the first sample.

By comparison with single sampling, multiple sampling offers a saving in the average number of items tested for given α and β risks. When lots are very good or very poor, decisions are taken after the first sampling; but when the decision is more difficult, it is supported by the results of a further sample. In a production setting this leads to a **variable** sampling size and test load and to more complex administration. The **average** sample size per lot under multiple sampling is a function of the actual percentage defective, and assumes a maximum value less than the combined sample number per lot (see Figure 8.6).

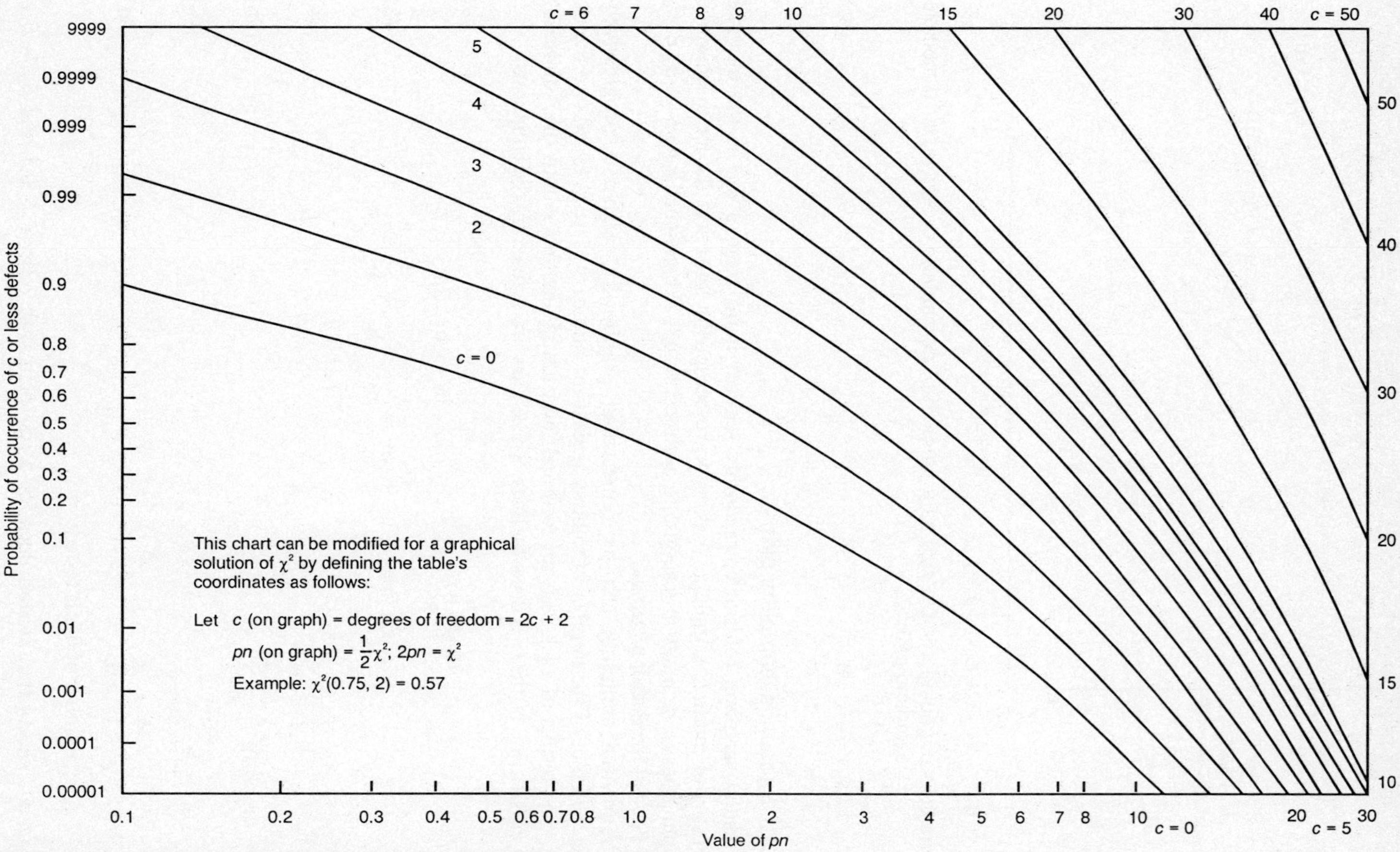
c = 6
7
8
9
10
15
20
30
40
c = 50
9999
0.9999
0.999
0.99
0.9
0.8
0.7
0.6
0.5
0.4
0.3
0.2
0.1
0.01
0.001
0.0001
0.00001
Probability of occurrence of c or less defects
5
4
3
2
c = 0
50
40
30
20
15
10
This chart can be modified for a graphical solution of χ² by defining the table's coordinates as follows:
Let c (on graph) = degrees of freedom = 2c + 2
pn (on graph) = ½χ²; 2pn = χ²
Example: χ²(0.75, 2) = 0.57
0.1
0.2
0.3
0.4
0.5
0.6
0.7
0.8
1.0
2
3
4
5
6
7
8
10
c = 0
20
c = 5
30
Value of pn

Figure 8.5 Cumulative probability curves—Poisson distribution.

Source: W. F. Dodge & W. G. Romig, Sampling Inspection Tables. New York: John Wiley & Sons Inc., 1944.

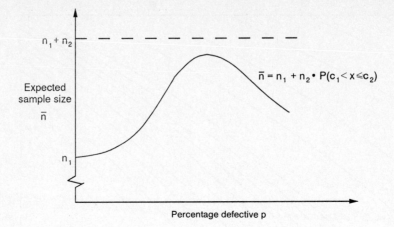

Figure 8.6 Expected sample size as a function of percentage defective, for a multiple sampling plan.

8.4.5.3 Sequential sampling plans

Sequential sampling (of single, randomly selected items from a lot) is the ultimate extension of multiple sampling. A decision to accept the lot, reject the lot or continue sampling has to be made in response to each test result. Pre-assigned values of α, β, θ_0 and θ_1 yield the decision criteria.

Sequential sampling can be represented graphically, within a progressively developed plot of the number of failures against the numbers of items tested (see Figure 8.7). The results of each test are recorded, and testing continues until the "staircase" plot crosses one of the two decision lines. These lines are defined by the equations

$$\text{accept line} = h_1 + sn = \frac{-a_1}{b_2 - b_1} + \frac{b_2 n}{b_2 - b_1} \tag{7}$$

$$\text{reject line} = h_2 + sn = \frac{-a_2}{b_2 - b_1} + \frac{b_2 n}{b_2 - b_1} \tag{8}$$

where $a_1 = \ln \dfrac{\beta}{1 - \alpha}$ $\hspace{4cm}$ (9a)

$a_2 = \ln \dfrac{1 - \beta}{\alpha}$ $\hspace{4cm}$ (9b)

$b_1 = \ln \dfrac{\theta_1}{\theta_0}$ $\hspace{4.5cm}$ (9c)

$b_2 = \ln \dfrac{1 - \theta_1}{1 - \theta_0}$ $\hspace{4cm}$ (9d)

n = number of items tested.

Figure 8.7 shows the division lines for a test plan with $\alpha = \beta = 0.1$ and $\theta_0 = 0.1$, $\theta_1 = 0.2$.

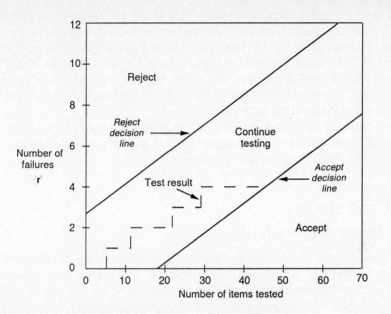

Figure 8.7 Graphical representation of sequential acceptance test.

The mean (or expected) sample size $\bar{n}$ is a function of the actual percentage defective. Together with the probability of acceptance P_{acc}, it can be found readily for five values of the percentage defective, including the two values that define the plan. Specifically,

$$\text{if } \theta = 0 \qquad P_{acc} = 1 \qquad \bar{n} = \frac{-h_1}{s} \qquad (10a)$$

$$\theta = \theta_0 \qquad P_{acc} = 1 - \alpha \qquad \bar{n} = \frac{(1 - \alpha)h_1 - \alpha h_2}{s - \theta_0} \qquad (10b)$$

$$\theta = s \qquad P_{acc} = \frac{h_2}{h_1 + h_2} \qquad \bar{n} = \frac{h_1 h_2}{s(1 - s)} \qquad (10c)$$

$$\theta = \theta_1 \qquad P_{acc} = \beta \qquad \bar{n} = \frac{(1 - \beta)h_2 - \beta h_1}{\beta - s} \qquad (10d)$$

$$\theta = 1 \qquad P_{acc} = 0 \qquad \bar{n} = \frac{h_2}{1 - s} \qquad (10e)$$

In the particular case where α and β are set equal and

$$h_1 = h_2 = h$$

the operating characteristic is defined by the equation

$$\theta = \left[\left(\frac{P_{acc}}{1 - P_{acc}} \right)^{s/h} - 1 \right] \left[\left(\frac{P_{acc}}{1 - P_{acc}} \right)^{1/h} - 1 \right]^{-1} \qquad (11)$$

Sequential sampling is more efficient, in terms of required sample size, than single or multiple sampling. For given risks α and β and discrimination ratio $\theta_1 : \theta_0$, the mean sample size is reduced by approximately 50% from that for a single sample plan. Consequently, sequential sampling is favored when the cost of testing is high or the number of items available for tests must be minimized. In some instances it is appropriate to set limits on the latter, thereby truncating the "continue testing" region of Figure 8.7 and modifying to some extent the statistical interpretation of the test plan. The overhead associated with scheduling and coordinating sequential tests is the principal disadvantage of the method.

The method of sequential sampling can be readily extended to tests involving the measurement of variables. An example will be given in the next section, where reliability demonstration testing will be discussed.

8.5 Case studies

8.5.1 System Test Plan

The outline in Table 8.1 of a System Test Plan is applicable to a broad class of systems, such as radars and navigational aids intended for international ground-based deployment. It illustrates the range of issues to be considered in this "high-level" project document, which would be issued under configuration management.

Table 8.1 Expanded Table of Contents for System Test Plan

---

1. INTRODUCTION

   1.1 *General*
   Background information for authorized users of the plan.

   1.2 *Purpose*
   The purpose of the integration, test and evaluation program.

   1.3 *Application of this plan*
   Systems covered by the plan.

   1.4 *Amendments*
   Provision for approved amendments to the plan under configuration control procedures.

   1.5 *Definitions*
   List of terms and words used in the plan with a project-specific meaning.

2. REFERENCED DOCUMENTS

   2.1 *Project specifications and publications*

   2.2 *Other documents*
   List of all relevant documents

3. REQUIREMENTS

   3.1 *General*
   Aspects such as acceptable verification methods, use of standard procedures, quality control and quality assurance requirements.

   3.2 *Test sequence*
   Delineation of sequential steps in the test program, illustrated by figures.

   3.3 *Equipment tests (stage I)*
   Tests on individual system elements, conducted by equipment suppliers.

   3.4 *Subsystem tests (stage II)*
   Tests on subassemblies and assemblies, conducted by their suppliers.

3.5 *Integrated system tests (stage III)*

   3.5.1 **System interface tests**
   Tests for compatibility of integrated subsystems across all interfaces, conducted by system test engineers.

   3.5.2 **System operability tests**
   Tests of the integrated system for functional performance under specified conditions, conducted by system test engineers.

3.6 *Field tests (stage IV)*

   3.6.1 **Field trials**
   Antenna range tests
   Field tests of performance, operability and compatibility (at selected sites).

   3.6.2 **Flight trials**
   Verification of siting.
   In-situ network integration.
   Final commissioning tests, prior to customer acceptance, conducted by independent authority.

3.7 *Service tests*
   Assessment of system additions and upgrades, under customer supervision.
   Periodic recalibration to confirm critical performance parameters.

3.8 *Analysis and inspection*
   Itemization of all system requirements that can be satisfied entirely by analysis or inspection.

3.9 *Factory tests*
   Itemization of all system requirements that can be met during factory tests, with due regard for combined performance, environmental and reliability testing.

3.10 *Reliability and maintainability demonstration*

   3.10.1 **Reliability demonstration**
   Execution of the reliability demonstration plan in conjunction with the test program.

   3.10.2 **Maintainability demonstration**
   Execution of the maintainability demonstration plan in conjunction with the test program.

3.11 *Integrated logistics support*
   Links with logistic engineering.

3.12 *Test index*
   Index of all factory and field tests, including description of item(s) under test, description of test, test identification details and responsible section.

8.5.2 System test specification

Table 8.2 is an edited version of a test specification for the generator protection systems (identified as "protection cubicles") to be installed in a new power station. It shows how the contractual test requirements divide into three categories: type tests, works tests and site tests. The type tests are concerned with general requirements; the works tests are concerned with stand-alone performance requirements; and the site tests are concerned with installed performance requirements.

   Two sets of standards are referenced: IEC (International Electrotechnical Commission) and AS (Australian Standard) publications.

Table 8.2 Extracts from a system test specification

.01 GENERAL

All equipment shall be subject to inspection and to testing in accordance with the appropriate standards as specified in the system specification. Where a standard provides for agreement on testing procedures or test levels between the "manufacturer" and the "purchaser", the Contractor shall submit proposed details of such tests to the Principal (i.e. Principal Contractor) for approval.

.02 TYPE TESTS

(a) *General*

Type tests shall be carried out in accordance with the appropriate parts of IEC Standard 255 to suit the purpose and application of the relay types and circuitry contained within the protection cubicles. It will be assumed that each reference to a protection cubicle is a reference to each separable portion of the cubicle(s) housing either the X or the Y protection equipment for each unit.

The specific relays supplied under this contract need not be subjected to these tests provided that test certificates are supplied declaring that these tests have been performed on the relay type and full details forwarded with the tender.

(b) *Test details*

The **type tests** shall include the following:

(i) **High frequency disturbance tests** (static relays only) in accordance with IEC 255—4, Appendix E, clause 5.

(ii) **Impulse voltage tests** with peak value of 5 kV and 1.2/50 ms waveform in accordance with IEC 255—5, clause 8.

(iii) **Environmental tests** on each relay in accordance with AS 1099—1980: *Basic Environmental Testing Procedures for Electrotechnology.*

There shall be no deterioration in the performance of the relay when subjected to the following type tests:

(1) **Dry heat test** (AS 1099, Part 2, Section 3, Test BC). In this test the relay shall be subjected to a temperature of 55°C for 2 h, and 40°C for 16 h, under load as a heat-dissipating specimen.

(2) **Damp heat test**. The relay shall be subjected to a temperature and relative humidity of $40 + 2°C$ and $90\% + 2\% - 3\%$ respectively for a period of 21 days. See *Test Ca* ref. AS 1099.2 Ca—1971.

(3) **Accelerated damp heat test.** The relay shall be tested during a 24 h cycle to observe the effects of high humidity when combined with wide temperature changes and resulting condensation on the components. See *Test Da*, ref. AS 1099.2 Da—1971.

(4) **Shock test**. This test is to simulate the effects of relatively infrequent nonrepetitive shocks likely to be encountered by the equipment in service or during transportation and to establish the relay's structural integrity. The tenderer is to advise what shock pulse the equipment offered can withstand without electrical or mechanical damage when mounted on a sheet steel control panel and when packed for shipment. See *Test Ea*, ref. AS 1099.2 Ea—1971.

(5) **Vibration tests.** The relays may be subjected to vibration. The tenderer shall advise to what degree the relay has been tested, in regard to vibration and its effect on the mechanical and electrical soundness of the relay.

.03 WORKS TESTS

(a) *General*

Following the completion of each protection cubicle, the Principal requires routine tests to be carried out on the cubicle and components in accordance with the appropriate standards in the presence of, and to the approval of, the Superintendent or his nominated representative.

These routine tests shall be carried out on each fully assembled protection cubicle within the premises of the manufacturer.

These tests shall be designed to fully prove the compliance of the protection system with the specification as far as the cubicle terminals, in order to minimize onsite commissioning time.

(b) *Test details*

The tests shall include the following:

(i) **Dielectric tests** on all circuits. The test voltage shall be 2 kV at 50 Hz for 1 minute for an insulation level of 500 V as given by the relevant parts of IEC 255—5 and AS 1138—1974.

(ii) **High-frequency disturbance tests** on all circuits, in accordance with IEC 255—4 Appendix E, clause 5.

(iii) **Impulse voltage tests** on all circuits with a peak value of 5 kV and 1.2/50 μs waveform in accordance with IEC 255—5, clause 8.

(iv) Each relay shall be tested to determine the **operating value**(s) (i.e. point on curve) of the various characteristic quantities (at least six) defining the relay characteristic. The test shall be in accordance with the reference conditions of IEC Standard 255 except that the setting(s) of the characteristic quantity shall be as stated by the Principal. The setting value(s) of the characteristic quantity will be in the region of the proposed setting value(s) of the relay. Under these test conditions the error in the operating values of the characteristic quantity, for at least six values of the characteristic quantity defining the relay characteristic, shall be determined and the error for each value shall not vary from the defined characteristic of the relay by more than $\pm 5\%$, unless otherwise specified.

Calibration checks shall be carried out on all other setting values of the relay's characteristic quantity at a sufficient number of operating values to satisfactorily determine the variation of error between the operating values. These tests shall be carried out under the reference conditions of IEC Standard 255 where applicable.

(v) With the required voltage and current values injected at the cubicle terminals, a functional test shall be performed to check the individual operation of every relay and its subsequent initiation of all **indications, alarms and trip outputs**.

(vi) Appropriate tests to the approval of the Superintendent shall be carried out to determine the **accuracy of the test equipment** in each cubicle. All operating sequences of the test equipment shall also be checked.

(c) *Test procedures*

All proposed testing procedures shall be submitted for approval, and approval obtained from the Superintendent, prior to the commencement of the tests.

(d) *Test certification*

Test certificates covering all the works tests referred to in this section shall be submitted for approval prior to the dispatch from the works of each protection cubicle.

.04 SITE TESTING AND COMMISSIONING

(a) *General*

(i) On arrival at site and during the course of erection all items of equipment shall be inspected by the Contractor so as to ensure that there shall be no delay in commissioning arising from defects, transport damage and/or deterioration of components.

(ii) Upon completion of the following:

(1) the erection of each completed protection cubicle,

(2) the erection of the current and voltage transformers and plant control room equipment that will interface with the protection cubicle in (1) above.

(3) the completion of the cabling associated with each protection cubicle in (1) above and its CTs, VTs, power supplies and plant control room equipment,

and in accordance with the program dates, the Contractor shall test and commission the whole system. This whole system comprises the Contractor's own equipment (in (1) of this section) and the equipment supplied, installed and tested by others (in (2) and (3) of this section).

(iii) Up to the time of taking over of each separable portion of the contract, the Principal reserves the right of inspection and rejection of any or all items not complying with the specification whether on account of faulty design, material or workmanship.

(iv) During the commissioning tests the Contractor's commissioning personnel shall work closely with, and cooperate fully with the Principal's personnel with the object of enabling the Principal's personnel to become fully conversant with the equipment.

(v) During the course of the commissioning tests the Contractor shall have suitably qualified labor readily available to rectify any defects or errors in any portion of the equipment that he has supplied and installed. This work shall be done to the approval of the Superintendent and at no cost to the Principal.

(vi) Final proving of each complete protection system shall be by primary injection.

The main alternator may be used for this purpose, in which case this testing procedure shall be subject to agreement between the Principal and the Contractor, and the execution of the procedure shall be fully subject to the supervision of the Principal with respect to operation, isolation and safety precautions. See also clause .04(b)(ii)(3).

(vii) Any or all of the tests may be witnessed by the Superintendent and the Contractor shall give him sufficient notice and afford him such facilities as is necessary for him to properly witness the tests.

(b) *Test details*

(i) The Contractor shall satisfy himself of the **compatibility** of all equipment and cabling interfacing with the Contractor's protection cubicles and necessary for the correct functioning of the protection system as a whole.

In this regard the Contractor shall:

(1) check all **interconnecting cabling** between the protection cubicles and the Principal's equipment and,

(2) check all of the Principal's equipment interfacing with the protection cubicles.

Information will be supplied to the Contractor as per .04(c).

(ii) The commissioning tests shall include:

(1) **Dielectric tests** as specified in clause .03(b)(i).

(2) **Functional checking** of the operation of each protection relay at a setting that will be provided by the Principal.

(3) Demonstration of the functional operation of each protection type from its respective source element(s) to the protection cubicles and to the plant control room. This shall include primary injection testing of the current transformers and execution of the appropriate tripping functions.

(4) Functional checking and demonstration of all testing routines of the test equipment contained in each cubicle.

(5) Functional checking and demonstration of all features of the complete protection system not covered elsewhere in this specification.

Any commissioning tests not covered by the above in (i) and (ii) but deemed necessary or advisable by the Contractor in order that he may assure himself of the correct and efficient functioning of the complete protection system shall be detailed and allowed for in the tender.

(iii) If on completion of the erection of each completed cubicle but prior to the termination by the Principal of the cabling to the cubicle, the Contractor deems that certain precommissioning tests should be carried out by the Contractor on his cubicles, then he shall allow the required time for these precommissioning tests in his program.

(c) *Information provided by the Principal*

(i) Prior to the commencement of the commissioning tests referred to in (b), the Principal will supply to the Contractor test results and drawings pertaining to the associated cabling, CT's and VT's.

(ii) Relay settings referred to in .04(b)(ii)(2).

(d) *Information to be provided by the Contractor*

(i) Six months prior to the commencement of the commissioning tests in (b) and the precommissioning tests in (c) (if any), the Contractor shall submit for approval commissioning information, which shall be provided in the following form:

(1) **Principles of testing**

This shall set out the testing and commissioning principles and objectives, relative to both initial completion and proving tests and station commissioning work. It will schedule the necessary tests and commissioning procedures.

The principles of testing shall give sufficient information about the design, function and performance of the plant and systems to facilitate approval of the testing and commissioning proposals by the Principal's commissioning engineer.

(2) **Plant completion test schedules**

These are based on the principles of testing and shall be in the form of instructions for use by the Principal's staff who will participate, assist or witness the tests.

The plant completion test schedules shall be provided in a format to the approval of the Superintendent.

(ii) Test certificates covering the commissioning tests shall be submitted before the issue of a certificate of practical completion.

8.5.3 System reliability demonstration/testing

Reliability testing is frequently an important aspect of the system verification process. It is of interest here because it:

- highlights the difficulties of verifying system lifecycle requirements even under operational conditions;
- occurs within an integrated test program which must be properly planned;
- develops bottom-up from the individual component to the system level;
- exploits statistical methods to provide assurance of system compliance with specification.

Above the level of individual components, testing for functional, environmental and reliability performance is normally constrained by inherent interdependencies.

At **system level**, the purpose of most reliability test plans is to demonstrate with prescribed confidence and at minimum cost that the system as designed and implemented meets specification. (Buried within this objective is the need for a precise definition of system failure—an issue examined further in Chapter 10.) This calls for test plans that free the final stages of testing from uncertainties about component reliability through tests at lower levels of integration, while providing for an assessment of the system under representative conditions.

The more obvious way to achieve the latter is to conduct the tests under operational conditions. In practice, however, cost pressures together with the complexities of program control and data collection, encourage the contractual recognition of standard tests in simulated environments. Under either conditions sequential sampling methods, applied to measurements of the time between system failures, allow efficient risk containment of the unknown **mean** (i.e. mean time before failure, MTBF). The time between failure is simply a continuous variable in such tests; its distribution is assumed to be exponential. For electronic and electrical systems the most widely accepted contractual standard for reliability testing is MIL-STD-785.

From the separate viewpoint of reliability engineering, this approach to testing is often criticized for its focus on failure times, as distinct from failure modes and their containment. Of itself it neglects weaknesses in design or implementation that may cause systematic failures (and may invalidate the statistical modeling assumptions and test inferences), and it provides no engineering direction if test results indicate a need for design improvements.

Consequently, the method must be recognized and employed for what it can offer in the **context of verification**, namely **efficiency** and **confidence**, subject to the validity of the test conditions and assumptions. While an estimate of achieved system reliability (under the conditions of test) can be made from the test results, it is worth remembering that the estimate itself would vary if the test program were rerun on another (or even the same) set of systems.

Reliability demonstration under MIL-STD-785 requires definition of the following parameters:

- target or design MTBF (θ_0), normally specified under the Reliability Development Plan to be greater than the MTBF required under the system specification;
- minimum acceptable MTBF (θ_1), corresponding to the contractual requirement that must be demonstrated—normally this value is the same as that given in the system specification;
- risk factors α and β, establishing respectively the probabilities that the system will be rejected though its actual MTBF exceeds θ_0, or accepted though its actual MTBF is less than θ_1.

The ratio $\theta_0 : \theta_1$ establishes the discrimination ratio, which, together with the risk factors, defines the acceptance and rejection criteria for sequential sample testing, as described in Section 8.4.5.3. With (preferably) several systems under test, the various times to failure are measured and the number of failures plotted as a function of the total test time T normalized in terms of θ_1). Accept and reject lines under standard plans (T_{acc} and T_{rej} respectively) are given by

$$T_{acc} = \frac{\ln(\theta_0/\theta_1)}{1/\theta_1 - 1/\theta_0} (r) - \frac{\ln[\beta/(1-\alpha)]}{1/\theta_1 - 1/\theta_0}$$

$$T_{rej} = \frac{\ln(\theta_0/\theta_1)}{1/\theta_1 - 1/\theta_0} (r) - \ln \frac{\ln[(1-\beta)/\alpha]}{1/\theta_1 - 1/\theta_0}$$

where r is the number of failures and T_{acc} and T_{rej} are normalized with respect to θ_1

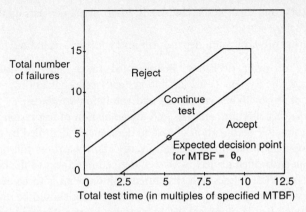

Figure 8.8 Total test time (in multiples of specified MTBF) for test plan III (MIL-HDBK-785). Decision risks: 10%; discrimination ratio: 2:1.

The procedure allows for system repair on failure, and return to testing. One of the standard plans under MIL-STD-785 is shown graphically by Figure 8.8. The operating characteristic and expected total test time (or number of failures) can be obtained from the statistical model for various values of actual MTBF (θ), as previously illustrated in Section 8.4.5.3.

For example, with $\theta = \theta_0$

$$P_{acc} = 1 - \alpha$$

$$\bar{r} = \frac{(1 - \alpha)\ln[\beta/(1 - \alpha)] + \alpha\ln[(1 - \beta)/\alpha]}{\ln(\theta_0/\theta_1) - (\theta_0/\theta_1 - 1)}$$

The truncation shown in Figure 8.8 affects the actual values of α and β, effectively reducing both risks.

If the system repair is achieved with design or implementation modifications, essentially the same procedure allows the demonstration of **reliability growth** through statistical analysis using the **method of least squares** and **regression theory** (Arsenault & Roberts). In effect, the test program allows inferences to be drawn about the (future) reliability of mature systems by analyzing the trends observed during tests of (currently) immature systems. These strategies are particularly pertinent to complex, newly developed systems (e.g. for aerospace or military applications), where the cost of reliability testing dictates containment of the total test time. Details of these methods, together with the standard environmental conditions to be considered in the reliability assessment of different classes of systems, are given in MIL-STD-785.

8.6 Summary

1. Verification ensures that the **lifecycle requirements** of the system specification are met. It is achieved after implementation through a **planned bottom-up process** of testing, evaluation, rectifying and documenting performance at each stage of system integration. Typically, this **sequential process** begins with laboratory tests, progresses

through factory and field tests, and ends after inservice tests under operational conditions.

2. The effectiveness of verification depends on many factors, including:

- the quality of **prior decisions, documented** (inter alia) in quality assurance requirements, the System Test Plan and system test specifications;
- the choice of **test methods, conditions and emulation interfaces;**
- the coordinated **collection, analysis and dissemination of test results;**
- the earliest possible **rectification** of system deficiencies identified by testing.

3. In the development of complex systems, verification warrants full-time engineering attention as an **embedded project**, with well-defined interfaces to the host project.

4. The **System Test Plan** defines in a structured fashion what is to be achieved, what is to be tested, who is responsible, when and where testing should be done, and how it should be done. Its development is an exercise in **top-down planning**, requiring continuous judgements about the validity and economy of the planned tests.

5. Standard methods of verification include experiment, analysis and inspection. **Sampling** procedures are widely used to improve the economy of testing.

6. Verification links engineering design to the specialized disciplines of **quality assurance** and **quality control**.

8.7 Statistical models used in sampling

8.7.1 Binomial

Sampling methods based on the binomial model are constrained by the following conditions:

1. Each item can be classified into one of two categories, say, effective or defective.
2. The probability p of obtaining a defective item remains the same for every item selected.
3. Each item is selected independently.
4. A fixed number of items, say n, is drawn for testing.

Under these conditions, the probability of drawing exactly x defectives (where $x \leq n$) is given by

$$P(x) = \binom{n}{x} p^x (1 - p)^{n-x} \qquad (x = 0, 1, 2, \ldots n) \qquad (1)$$

In effect, use of the binomial distribution assumes that samples are drawn at random from an infinite population. While all production batches in engineering are of finite size, the binomial model remains the most widely used. To some extent it can be more appropriate to the producer (who may be concerned with a "continuous" process) than to the consumer (who may purchase a batch of quite limited size).

8.7.2 Hypergeometric

The hypergeometric distribution affords the correct model when the percentage defective changes with the removal from the population of samples. It requires the following conditions to apply:

1. The sampling occurs from a batch (or population) of N items.
2. A simple random sample of n items is drawn without replacement.
3. Out of the N items, k are defective.

The probability of drawing exactly x defectives is then given by

$$p(x) = \left[\frac{\binom{k}{x} \binom{N-k}{n-x}}{\binom{N}{n}} \right] \left(\begin{array}{l} a \leq x \leq b \\ a = \max[0, n - (N - k)] \\ b = \max[k, n] \end{array} \right) \qquad (2)$$

(This result is easily found by considering the number of ways that x defectives, (n − x) effectives and n items can be drawn from their respective host numbers of k, (N − 1) and N.)

The use of equations 1 and 2 above in the extended formulae for sampling plans is facilitated by tables or statistical calculators. However, it is advantageous to know when the simpler binomial result is adequate (e.g. when n < N/10) or when both results can be approximated by more readily utilized formulae.

8.7.3 Poisson

The Poisson distribution is often used to compute the probabilities of defects per unit. It produces a somewhat less "square-shouldered" operating characteristic and therefore yields a conservative approximation in terms of required α and β risks than does the hypergeometric distribution. The strict conditions applying to its use are as follows:

1. The number of defects that occur in one "unit" is independent of the number occurring in any other.
2. For small units, the probability of a defect is proportional to the unit size.
3. The probability of two or more defects' occurring in a "very small unit" is so small that it can be neglected.

The probability of finding exactly x defects per unit is then given by

$$p(x) = \frac{(\exp)^{-\mu} \mu^x}{x!} \qquad (x = 0, 1, 2, \ldots) \qquad (3)$$

where μ is the mean number of defects in units of the size being examined.

In general, it is very difficult to confirm that the above conditions apply in many situations where Poisson statistics are used. However, the model is used for inspecting materials measured by area or volume (e.g. silicon crystals for integrated circuit fabrication, where defects in lattice structure affect yield).

8.7.4 Normal

When the normal or logarithmic distributions are required, the **sample** characteristics (i.e. mean and variance) and their probability distributions become important. For repeated samples drawn from a normal distribution, the familiar results for sets of measured variables x_i $(1 \leq i \leq n)$ are given by the following:

Mean value of sample means
$$= E(\bar{x}) \tag{4}$$
$$= \text{population mean } \mu$$

Standard deviation of sample means
$$= \sigma / \sqrt{n} \tag{5}$$
$$= \frac{\text{population standard deviation}}{\sqrt{n}}$$

with sample means normally distributed.

Mean value of sample variances
$$= E(s^2)$$
$$= \frac{n-1}{n} \sigma^2 \tag{6}$$

Standard deviation of sample variances $= \sigma^2 \sqrt{\dfrac{2}{n-1}}$ \hfill (7)

Unbiased estimate of population
variance afforded by sample $= \dfrac{n}{n-1} s^2$

where the variable $s\sqrt{n}/\sigma$ is distributed like χ with $(n-1)$ degrees of freedom.

If a population is **non-normal** but exhibits a mean μ and standard deviation σ, equations 4, 5, and 6 hold exactly, and the distribution of sample means approaches normality as n increases, as assured by the central limit theorem. Equation 7 is then approximately true, and the appropriate chi distribution is widely assumed for the variable $s\sqrt{n}/\sigma$ (σ being the standard deviation of the non-normal population).

For all the distributions mentioned above, where **estimates** are sought for population parameters (e.g. the percentage defective under a binomial distribution or the standard deviation under a normal distribution), consideration should be given to the **type** of information required. The choice lies between an unbiased point estimate or an interval estimate affording a prescribed degree of confidence. Details are found in the referenced textbooks and elsewhere. In all cases, however, it is necessary to know or assume the type of population (i.e. normal, logarithmic, binomial or whatever).

8.8 Short questions

1. What are the principal issues to be considered in the development of a System Test Plan?
2. Verification can be described as a sequence of tests. What distinguishes each step in the sequence, and why is testing sequential? What are the main methods of verifying system performance?
3. What is the rationale for using sampling techniques during the verification phase?
4. Describe the terms "producer's risk" and "consumer's risk" in relation to a sampling test procedure and its operating characteristic. Give three typical system requirements that are normally verified using sampling methods.
5. What are the principal advantages and disadvantages of sequential-sampling test plans with respect to single-sample test plans?

8.9 Problems

1. *Sequential testing*: A complex software algorithm transforms a set of input parameters into a set of output parameters. To test this algorithm a known good algorithm (but one that is perhaps much slower) will be used. A set of random input values is generated and applied to both algorithms, and the results are compared. Let the probability of detecting and identifying an error by one such test run be denoted by P, and let the probable number of initial errors be denoted by N_0. Both P and N_0 are of course initially unknown. Once identified, it is assumed that an error is automatically rectified.

   (a) Propose a sequential test and rectification procedure that will allow it to be stated that the probability of the algorithm not being error-free is less than e ($e \ll 1$).

   (b) Demonstrate the use of the test procedure for the case where the numbers of errors found and rectified in the first four runs were

$$n_1 = 4, \quad n_2 = 2, \quad n_3 = 1, \quad n_4 = 1$$

   and it is demanded that $e = 0.01$.

   (c) Discuss the limitation inherent in a sequential test procedure of this kind.

2. *Satellite testing*: In the design and testing of satellites prior to launch, each of the special conditions associated with either their deployment or their subsequent operation in space must be thoroughly assessed to minimize the risk of premature failure.

   (a) List six or more distinctive physical conditions that have to be taken into account in designing satellites (i.e. conditions linked with the operating environment or the transport process from the earth's surface).

   (b) Assuming that each of these conditions has been recognized in preparing the system specification for a new satellite, what assurances of compliance can be provided under a system test plan? (For each condition, summarize how the resulting requirements of the system specification should be tested for maximum validity.) Clearly identify any limitations that must be accepted as part of the test plan.

   (c) During an extended "end-to-end" performance evaluation of a satellite prior to launch, a technician records a brief burst (say 8–10 seconds) of high error-rate signals on the satellite transmissions. This isolated transient effect occurs during environmental cycling. What criterion should dictate the response to this observation? Briefly discuss the basic strategy that should be followed in the application of this criterion and the tools available to support that process.

8.10 References

Arsenault, J.E. & Roberts, J.A., *Reliability and Maintainability of Electronic Systems*, Computer Science Press.

Blanchard, B.S. & Fabrycky, W.J., *Systems Engineering and Analysis*, Englewood Cliffs, N.J.: Prentice Hall, 1990.

Bowker, A.H. & Lieberman, G.J., *Engineering Statistics*, Englewood Cliffs, N.J.: Prentice Hall, 1977.

Carter, D.E., "The role of test in design creativity", *Computer Aided Engineering Journal*, vol. 4, no. 3, June 1987, pp. 137–9.

MIL-STD-105, *Sampling Procedures and Table for Inspection by Attributes*, Washington DC: US Department of Defense, 1976.

MIL-STD-785, Military Standard, *Reliability Testing for Engineering Development, Qualification and Production*, Washington DC: US Department of Defense, 1967.

O'Connor, P.D.T., *Practical Reliability Engineering*, New York: Wiley, 1985.

Von Alven, W.H. (ed.), *Reliability Engineering*, Englewood Cliffs, N.J.: Prentice Hall, 1964.

Wetherill, G.B., *Sampling Inspection and Quality Control*, London: Chapman & Hall, 1977.

System
characteristics

CHAPTER 9

Performance

9.1 Choice of parameter set

9.1.1 Level of detail

Every system is designed to produce something: a service, a function or a product (goods). The characterization of that production process will be called the system **performance**.

Note that this is by no means a complete characterization of a system. Firstly, it only describes the system "as designed"; it does not consider the behavior of the system over its whole lifecycle. Secondly, it does not describe how that performance is supported and maintained; that is, the whole area of **logistic support** and system **operation** is neglected. Some of the important supporting issues, such as reliability and maintainability, will be treated in subsequent chapters; the all-encompassing system characterization in terms of cost-effectiveness will be treated separately in Chapter 14.

The definition of performance is tied to the functionality of the system. In the same way that this functionality can be broken down into more and more detailed subfunctions (characterizing smaller and smaller system elements), the definition of the system performance can be more or less detailed. Three points are important in this regard:

1. The set of parameters characterizing system performance should be developed and expanded in parallel with or as part of the system design, and should be subjected to the same formal traceability requirements.
2. The definition of system performance used in any particular case will depend on what aspects of system performance are being studied, but the parameter(s) used must always be a subset of the full definition (to the extent that it has been developed at that point in time). That is, a new definition of system performance can

be developed for a special purpose by simplifying or contracting the full definition, but no new performance criteria can be introduced. Doing so would really be a "hidden" way of changing the project objectives.

3. The definition must be compatible with the data available. It is useless to come up with a definition that, although formally correct, requires data that simply are not available at that stage of the engineering process, neither as facts nor as estimates.

9.1.2 Mutual independence

Choosing a set of parameters to characterize system performance is analogous to choosing a coordinate system to span an n-dimensional space. There are many choices of coordinate systems, but they all consist of n mutually independent coordinates. In choosing (n + 1) coordinates there would exist a relation between all or some of them; it would not be possible to specify values for all (n + 1) independently.

In the case of a system the situation is much more complex. For a start it is generally impossible to assign a particular "dimensionality" to a system; it is not so much a characteristic of the system as a function of the level of detail wanted in describing system performance. So it is impossible to say what the correct (i.e. minimum) number of parameters should be in order to have an independent set. In addition, the parameters generally must not be defined in an arbitrary fashion (e.g. by a linear transformation to an orthogonal set, as is done with coordinates). As discussed in Section 9.2, certain requirements are placed on any parameter definition. Finally, the relationships between the system parameters are often far from being linear or even expressible as simple functions. They may be non-linear, discontinuous (e.g. step functions) or statistical in nature.

As a result of all this, system performance parameters cannot be treated with the same mathematical rigor as a coordinate system. However, that does not mean that no attempt should be made to define the parameters so that they are as independent as possible, as the benefits of doing so are still very real:

1. The allocation process (requirements flowdown) is simplified; the number of system elements influencing the value of a parameter is reduced.
2. The design process is easier to manage, as responsibility for the achievement of a particular parameter value can be limited to a smaller group of engineers (ideally one).

9.1.3 Completeness

The question of the **completeness** of the set of parameters describing system performance can be viewed from three points of view:

1. The set must provide an adequate basis for the next step in the design process.
2. It must be possible to optimize system performance, as described by the set, in a self-consistent manner.
3. The set must cover (or correspond to) all aspects of user requirements.

The latter requirement is really a reflection of two separate, more precise requirements. On the one hand, any change whatsoever in user requirements must result in a change in one or more system performance parameter values. If not, the system is not an adequate solution to the problem of satisfying user requirements, and there is a lack of traceability. On the other hand, a change in one or more system-performance parameter values must result in a system that looks different to the user(s). A more precise way to express this is to say that, if the set of user requirements is viewed as a point in a user requirements space and the corresponding values of the system performance parameters are viewed as a point in a system performance space, then the function that maps the former onto the latter, as well as its inverse, must both be one-to-one.

The aspect of completeness expressed in point 2 above is related to an optimization criterion that includes one (or possibly more) additional parameter(s). In a spacecraft this may be weight, but in most systems it is cost. The second requirement then says that it must be possible to express the cost of the system as a function of the system performance parameters, and that this function must be single-valued (i.e. the inverse must be one-to-one).

Of course, there may be several such functions, each representing a different **solution** to the system design problem at that stage in the process (e.g. different technologies or different architectures). In that case the parameter set must be adequate for all the solutions. The performance parameters are a reflection of **user** requirements (albeit transformed into engineering terms); they must not be chosen with any particular solution in mind.

9.2 Definition requirements

9.2.1 Assumptions

It lies in the nature of the systems engineering methodology as a top-down process that, in order to construct useful definitions of performance parameters, it is necessary to make some assumptions, particularly in the early part of the process. For example, a common type of assumption consists of assumptions about the size, importance or influence of the terms in an expansion of a given parameter. The performance parameter is defined to be only one or a few of these terms, ignoring all the others. In this way a simple useful definition is obtained for which data can be obtained at that stage of the design process.

Another aspect of the definition of performance parameters, which may be viewed as a form of assumption, arises from the separation of the factors that influence performance into controllable and uncontrollable factors. **Uncontrollable** factors are typically interest rates, labor rates, cost of materials and factors related to the use of the system (e.g. cost of a system failure). These factors must be given, and their values may themselves be based on fact or on assumptions. The latter is not the issue here. It is rather that, by assuming certain factors to be uncontrollable (i.e. design-independent), the degree of complexity in the definition of performance parameters can be reduced.

An example of this occurs in the definition of the performance (e.g. speed or reaction time) of a man-machine system. By making the human performance an uncontrollable factor (or constant), the performance parameter will be time-independent and a function only of certain features of the man-machine interface (MMI) design. However, in reality the human performance is not uncontrollable. Due to the adaptability of human behavior there is a learning process, which can be influenced by the MMI design. The performance parameter is thus a time-dependent function of both MMI design features and human characteristics. The simplification realized by making the human performance an uncontrollable factor is clearly very considerable; the extent to which it is valid or useful must be investigated for each individual case.

9.2.2 Standards

When defining performance parameters for a particular system, the engineer must take into account that there exist a number of standards concerned with the definition and measurement of performance parameters. Clearly, there is no need to reinvent the wheel; whenever it is possible and useful to adopt an existing definition, this should be done. In particular, the definition should preferably be a specialization or further refinement of an existing (general) definition, and the vocabulary used and the test methods called upon for verification should conform to accepted standards.

A number of organizations produce and administer standards, and normally first preference should be given to **national standards**, to the extent that they are applicable. Such national standards may be subdivided into three groups:

- general standards applicable to the whole technical community, dealing mainly with matters relating to documentation (e.g. units, vocabulary, symbols, drawing practices);
- standards pertaining to general classifications and test procedures;
- standards containing performance requirements for specific plant and equipment.

Of these, it is only in the last group that special national requirements are important, to cater for dominant user groups (e.g. agriculture), climatic conditions (e.g. humidity, fungal growth) or legislation (e.g. safety, environment).

In the first two groups the national standards are often nothing but a reissue of **international standards**, sometimes with minor modifications to suit local tradition. When dealing with suppliers or customers in other countries, it is very useful to be aware of this connection and to refer to it.

The International Standards Organization (ISO) has its main office in Geneva, Switzerland, and as part of this organization the International Electrotechnical Commission (IEC) is responsible for issuing standards in the field of electrical engineering. Within the IEC there are many technical committees, in which experts from all over the world cooperate in working out the content and wording of the standards in draft form. These drafts are then submitted to the national electrotechnical committees, which have to give their approval, approval with comments or rejection within 6 months. The IEC standards are consecutively numbered, and a listing is published yearly.

The electromagnetic interference generated by electrical equipment in the frequency domain above 150 kHz has been the object of international cooperation for a long time, and the International Special Committee on Radio Interference (CISPR) has issued a number of standards.

Of the various national standards, some have gained importance beyond their national boundaries. Among these are the German Industrial Standards (DIN) and British Standards (BS), which cover practically all aspects of engineering activity. In the USA there are a number of bodies that issue standards, such as the Society of Automotive Engineers (SAE), the American Petroleum Institute (API) and the Institute of Electrical and Electronics Engineers (IEEE); the national body is called the American Standards Association (ASA).

Of particular importance are the US military standards (MIL-STD), for several reasons:

1. They are very extensive; and although they cover a lot of material that is of interest only to the military, they give an excellent coverage of the electronics field.
2. They are concerned with a quality level above that used in entertainment or even commercial electronics and are therefore increasingly being adopted for industrial applications.
3. They include a wealth of didactic material, mostly in the form of handbooks (MIL-HDBK), which is very useful and can be highly recommended as educational material. (The same is true also of many technical manuals for US military equipment.)

The MIL standards are subdivided into specifications for particular components and materials (e.g. MIL-F-495, *Finish, Chemical, Black, for Copper Alloys*), where the letter in the designation corresponds to the first letter of the test, and into the actual standards (e.g. MIL-STD-781B, *Reliability Tests: Exponential Distribution*). The various handbooks give information about the application of the standards (e.g. MIL-HDBK-217D, *Reliability Prediction of Electronic Equipment*).

9.2.3 Testability

It is not enough for a system performance parameter to be defined with mathematical precision; it must also be defined in such a way that it can be **tested** (or verified). In other words, the definition must be **operational**.

To ensure that this requirement is met, development of the test philosophy and test requirements must be started early in the engineering process, and then continued in parallel with all the other aspects of the system (e.g. performance, reliability, maintainability). Only in this manner is it possible to ensure that **all** system requirements are taken into account at each step or level of the design process, and that the solution converges towards one in which the **overall** cost-effectiveness is optimized. Only by ensuring the testability of the performance parameters will it be possible to close the loop at the end of the design phase, and to demonstrate (on paper) that the original user requirements (as spelled out in the project definition) will be satisfied by a system produced according to the specifications and tested according to the test requirements.

9.2.4 Technical performance measurement

Closely related to the definition and testability of performance parameters is the activity of actually determining values for such parameters during the course of the systems engineering process. This activity, called **technical performance measurement (TPM)** is not an actual measurement, as the system does not exist before the very end of the process. (When existing equipments or subsystems are to be incorporated into a new system, actual measurements can form a part of TPM.) Rather, TPM is the continuing prediction and demonstration of the degree of anticipated or actual achievement of selected technical objectives.

TPM includes the analysis of any differences between the achievement to date, the current estimate and the specification requirement. **Achievement to date** is the value of a technical parameter estimated or measured in a particular test and/or analysis. **Current estimate** is the value of a technical parameter predicted to be achieved at the end of the project within existing resources.

Selection of the key technical parameters to be monitored is guided by two criteria:

1. The parameters must be **critical** to the achievement of the overall system performance.
2. A well-defined (unambiguous) **measure** of the value of each parameter must be available throughout the design process (e.g. as the output of a model, such as a reliability model).

The number of parameters selected should initially (i.e. in the analysis phase) be kept small, as the total number of parameters that will need to be measured towards the end of the design process to support the key parameters may be ten times greater. Typical key parameters are:

> weight
> power
> computer throughput
> computer memory size
> processing time
> response time
> availability/reliability
> communications capacity (e.g. bandwidth, number of links).

Tracking the value of the parameters throughout the design process is a means of verifying that that process is on the right track. However, the parameters chosen may be affected by certain parts of the system only. So tracking a particular parameter may involve following the design process of certain system elements only or, in other words, of monitoring the output of certain work packages only. These are identified for each parameter using the work breakdown structure (WBS) and the requirements flowdown (or budgeting) process. The result is that each key parameter is associated with a subset of the WBS, sometimes called the **TPM tiered dependency tree**.

The TPM process involves a number of activities:

> planned parameter profile development
> TPM parameter status tracking and forecast
> TPM status reporting

The concept of a **parameter profile** arises from the fact that many (but not all) TPM parameters reach their final value by means of a development process. For instance, the real-time budget of a cyclic process-control system will be used up as more and more control processes are added in; the weight budget on a spacecraft payload will be used up as the design progresses (or may decrease towards the budgeted value, depending on design philosophy); and so on. The end value is restrained by the value defined in the appropriate specification; the planned approach to this end value is the parameter profile. However, there are of course TPM parameters from which no change is expected; the end value is achieved as the result of a single task in the design process.

The major effort in TPM is directed towards tracking the status of the TPM parameters. Status may be determined by design analysis, simulation or any type of testing. The accumulation of performance-parameter status data over time constitutes the **achieved parameter profile**. Many parameter profiles for lower level elements may be constructed directly from the status data; others, for higher level elements, can be derived through appropriate summation models (as used in the budgeting process) from parameter values of lower level elements.

TPM status reporting is based on a comparison between the planned and achieved parameter profiles. Any excessive deviation is quickly identified. By injecting the resulting TPM report into the regular systems engineering review process, remedial action, in the form of an increase or redirection of the program effort, may be initiated in order to overcome or minimize the effect of any weakness that may develop in the planned program.

9.3 Case studies

9.3.1 Tactical communications system

9.3.1.1 Definition of the grade of service

The next step in the characterization and modeling of tactical networks is to define the system performance. From the user's point of view, this performance will be judged by several characteristics, such as intelligibility, speed of obtaining a connection, availability of trunk capacity, available services (e.g. directory service) and so on. Somehow these characteristics must be reflected in precisely defined system performance parameters. However, in the spirit of the top-down design process, the first thing to look for is a measure of system performance that is as simple as possible, preferably a single parameter.

Following on from Section 6.5.2.2, the engineer could try to calculate a **grade of service** (GoS) by matching traffic against network capacity. Let GoS be defined as *the probability of establishing a connection on the first attempt between any two subscribers, given that the called subscriber is not busy*. This is quite a common definition for a telephone system, and it reflects a characteristic of the service that is both of major importance and determined mainly by the high-level system design. A characteristic such as intelligibility is determined more by the detailed design of the system, as reflected in the equipment specifications.

9.3.1.2 Traffic

The simplest description of the traffic arises by assuming it to be generated in a random but uniform manner, such that all links are on average equally loaded, and then to operate only with this average value. Let each subscriber generate on average g calls per hour, and let each call have duration h. Then the traffic generated by each subscriber s is given by

$$s = gh$$

The next step up in complexity (and thereby realism) in the description of the traffic is to assume that the traffic is still spatially uniform (i.e. equally loaded on all links), but to take into account the fluctuations of the traffic in time. With the calls being placed at random, the probability of having x calls on a link is equivalent to the probability of having x calls initiated in a time interval h, and this is given by a Poisson distribution $P(x)$ with

$$P(x) = \frac{e^{-q} q^x}{x!}$$

where q is the average traffic density per link, in erlang.

Referring back to Section 6.5.2.2, q is determined as follows:

$$q = \frac{UsL}{M} = \frac{29s}{C_s} \approx 230s$$

where U = number of subscribers
L = average connection length
M = number of links
C = traffic capacity of the network
C_s = subscriber capacity.

9.3.1.3 Evaluation of the grade of service

Using the average traffic model, GoS is unity as long as $s \leq C_s$; for $s > C_s$, GoS is simply C_s/s, as shown in Figure 9.1.

Using the Poisson model, GoS is given by

$$GoS = 1 - B(29,q)$$

where

$$B(c,q) = (q^c/c!) \sum_{i=0}^{c} (q^i/i!)$$

is the Erlang B formula. This formula gives the probability of not being able to establish a call between two subscribers connected by c circuits in parallel, in the presence of total traffic of q erlang. This result also is shown in Figure 9.1.

While at first glance the simple model may seem an acceptable approximation, it is important to remember that it is normally the region $0.9 \leq GoS \leq 0.999$ that is of particular interest. Expanding this region on a logarithmic scale, as shown in Figure 9.2, it becomes apparent that the use of averages (i.e. neglecting the fluctuations) could lead to significant underdimensioning of the network.

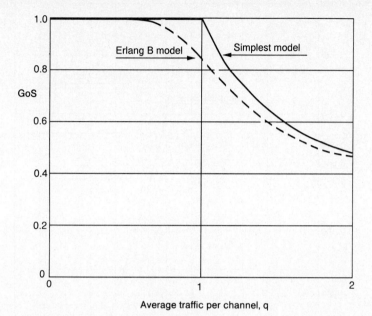

Figure 9.1 Grade of service for models of the tactical communications system. The simplest model characterizes the traffic by its average value only; the Erlang B model takes the fluctuations arising from random call generation into account.

9.3.2 Combustion optimization system

9.3.2.1 Performance definition

The purpose of the combustion optimization system should now be clear: to effect cost savings by minimizing flue losses, and to do this by measuring the oxygen content in the flue and using feed-back control to optimize the fuel/air ratio consistent with acceptable pollution levels. The system was introduced in Section 1.4.2, and an economic model was developed in Section 6.5.1. However, the latter implicitly assumed ideal operation of the system and was concerned only with determining the economic viability of the optimization idea. To proceed with the design and to be able to allocate parameter values as design requirements, a definition of system performance is first needed.

From the purpose of the system, it follows that the degree to which the system is able to reduce the flue losses must be an overall performance parameter. Call this parameter the **reduction factor**, and let it be denoted by $(1 - q)$. Then $q = 0$ corresponds to the case where the system maintains the oxygen content of the flue equal to the optimum value at all power levels, as given by the full curve in Figure 1.7. The value $q = 1$ corresponds to the performance available without optimization.

Simplifying the description of the system performance to its static (or average) behavior, the reduction factor is simply related to the **accuracy** of the system. It is only the presence of **errors** that prevents the system from keeping the combustion operating point on the ideal curve. These errors are essentially twofold: the acquisition error and the deadband error.

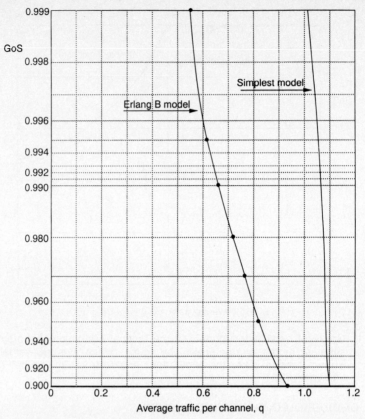

Figure 9.2 Grade of service for the same two models as in Figure 9.1, but focused on the range $0.9 \leqslant \text{GoS} \leqslant 0.999$.

The acquisition error is associated with the determination of the two input values, oxygen level and power level, both of which are needed for the electronics to calculate the appropriate correction. That is, the measured oxygen level x' will differ from the true value x by an error e_m, such that

$$x' = x(1 \pm e_m)$$

Similarly, the calculated set-point value x_0' will differ from the true value x_0, by an error e_0, such that

$$x_0' = x_0(1 \pm e_0)$$

In the worst case the total error will be the sum of the two errors,

$$e = e_m + e_0$$

and the acquisition error can then be written as $ex_0/\triangle x$, where $\triangle x$ is the distance between the full and dotted curves in Figure 1.6.

The **deadband error** results from the operation of the servo mechanism that performs the physical (or mechanical) correction of the air valve position. If that deadband is expressed in terms of the effect it has on the oxygen content in the flue, and this equivalent change is denoted by d, then the deadband error can be written as $d/(\triangle x u)$, where u is the control loop gain. As a consequence of expressing the errors relative to $\triangle x$, the reduction factor is given by

$$q = \frac{x_0}{\triangle x} e + \frac{d}{\triangle x} \cdot \frac{1}{u}$$

That is, the system performance parameter q is now expressed in terms of the three subsystem parameters e, d and u; and once a requirement has been determined for q, this can be allocated to e, d and u. The appropriate value of q can be determined only by taking the associated **cost** into account, and this will be discussed in Chapter 14.

9.4 Summary

1. The **performance** of a system is a measure of how well it produces the required output (e.g. service, product). This meaning is restricted; it does not include any of the operating and supporting functions.
2. With this definition, performance becomes a measure of the "prime equipment" design.
3. To describe performance, it is necessary to **choose** a set of performance parameters. This choice should be governed by **independence** and **completeness**.
4. The choice of performance parameters will often include **assumptions**. Two types of assumptions are those concerning the importance of terms or factors (i.e. which ones can be neglected) and those concerning the controllability of factors. Uncontrollable factors become constants.
5. Where possible, parameters should conform to accepted **standards**, and each parameter must be **testable**.
6. **Technical performance measurement (TPM)** introduces the time factor into performance requirements. It is not adequate to specify an end requirement; it is also necessary to specify how this is to be approached as a function of the design effort expended.

9.5 Short questions

1. Subdivide a system into major parts so that the performance concept applies to only one of them. Illustrate this by an example.
2. What is meant by "independence" of the performance parameters?
3. Discuss two separate requirements placed on the completeness of the set of performance parameters.
4. Name two requirements placed on the definition of any performance parameter.

9.6 Problems

Power generation: An isolated power grid is fed by N identical turbogenerators, each with the following characteristics:

maximum power output: S_0
operational availability: A

The demand on the grid is a rapidly varying stochastic variable, and is equally likely to have a value anywhere in the range S_1 to S_2.

(a) Develop a reasonable definition of the grade of service (GoS) for this general type of power supply.

(b) Calculate the value of the GoS for the special case:

$$N = 4 \qquad S_0 = 350 \text{ MW}$$
$$A = 0.95 \qquad S_1 = 600 \text{ MW}$$
$$S_2 = 1200 \text{ MW}$$

9.7 References

MIL-F-495, *Finish, Chemical, Black, for Copper Alloys*, Washington DC: US Department of Defense, 1965.

MIL-STD-781B, *Reliability Tests: Exponential Distributor*, Washington DC: US Department of Defense, 1967.

MIL-HDBK-217D, *Reliability Prediction of Electronic Equipment*, Washington DC: US Department of Defense 1982.

Systems Engineering Management Guide, Fort Belvoir VA: Defense Systems Management College, US Department of Defense, 1990.

CHAPTER 10

Reliability

10.1 Systems and components

10.1.1 Definition

The **reliability** of a system is a characteristic deriving from its design and identified with the frequency of operational failures and their effects on performance. Quantitatively, the reliability R(t) of a system is defined as *the probability that the system will perform satisfactorily for a given period of time t, when used under specified conditions*—a definition valid also for system elements and their individual components.

Essentially, this definition allows performance uncertainties associated with random failures to be described quantitatively by means of a **probabilistic** and **time-dependent** function, which is distinctive as a system characteristic of near universal interest, and yet **specific** in its meaning for any particular system. Specificity arises from the need to define satisfactory performance adequately—in particular, to be precise about what constitutes failure. (For complex systems, failure is typically a condition requiring engineering judgement, being neither self-evident nor clearcut!)

Additionally, the definition recognizes that failure is frequently sensitive to changing **external conditions** and can be interpreted consistently only if these are suitably constrained. The specification and maintenance of these conditions can again be very challenging at a system level, with environmental and operational stresses (including those due to human involvement at every stage of the system lifecycle) to be considered.

For systems or components commencing their useful life with satisfactory performance, reliability as defined above has an initial value (at t = 0) of unity, together with a steady state value (as t → ∞) of zero, regardless of design or operational measures. ("One-time" systems or components rapidly consumed by normal use must be treated in the limit as a special case.) In practice, intended lifetimes are finite, and the engineering process is concerned with the achievement of high reliability over a specified period—that is, with probabilities of success approaching unity for a given time. In most cases this is best done by examining the corresponding small probabilities of failure and by containing them through effective engineering **design**. At the system level, this typically requires recourse to redundancy and maintenance, among other measures.

For those accustomed to deterministic performance parameters, the idea of engineering a statistically measurable outcome may be new; yet it captures the very essence and significance of reliability engineering. The extensive literature on reliability engineering emphasizes its importance as a specialist discipline that has contributed enormously to the effectiveness of modern systems and associated user satisfaction.

10.1.2 Relationship between component and system failure

References to reliability, and the failures that lead to a lessening of reliability, are usually associated with the random failures of single components that are in mind, and this is the case that will be considered in this chapter. There are of course many cases where larger entities, such as whole subsystems, may fail at once due to some external cause (e.g. voltage transients, fire, sabotage), but this aspect of system performance is better treated as **vulnerability** to particular threats (although there is no sharp division between the two). Thus, system failure is always attributable to **component failure**, and the theory of component failure provides one part of the foundation of system reliability theory.

The other part is peculiar to systems—that is, to the fact that system parameters are not generally just the addition of component parameters, but a complex function of those parameters, due to the interaction of the components. In particular, a single component failure may often not lead to system failure; the system makes a transition to another **state** (see Section 5.2). It is important to realize that with regard to reliability a state representing system failure is always a **terminal state**. This is reflected in the theory of system reliability in the structure of the **reliability block diagram**. This is a graphical description of how the components contribute to maintaining the system performance and will be discussed in some detail in Section 10.2. Before that, however, it is appropriate to review the theory of component reliability.

10.1.3 Probability functions

Reliability can be expressed mathematically in terms of failure density functions or hazard (i.e. failure) rates, noting that the probability of failure or unreliability $Q(t)$ is the complement of the probability of success; that is,

$$R(t) + Q(t) = 1$$

Consider now a population of N identical components, all operational at time $t = 0$, and let $n(t)$ be the number of components still operational at time t, as individual components fail at random. Then the **failure density function** $f(t)$, which measures, as a function of time, the rate at which failures occur in the original population, is defined in terms of n and N as

$$f(t) = \frac{n(t) - n(t + dt)}{Ndt}$$

in the limit as $N \to \infty$. As $n(t)/N$ is simply the probability of survival (or success) after an elapsed time t if N is large, this can be rewritten in terms of $R(t)$ as

$$f(t) \, dt = R(t) - R(t + dt)$$

or in the limit as

$$f(t) = -dR(t)/dt$$

The function $F(t)$, defined by

$$F(t) = \int_0^t f(t') \, dt'$$

is not only the **failure distribution function** but also the unreliability $Q(t)$.

The **hazard rate** $z(t)$ measures, as a function of time, the rate at which failures occur within a surviving population. (The term "failure rate" is widely applied to this function, although, strictly, only repairable items should be so characterized.) In terms of the same variables n and N, the hazard rate is defined as

$$z(t) = \frac{n(t) - n(t + dt)}{n(t) \, dt}$$

in the limit as $N \to \infty$. Dividing through by N, this becomes in the limit

$$z(t) = -\frac{dR(t)}{dt} \cdot \frac{1}{R(t)}$$

or

$$-\int_0^t z(t') \, dt' = \ln R(t)$$

and

$$R(t) = \exp\left(-\int_0^t z(t') \, dt'\right)$$

As a consequence of the above definitions there exists a simple relationship between the three functions, namely

$$z(t) \, R(t) = f(t)$$

A common measure of the reliability performance of a component is the **mean time to failure** (MTTF), where

$$MTTF = \int_0^\infty t \, f(t) \, dt$$

By using the relation $f(t) = -dR(t)/dt$ and integrating by parts, the MTTF can also be expressed as

$$MTTF = \int_0^\infty R(t)\, dt$$

For repairable items exhibiting a constant failure rate, including systems for which the population is typically small, the **mean time between failures** (MTBF) is another probability function used in reliability engineering. Under special conditions, MTTF and MTBF can have equal values, although their physical bases are quite distinct.

10.1.4 Reliability data and models

In principle, the reliability of a particular component type can be determined by operating a very large population of such components until they have all failed, noting their times of failure, and then presenting the data as a table or a graphic representation. This is impractical for two reasons:

1. The testing effort would be enormous; each time a slightly different component (e.g. a new resistance value) was designed, new tests would have to be made.
2. Numerical data in the form of tables or graphs are not that easy to use; it is much more convenient to characterize the data by one or two parameter values.

The solution to these problems lies in realizing that there is a **physical mechanism that leads to failure** (e.g. ion migration, mechanical wear, fatigue due to vibration) and that this mechanism is **common to large classes of components**. For each class it is possible to construct a **model** of the failure mechanism that leads to an analytic expression for the failure probability density that depends on only one or a few parameters, and then to determine the value of these parameters for each component type within the class by fitting the expression to the available failure data.

Without going into the physical details of models, it is adequate for the present purposes simply to note that experience has shown that the behavior of the failure probability density can, for most component types, be represented by one of a small set of functions, where time is the continuous variable. Their statistical properties are summarized below:

1. *Exponential model*:

$$f(t) = \lambda \exp(-\lambda t)$$
$$R(t) = \exp(-\lambda t)$$
$$z(t) = \lambda$$
$$MTTF = 1/\lambda$$

The most significant characteristic of the exponential law is the (time) constant failure rate, given by the parameter λ. This behavior is true of most semiconductor components over most of their lifetime, and is otherwise applicable where ageing is not a factor in failure rate statistics.

2. *Rayleigh model*:

$$f(t) = (t/\sigma^2) \exp(-t/2\sigma^2)$$
$$R(t) = \exp(-t/2\sigma^2)$$
$$z(t) = t/\sigma^2$$
$$\text{MTTF} = \sigma(\pi/2)^{\frac{1}{2}}$$

The linearly increasing failure rate is typical of components subject to wear, corrosion and similar ageing processes. The parameter σ determines the shape of the failure density function.

3. *Gamma model*:

$$f(t) = \frac{\lambda (\lambda t)^{k-1}}{(k - 1)!} e^{-\lambda t}$$

$$R(t) = 1 - \lambda^k (k - 1)! \int_0^t [\tau^{k-1} \exp(-\lambda \tau)]dt$$

$$z(t) = \frac{\lambda (\lambda t)^{k-1}}{(k - 1)!} \left[\sum_{i=0}^{k-1} \frac{\lambda t}{i!} \right]^{-1}$$

$$\text{MTTF} = k/\lambda$$

The gamma law is appropriate to components subject to wear and to redundant systems where there is a renewal process based on component replacement. The parameters λ and k determine the scale and shape respectively of the failure density function.

4. *Weibull model*:

$$f(t) = \alpha\beta t^{\alpha-1} \exp(-\beta t^\alpha)$$
$$R(t) = \exp(-\beta t^\alpha)$$
$$z(t) = \alpha\beta t^{\alpha-1}$$
$$\text{MTTF} = \frac{\alpha\Gamma(1/\alpha + 1)}{\beta}$$

Here Γ is the gamma function, defined by

$$\Gamma(x) = \int_0^\infty \exp(-t)t^{x-1} dt$$

The Weibull law includes all the previously discussed models. In particular, for $\alpha = 1$, the Weibull law changes to an exponential law. The parameter α determines the shape of the failure density function, and β its scale.

Presenting reliability data for a component or component type consists of specifying the appropriate density function and the values of the parameters characterizing this function. However, for most components more than one distribution and/or more than one set of parameter values are required to describe the reliability behavior of the component over its complete lifetime. For example, one may be required to model the infant mortality in the early part of the component life, then another to model the main part of its life, and finally a third to model the wearout or end-of-life period.

Constructing such a complex model for a particular component type can be justified only in special cases; in general, simplification through approximation is used. For electronic systems this works particularly well, because even the smallest (i.e. lowest level) elements of which a system would be composed contain many individual components (i.e. devices); so the peculiarities of the individual component types become blurred. Two popular methods of approximation or estimation will be discussed in the next section.

10.1.5 Reliability estimation

A number of factors influence the reliability of components and thereby systems, including:

- the quality of the parts used (e.g. electronic components can be screened for "high rel" parts);
- the electrical stress on parts (e.g. components operating at rated levels fail more frequently than those operating at reduced levels);
- the environmental stress on parts (especially temperature, but not exclusively).

Reliability data for electronic components, taking into account the above factors, are contained in MIL-HDBK-217; similar data for other types of components (e.g. circuit breakers, power transformers, cables) can be found scattered throughout the technical literature.

However, MIL-HDBK-217 also contains two methods (or models) for applying these data: the **parts count method** and the **parts stress analysis method**. Both methods apply to modules and equipment where it may be assumed that the item fails when any of its parts fail; that is, **the item failure rate is equal to the sum of the failure rates of its parts**.

10.1.5.1 Parts count method

The information needed to apply this method is:

- generic part types (including complexity for microelectronics) and quantities;
- part quality levels;
- equipment environment.

The general expression for equipment failure rate with this method is

$$\lambda = \sum_{i=1}^{n} N_i (\lambda_G \, \pi_Q)_i$$

where, for a given environment,

λ = total equipment failure rate (per 10^6 h)
λ_G = generic failure rate for the ith generic part (per 10^6 h)
π_Q = quality factor for the ith generic part
N_i = quantity of the ith generic part
n = number of different generic-part categories.

Information to compute equipment failure rates using the equation above is given in MIL-HDBK-217 and reproduced in Section 10.11 as Tables 10.1 to 10.14. It applies if the entire equipment is being used in one environment. If the equipment comprises several units operating in different environments (e.g. avionics with units in airborne inhabited (A_I) and uninhabited (A_U) environments), then the equation above should be applied to the portions of the equipment in each environment. These "equipment or environment" failure rates should be added to determine total equipment failure rate. Environmental symbols are as defined in Table 10.12.

The quality factors to be used with each part type are shown with the applicable tables and are not necessarily the same values as are used for stress analysis. Multiquality levels are presented for microelectronics, discrete semiconductors, and established reliability (ER) resistors and capacitors. The λ_G values for the remaining parts apply, providing that they are procured in accordance with the applicable parts specification, and for these parts $\pi_Q = 1$. Microelectronic devices have an additional multiplying factor π_L—a learning factor as defined in Table 10.4.

10.1.5.2 Parts stress-analysis method

This method, requiring detailed parts list and parts stresses, can be used for trade-off studies of part selection, part quality and stresses. (Procurement costs are usually included in this process.) The most commonly used source for the application of this method (i.e. paragraph 2.0 of MIL-HDBK-217) is based on the use of large-scale data collection efforts to obtain the relationships (i.e. models) between engineering parameters (e.g. temperature, stress) and reliability variables (e.g. part quality, failure rate).

Part failure models vary with different part types, but their general form is

$$K_i = \lambda_B \ \ \pi_E \ \ \pi_A \ \ \pi_Q \ \dots \ \pi_N$$

where λ_B = base failure rate
π_E = environmental adjustment factor
π_A = application adjustment factor
π_Q = quality adjustment factor
π_N = additional adjustment factors.

1. *Base failure rate* (λ_B) is obtained from reduced part-test data for each generic part type. The data are generally presented in the form of failure rate against normalized stress and temperature factors. These values of applied stress relative to the rated stress represent the variables over which design control can be exercised and that influence part reliability.
2. *Environmental adjustment factor* (π_E) accounts for the influence of environments other than temperature and is related to operating condition (e.g. vibration, humidity). The environments are defined in Table 10.12.
3. *Application adjustment factor* (π_A) depends on the application of the part and takes secondary stress factors into account.
4. *Quality adjustment factor* (π_Q) is used to account for the degree of manufacturing control with which the part was fabricated and tested before shipment to the user. Table 10.14 identifies those parts with multilevel quality specifications, while Tables

10.3, 10.6 and 10.9 identify quality factors for microcircuits, discrete semiconductors, and established reliability (ER) capacitors and resistors, respectively.

The detail involved in generating a parts stress analysis and the volume of design parameters under various assumptions warrant the use of a computer program with stored information on all generic parts. Such programs (e.g. Electromagnetic Sciences "Predictor") are generally available.

10.1.6 Software reliability

10.1.6.1 Overview

So far it has been implicitly assumed that a system failure can be traced to the failure of one or more hardware components. That is, if the system contains software, which almost all electronic systems do, it has been assumed that this software is free of errors and therefore does not contribute to system failure. This may be approximately true in some special cases (e.g. programmable logic controllers), but in general software is a significant contributor to system failure. Then why has software reliability not been treated on an equal footing with hardware reliability?

The reason is primarily to be found in the **differences** between the two, some of which are listed in Table 10.15. As a consequence there has been a tendency to regard bad production as the cause of software failure. In other words, the software itself does not fail; it is "wrong" from the very beginning. In the same way a bolt may go through its production process without getting threads cut into it; in such a case it would not be said that the bolt has failed or is unreliable.

If this line of thought is pursued to its logical conclusion, it may seem that testing is the answer to improving software quality. This is certainly one approach vigorously pursued in the context of systems engineering, as was discussed in Chapter 8. However, it is easy to show, even for relatively simple programs, that exhaustive testing is an impossibility. It is therefore necessary to accept the situation that operational software contains errors that manifest themselves as system failures from time to time.

To take this fact into account in the systems engineering process, it is most beneficial to think of system failure due to software errors as a reliability problem, and to develop the concept of software reliability in a manner similar to that of hardware reliability. This approach is supported by focusing on the **similarities** between the two rather than on their differences. Some important similarities are as follows:

1. Both are functions of **complexity**. While this is not an easily measurable parameter, it clearly points the way for software reliability to be developed using methods similar to those used to handle complexity in hardware reliability design.
2. The assumption of **constant failure rate**, so commonly used in hardware reliability theory, implies latent causes of failure (e.g. small variations in materials or in the manufacturing process) that manifest themselves at random times. Exactly the same situation is encountered with software.
3. Hardware reliability can be improved by identifying critical failure modes (see Section 10.6) and eliminating them by design modifications. This is similar to debugging software. Both hardware and software thus allow **reliability growth**.

10.1.6.2 Concept development

System failures due to software occur because an **error** in the software becomes apparent. The error is present in the software from the moment it is produced, but it only becomes **effective** when a particular combination of data is processed. Software errors are errors in the code as it exists at any one time, and they can arise from a number of sources:

1. The requirements specification was incorrect and the software has consequently been designed to "do the wrong thing". Here there is obviously a grey area between error and inadequate performance; and as always with reliability, there is a need to be specific in defining what constitutes a system failure.
2. The design was incorrect; for example, the specification was misinterpreted, certain conditions were overlooked, or the design practice was just poor.
3. The coding (i.e. implementation) was incorrect; for example, an inverted IF statement or incorrect statement label was used.

The source of an error is irrelevant to the concept of software reliability, but it is obviously of great importance when considering how to avoid errors.

In any case, when it goes into service at time $t = 0$, any particular software package contains an unknown number of errors N_0. During the service (i.e. operation) of the software, there is a certain probability that a particular error will become effective and cause a system failure. This probability may be different for each error, be characteristic of certain grouping of errors or be the same for all errors.

When an error causes a system failure, it can be identified and eliminated through a modification to the code, or it can be left unchanged and the system simply restarted. The latter response is the only course of action when the source code cannot be accessed, as is often the case for standard software packages. The former response, and any other modification, can be viewed as an action leading to one of two outcomes: either the software remains essentially unchanged, with only the error removed; or the changed software has new errors and new failure rates, as a result of the redesign to remove the error.

Depending on the choices or assumptions made above, the result is a different **model** of software reliability. The two simplest models will be discussed briefly in the next section. However, no matter which model is chosen, the concept of software reliability, defined simply in terms of its effect on system performance, is no different from that of hardware reliability; and therefore, at a system level, the two can be treated as indistinguishable.

10.1.6.3 The two simplest models

The very simplest model arises from assuming that no attempt is made to correct the software after a fault has occurred. The system is just restarted, and the number of errors in the code remains constant and equal to N_0. In this case it is not necessary to make any assumptions about the probability of individual errors becoming effective. The errors are indistinguishable; all that can be measured is their combined effect on the system, which is expressed directly in terms of the **failure rate** λ of the software. The failure rate is in this case constant by definition; the parameter N_0 is irrelevant, and the model has no internal structure.

The simplest model applies to cases where the source code cannot be accessed and where also it is acceptable to leave the error in place. The system is simply restarted and the risk that the error will cause a further failure at some future date is accepted. Examples of this are the many smaller programs produced for personal computers.

However, in most larger systems the consequences of a failure are so serious that, once an error has been detected, it is well worth while investigating and eliminating it. Assuming that an error is eliminated without causing any new errors, the simplest model that takes this into account is the following.

Let the number of errors N initially equal N_0, and let each error have the same probability p of detection (and hence elimination) per unit time. Then

$$dN/dt = -pN$$
$$N = N_0 e^{-pt}$$
$$\lambda = pN = pN_0 e^{-pt}$$

Under this model the failure rate λ is decreasing with time in an exponential fashion and contributing to the **reliability growth** of the system.

The assumption that p is the same for each error can of course be only an approximation. Any one error becomes effective for a particular subset of all possible input data values. These subsets are not the same size, nor is the probability of the input data being in their vicinity the same. Consequently, the probability p of the input data intersecting such a subset varies, often greatly, and the errors with a high p value are eliminated first. The mean value of p over all remaining errors decreases as N decreasess; and by postulating various functional relationships between p and N, a whole class of more complex models can be generated.

10.2 Analysis of system reliability

10.2.1 Reliability as a system performance parameter

It should by now be clear that the concept of reliability is inseparable from a definition of failure. However, while this may be fairly straightforward in the case of a single part or a simple combination of parts, the definition of failure for a system is much more complicated. For example, if a single telephone is out of order, this does not indicate that the telephone system has failed. On the other hand, if all telephones have failed, this certainly indicates that the system has failed. Where, between these two extremes, does the point of failure lie?

The way out of this difficulty lies in considering the functionality of the system— that is, the **service** provided by the system. From the user's viewpoint, there is an ideal service that the user would like to receive. In reality the system provides a service somewhat less than the ideal, or a service that has a certain probability of reaching the ideal level. This is called the **grade of service** (GoS), and is a number between 0 and 1. GoS depends on the system design (even a fully intact system usually provides GoS < 1) and on the state of the system elements. As elements fail, GoS is reduced; the steady state value is reached when replacements or repairs equal element failures. System failure is **defined** to occur when GoS falls below a certain limit.

The most general measure of performance, of which GoS is a special case, is of course the **system effectiveness**—that is, the probability of the system's fulfilling its mission under specified conditions (see Chapter 9). If this probability falls below a certain limit, the system has failed. Such a definition clearly provides a relationship between element reliabilities and system reliability, but the term "fulfilling its mission" is in general not any simpler to define than GoS. Only in special cases, such as the success of a missile, is the meaning intuitively clear. In most cases it becomes necessary to approach the problem by building a **reliability model** that encompasses the system effectiveness model. In different words, the effectiveness model must be extended to take element reliability into account.

10.2.2 System reliability models

Ideally, a system reliability model should be an entity formed from a number of elements, each characterized solely in terms of its reliability parameters (i.e. not in terms of any performance parameters). These elements are in one of only two states: operating or failed. The output of the model is the reliability state of the system.

Such a model can be described in terms of Boolean logic. Its arithmetic expressions describe the relationships among the elements of the model (i.e. the **structure** of the model). As usual, this structure can also be represented in graphical form, in which case an appropriate **reliability block diagram** must be developed. It is important to realize that the reliability block diagram, although it looks like a representation of the actual system, is very different in nature from the latter. Not only is the structure usually completely different, but also the **interactions** between the elements of the reliability block diagram are now logical (i.e. logic AND and OR), as illustrated in Section 3.1.2. The interactions between the elements of the actual system are signals in real time.

For systems with multiple performance objectives (or missions), different reliability block diagrams may be needed to highlight the logical relationships in each case among the elements of the system model.

10.2.3 Series/parallel structures

10.2.3.1 Series configuration

The most common configuration in reliability modeling is the **series configuration**. It applies to any system whose successful operation requires the correct functioning of **all** its elements—that is, to a system where any (random) element failure represents immediate system failure. The series reliability block diagram shown in Figure 10.1 conveys this dependence of the system on every element.

The series reliability block diagram finds wide application, as is suggested by a few simplified high-level models. For instance, to achieve successful point-to-point line communications, the transmission, interconnecting cable and receiver that constitute a

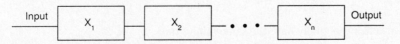

Figure 10.1 Series-configuration reliability block diagram.

three-element system model must all operate satisfactorily. Again, for the successful operation of a desktop computer, the input devices, processor, memory and display units must all be functioning satisfactorily. (While the same series logic might be applied to a more complex computer, the adequacy of the four-element system representation would have to be tested first.) Also, for submerged submarine operations, various subsystems are clearly critical to success, including the submarine's life support, drive, control and navigation subsystems, to name a selection. Depending on its mission at any particular time, others could be added (e.g. the underwater detection or weapons subsystem).

In each of these examples of series reliability structures, a performance objective has been given that permits the selection of suitable elements and a logical test of their part within the system. The rather intuitive way this has been handled in these examples of complex systems disguises the analysis that is more typically required. When the elements are themselves quite complex, the possibility of their operation with various grades of service must also be appreciated. An understanding therefore of what constitutes element failure is a prerequisite to determining overall system reliability. In particular, there is no requirement that each, or indeed any, of the elements in the examples above must perform its function ideally.

The probability that X_1 and X_2 and X_3 and . . . X_n will operate successfully is given by the general relation

$$R = P(X_1) \, P(X_2|X_1) \, P(X_3|X_1X_2) \ldots P(X_n|X_1 \ldots X_{n-1})$$

where $P(X_2|X_1)$ = conditional probability that X_2 is good, given that X_1 is good.

If element failure is **statistically independent**, the above expression reduces to

$$R = P(X_1) \, P(X_2) \, P(X_3) \ldots P(X_n)$$

or,

$$R = \prod_{i=1}^{n} P(X_i)$$

The sensitivity of series system reliability to individual element characteristics implies that the reliability of a series system is less than the reliability of its least reliable element.

If all the elements have failure rates independent of time, given by $z_i(t) = \lambda_i$ ($i = 1, 2, \ldots, n$), then

$$R(t) = \prod_{i=1}^{n} \exp(-\lambda_i t),$$

or

$$R(t) = \exp\left(-\sum_{i=1}^{n} \lambda_i t\right)$$

If the n elements actually have the same failure rates,

$$R(t) = \exp(-n\lambda t)$$

From Section 10.1.3, the mean time to failure (MTTF) is given by

$$\text{MTTF} = \int_0^\infty \exp\left(-\sum_{i=1}^{n} \lambda_i t\right) dt = \left(\sum_{i=1}^{n} \lambda_i\right)^{-1}$$

or, for n identical elements, by

$$\text{MTTF} = 1/n\lambda$$

10.2.3.2 Parallel configuration

The second commonly occurring configuration encountered in the reliability modeling of systems is the **parallel configuration**. It can be represented by the reliability block diagram shown in Figure 10.2, which affords a number of system performance options, including active or standby redundancy.

In the simplest example of a parallel configuration, with n units all initially performing the same operation, the system remains operational if **one or more** units are still operating. This corresponds to an **active redundant** system.

Examples of parallel or redundant configurations are multiple. Easily recognized examples include triplicated display consoles in control rooms, multiple engines on aircraft, switchable buffers on processors and free boards in computer memories. More subtle examples, requiring careful assessment of the level of redundancy, arise in power systems, communications systems, transport systems and so on. By introducing genuinely redundant units, system reliability is improved.

The parallel system fails only if all the units fail. With P(X) (the probability of element success) replaced by Q(X) to indicate the probability of element failure, the probability that an active redundant system will operate successfully is given by

$$R = 1 - \prod_{i=1}^{n} Q(X_i)$$

or

$$R = 1 - \prod_{i=1}^{n} [1 - P(X_i)]$$

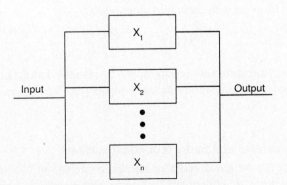

Figure 10.2 Parallel-configuration reliability block diagram.

The reliability of such a system is better than that of its most reliable element. However, a few numerical examples will serve to show that the overall improvement is quite modest, in the absence of further initiatives such as element maintenance following failure (see Section 10.3).

In the case of constant failure rate,

$$R(t) = 1 - \prod_{i=1}^{n} [1 - \exp(-\lambda_i t)]$$

and for identical elements

$$R(t) = 1 - [1 - \exp(-\lambda_i t)]^n$$

The calculation of the parallel system MTTF is quite demanding. However, for a constant failure rate it can be found as follows for an active redundant system:

$$MTTF = \int_0^\infty R(t)\, dt$$

$$= \int_0^\infty \left(1 - \prod_{i=1}^{n} [1 - R_i(t)]\right) dt$$

$$= \int_0^\infty \left(1 - \prod_{i=1}^{n} [1 - \exp(-\lambda_i t)]\right) dt$$

which, after multiplying through and integrating, becomes

$$MTTF = \sum_{i=1}^{m} \frac{1}{\lambda_i} - \sum_{\substack{i,j=1 \\ i<j}}^{m} \frac{1}{\lambda_i + \lambda_j} + \sum_{\substack{i,j,k=1 \\ i<j<k}}^{m} \frac{1}{\lambda_i + \lambda_j + \lambda_k} - \text{etc.}$$

For n = 2 this reduces to

$$MTTF = \frac{1}{\lambda_1} + \frac{1}{\lambda_2} - \frac{1}{\lambda_1 + \lambda_2}$$

and for n = 3 to

$$MTTF = \frac{1}{\lambda_1} + \frac{1}{\lambda_2} + \frac{1}{\lambda_3} - \left(\frac{1}{\lambda_1 + \lambda_2} + \frac{1}{\lambda_1 + \lambda_3} + \frac{1}{\lambda_2 + \lambda_3}\right) + \frac{1}{\lambda_1 + \lambda_2 + \lambda_3}$$

If all elements are identical with constant failure rate λ,

$$MTTF = \frac{1}{\lambda} \sum_{i=1}^{m} \frac{1}{i}$$

Corresponding (but different) results apply to **standby redundant configurations**. The results are given in the literature in many places (e.g. Von Alven 1964, O'Connor 1985).

10.2.3.3 Series/parallel and partly redundant systems

Most systems will not be just a series or parallel combination of elements, but some more complicated structure containing both series and parallel combinations. Also, in the case of the parallel combinations, more than one of the elements may have to be operable for the system to operate satisfactorily; that is, the system has **partial redundancy**.

Take the case where there are three components in parallel and at least two have to be operating for the combination to work properly. Let the reliabilities be given by R_1, R_2 and R_3, and the unreliabilities, $Q = 1 - R$, be denoted by Q_1, Q_2 and Q_3 respectively. Recalling the basic definition of reliability as a probability, the reliability of the combination is the sum of the probabilities of all the states that will satisfy the criterion of success; that is,

$$R_T = R_1R_2R_3 + R_1R_2Q_3 + R_1R_3Q_2 + R_2R_3Q_1$$
$$= R_1R_2 + R_1R_3 + R_2R_3 - 2R_1R_2R_3$$

In the case where $R_1 = R_2 = R_3 = R$,

$$R_T = R^3 + 3R^2Q$$
$$= 3R^2 - 2R^3$$

The calculation of the reliability of a more complex series/parallel combination is best illustrated by an example.

Consider the system shown in Figure 10.3(a), consisting of seven elements. The parallel branches are fully redundant, except for the combination of elements 4, 5 and 6, where at least two must be operating. The successive reductions of this system also are shown in Figure 10.3, and the system reliability is calculated as follows:

$$R_8 = 1 - (1 - R_1)(1 - R_2) \quad (\text{or } R_1R_2 + R_1Q_2 + R_2Q_1)$$
$$= R_1 + R_2 - R_1R_2$$
$$R_9 = R_4R_5 + R_4R_6 + R_5R_6 - 2R_4R_5R_6$$
$$R_{10} = R_3R_9$$
$$R_{11} = R_7 + R_{10} - R_7R_{10}$$
$$R_T = R_8R_{11}$$

If $r_i = 0.9$ for all i,

$$R_8 = 0.99 \quad R_9 = 0.972 \quad R_{10} = 0.8586 \quad R_{11} = 0.98568 \quad R_T = 0.9758232$$

10.2.4 Complex structures

10.2.4.1 General

The series and parallel configurations just discussed (with or without partial redundancy) form the basis for evaluating the reliability of more complex systems. In one way or another, an attempt is made to reduce the complex structure to a function of series and parallel structures. A couple of the more often used methods will be introduced in the following sections.

To illustrate these methods, the **simple bridge system** shown in Figure 10.4 will be used; this bridge structure is the simplest one that cannot directly be represented by a combination of series and parallel branches. The system is considered to be operational if there exists a **path** from the input to the output (i.e. at least one of the element combinations AB, DE, ACE or DCB is operating).

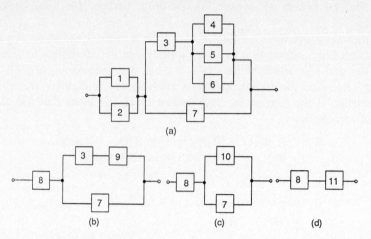

Figure 10.3 Reliability block diagram for a system consisting of seven elements, and its successive reductions.

10.2.4.2 Conditional probability method

The method consists essentially of introducing a hierarchical ordering of the possible operating states, using the states of one or more subsystems as the ordering criterion. In the case of the bridge network, the subsystem simply consists of the element C, and the possible operating states fall into two groups: one if C is operating, and one if C has failed:

C operating: AB, DE, AE, DB, ADB, ADE, BEA, BED, ABDE
C failed: AB, DE, ADB, ADE, BEA, BED, ABDE

The system reliability R(S) is then given by the sum of the probabilities of each state in the upper line times the probability of C operating, plus the sum of the probabilities of each state in the lower line times the probability of C having failed. Remember that the probability of the system being in state AB is $R(A)R(B)Q(D)Q(E)$.

The above method of evaluating system reliability is very straightforward and suitable for use with a computer. For manual calculations it is more convenient to reduce the two derived systems as shown in Figure 10.5, and then to use the expressions from Section 10.2.3. Let R(X) be the reliability of the derived system for C operating, and R(Y) the reliability of the derived system for C failed. Then

$$R(X) = [R(A) + R(D) - R(A)R(D)][R(B) + R(E) - R(B)R(E)]$$
$$R(Y) = R(A)R(B) + R(D)R(E) - R(A)R(B)R(D)R(E)$$
$$R(S) = R(X)R(C) + R(Y)[1 - R(C)]$$

If $R(i) = 0.9$ for all i,

$$R(X) = 0.9801 \quad R(Y) = 0.9639 \quad R(S) = 0.97848$$

For more complex systems the hierarchy can obviously have more levels, and the subsystems used for ordering can consist of several elements and can, in principle, themselves be complex.

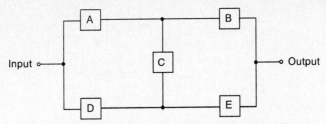

Figure 10.4 A simple bridge configuration, which cannot be directly reduced to a combination of series and parallel subconfigurations.

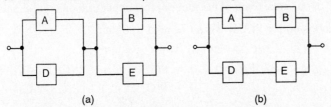

(a)   (b)

Figure 10.5 Two states of the bridge system from Figure 10.4, depending on whether element C is operational or failed.

10.2.4.3 Cut-set method

Cut-sets were introduced in Section 3.3.1. In the context of reliability analysis, a cut-set is defined as *a set of elements that, when all have failed, causes failure of the system but, when any one element has not failed, does not cause failure.*

The fact that all elements in a cut-set must fail for the system to fail means that the elements of the cut-set are effectively in **parallel**. Furthermore, as the system fails if any one cut-set fails, the cut-sets may be considered to be connected in series, as far as a reliability model is concerned.

In the case of the bridge system in Figure 10.4, the cut-sets are AD, BE, ACE and DCB. Consequently a reliability model with the structure shown in Figure 10.6 is obtained. However, it is important to note that the elements of this model are not independent; on the contrary, the same physical system element appears more than once. Therefore the simple expression for a serial connection of independent elements does not apply, and the expression that involves the conditional probability must be used, namely

$$P(A \text{ or } B) = P(A) + P(B) - P(A|B)$$

or, in the language of set theory,

$$P(A \cup B) = P(A) + P(B) - P(A \cap B)$$

For the structure in Figure 10.6, the unreliability Q_T (i.e. the probability that at least one of the cut-sets will occur) is given by

$$Q_T = P(C_1 \cup C_2 \cup C_3 \cup C_4)$$

Writing this as

$$Q_T = P[(C_1 \cup C_2) \cup (C_3 \cup C_4)]$$

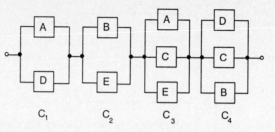

C_1 C_2 C_3 C_4

Figure 10.6   Reliability block diagram of the system from Figure 10.4, but now in terms of cut-sets. Each of the four parallel configurations represents a cut-set.

the above expression for $P(A \cup B)$ can be used to obtain

$$
\begin{aligned}
Q_T = {} & P(C_1) + P(C_2) + P(C_3) + P(C_4) \\
& - P(C_1 \cap C_2) - P(C_1 \cap C_3) - P(C_1 \cap C_4) \\
& - P(C_2 \cap C_3) - P(C_2 \cap C_4) - P(C_3 \cap C_4) \\
& + P(C_1 \cap C_2 \cap C_3) + P(C_1 \cap C_2 \cap C_4) \\
& + P(C_1 \cap C_3 \cap C_4) + P(C_2 \cap C_3 \cap C_4) \\
& - P(C_1 \cap C_2 \cap C_3 \cap C_4)
\end{aligned}
$$

with

$$
\begin{aligned}
P(C_1) &= Q_A Q_D \\
P(C_2) &= Q_B Q_E \\
P(C_3) &= Q_A Q_C Q_E \\
P(C_4) &= Q_B Q_C Q_D \\
P(C_1 \cap C_2) &= P(C_1)P(C_2) = Q_A Q_B Q_D Q_E \\
P(C_1 \cap C_3) &= P(C_1)P(C_3) = Q_A Q_C Q_D Q_E \\
P(C_1 \cap C_4) &= P(C_1)P(C_4) = Q_A Q_B Q_C Q_D \\
P(C_2 \cap C_3) &= P(C_2)P(C_3) = Q_A Q_B Q_C Q_E \\
P(C_2 \cap C_4) &= P(C_2)P(C_4) = Q_B Q_C Q_D Q_E \\
P(C_3 \cap C_4) &= P(C_3)P(C_4) = Q_A Q_B Q_C Q_D Q_E
\end{aligned}
$$

Each of the five third- and fourth-order terms in the expression for Q_T equals $Q_A Q_B Q_C Q_D Q_E$. Thus,

$$
\begin{aligned}
Q_T = {} & Q_A Q_D + Q_B Q_E + Q_A Q_C Q_E + Q_B Q_C Q_D \\
& - Q_A Q_B Q_D Q_E - Q_A Q_C Q_D Q_E - Q_A Q_B Q_C Q_D \\
& - Q_A Q_B Q_C Q_E - Q_B Q_C Q_D Q_E + 2\, Q_A Q_B Q_C Q_D Q_E
\end{aligned}
$$

If $Q_i = 0.1$ for all i (as before), $Q_T = 0.02152$, or $R_T = 0.97848$ (as before).

10.3 Influence of maintenance

10.3.1 Maintenance as part of the operating environment

So far in this chapter, reliability has been considered in its primary form. This means that an (infinitely) large collection of identical systems is observed, all known to be operating at time t = 0. The fraction operating at time t is the reliability R(t).

While this situation is applicable at component level (e.g. for resistors or diodes), it would not normally apply directly to systems, for the following reasons:

1. Systems are often one-off; at most there is a small number of similar ones.
2. A system is designed to operate satisfactorily for a certain lifetime, usually many years; but within this lifetime many component failures will occur, and there is a probability that one of these will lead to a system failure.
3. Even when a system failure occurs, the system is usually not abandoned or replaced; it is repaired.

After each component failure the system is restored to its operating state by carrying out corrective maintenance (i.e. repair), under the assumption that the failure rates of all system components remain constant over the lifetime of the system (i.e. have an exponential distribution).

10.3.2 Three-component systems

Consider a maintained system consisting of three main components, two of which are identical and operating redundantly in parallel, as shown in the reliability block diagram in Figure 10.7. The failure rates f_1 and f_2 and, for the redundant elements, the repair rate g_2 also are indicated. This system has been designed for a mission duration of 96 h; that is the reliability is the probability of its surviving 96 h without a system failure.

If no maintenance is carried out during the mission, the system reliability is immediately given by

$$R = R_1[1 - (1 - R_2)^2]$$
$$= 0.953 \, (0.76) = 0.724$$

and it is clearly the two identical elements that limit the reliability, despite being redundant.

With maintenance performed during the mission, however, the situation is quite different. Then the failure rate of the parallel combination is approximately given by the expression $2f_2^2/g_2$ which equals about 1.10^{-4} per hour. The total failure rate is therefore given by

$$f = f_1 + 2f_2^2/g_2$$
$$= 5.10^{-4} + 1.10^{-4}$$

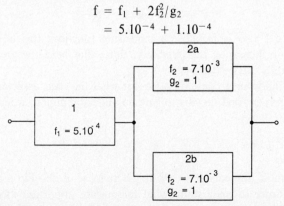

Figure 10.7 Reliability block diagram of a three-component system; f = failure rate, g = repair rate (both per hour).

So now it is the first component that is the weak link.

Thus, a procedure attempting to define the system reliability as a parameter independent of (or orthogonal to) maintenance, and asking how the reliability could be improved, might lead to the conclusion that the reliability of the two parallel components should be improved as a priority. However, this would be quite wrong, because the actual conditions under which the system is to operate include maintenance, and it is this operation that should be optimized with respect to cost-effectiveness. Consequently, to achieve an improvement in the reliability of this particular system, only the first component would initially be examined.

This example illustrates a recurring theme, namely the importance of **defining the system parameters for each particular system**. There is no such thing as a "generally valid" definition of a system parameter, unless it is so general as to be useless in an operational sense. And the definition must be complete in the sense of taking into account **all relevant aspects of the whole system**—not just the physical system, but the integrated logistic support subsystem as well.

10.4 Economics of reliability

10.4.1 Cost of reliability

10.4.1.1 General

A given level of system reliability does not come about by chance; it is the result of numerous concerted activities throughout all phases of the systems engineering process, as well as further activities during the system lifetime. As discussed later in this chapter, each phase of such a "reliability program" must be properly integrated with the engineering design process, under an appropriate plan.

Clearly, there is a cost associated with reliability, and in general both the cost function C(R) itself and its derivative dC/dR are monotonically increasing functions. This is fairly obvious, as the cost must go to infinity as the reliability approaches unity. A more significant representation of the cost function results from introducing the variable

$$P = -10 \log (1 - R)$$

The function dC/dP also is usually an increasing function; this is because, of the activities involved in increasing the reliability, those that have the greatest benefit-to-cost ratio are naturally carried out first.

It is therefore important to develop a feel for what the costs associated with various types of activities may be, and thereby to gain an understanding of where on the P-scale they fit.

10.4.1.2 Modeling

The simplest and most immediately rewarding activity is to develop a reliability model of the system, in order to understand what factors are important in determining the system reliability. There are in principle two ways of developing such a model: bottom-up and top-down.

The bottom-up approach looks at all possible component failures and analyses their effect on system reliability. This is essentially an extension of the **failure mode, effect and criticality analysis (FMECA)** procedure to systems.

The top-down approach starts with a system failure, as defined in the system reliability definition, and then breaks the causes of this failure down into finer and finer detail, so that each system failure is represented by a **fault tree**.

In either case the result will be some form of reliability block diagram, as discussed in Section 10.2. Developing such a model will typically cost between 10^{-4} and 10^{-3} of the total system cost, but the increase in P (without any further cost) is often in the range 3 to 6. The major effect of developing a reliability model is to eliminate errors in the system structure.

10.4.1.3 Element requirements

The elements of a system are generally equipments, and the reliability of individual equipments has a major and direct influence on system reliability. The required level of reliability is specified through the systems engineering process in the equipment specifications produced at the end of the design phase, but how that level of reliability is achieved is not normally within the jurisdiction of the systems engineer. Nevertheless, it is useful to recall some of the main actions employed to increase equipment reliability:

- choice of a more reliable technology (e.g. substitution of EEPROMs for floppy disks, use of nonflammable cable insulation);
- component screening (e.g. testing every component instead of testing on a sample basis);
- component derating, particularly with respect to operating voltage, thereby reducing the electrical stress on the components;
- use of worst case rather than mean environmental requirements, particularly for temperatures internal to the equipment.

By employing a suitable mix of these, a decrease in failure rate of almost an order of magnitude, say $\Delta P = 8$, can typically be effected but at a cost increase for the total system of something like 25%.

10.4.1.4 Equipment screening

By increased acceptance testing of equipment received for integration into the system, it is usually possible to effect an increase in P of 2 to 3, with a resultant increase in the total system cost of about 2.5%. However, this is a **transient** effect; after a time somewhat less than the equipment MTBF, the same improvement would be achieved anyway.

10.4.1.5 Redundancy

The duplication of equipment and/or subsystems obviously has a significant effect on the reliability, increasing P by something in the range 5 to 25. However, the cost penalty is correspondingly high; the total system cost increases by 50–150%.

10.4.1.6 Preventive maintenance

Increasing the level of preventive maintenance, particularly in the area of periodic inspections and testing, sometimes leads to a modest increase in reliability (e.g. $\Delta P = 2$), but this depends very much on the particular system and the degree to which a gradual deterioration can be detected. With completely random failures (i.e. a constant failure rate), preventive maintenance is useless.

10.4.2 Cost of failure

10.4.2.1 General

Knowledge of the means of increasing reliability and how much it will cost is only one-half of the information needed to optimize the system design. Clearly, there is also a need to measure the benefit of higher reliability; that is, there must be a function, say B(r), that expresses the savings, in dollars, of having a reliability corresponding to r. The optimum system design is then determined by the condition

$$dB(r)/dr = dC(r)/dr$$

However, while everyone will readily agree that "high reliability" is a worthwhile objective, it is often surprisingly difficult to get the users or owners of a system to put a dollar value on any particular value of (or increase in) reliability. Typically, users or owners are not willing to pay for reliability, because "with a little luck, a failure may not occur anyway". To gather the information required to make a rational decision, it is necessary to determine the cost of failure in a systematic fashion, first of all recognizing that there are several different factors that contribute to cost.

10.4.2.2 Cost of repair

Once a fault has occurred, it has to be repaired to bring the system back into its normal operating state. The costs of this repair consists of labor, consumables and spare parts. Note that all of these may relate not only to the part that failed, but also to any other action that has to be taken to bring the system up again (e.g. retuning, reloading software, testing and calibration). This is a periodically occurring cost, with the period being the MTBF.

The easiest way to calculate the benefit of increasing the MTBF from one value, say A, to another value, say B, is to introduce an average yearly repair cost. If C_R is the cost of repairing one fault, the equivalent yearly cost is C_R/MTBF, and the benefit is

$$\text{saving} = C_R \left(\frac{1}{A} - \frac{1}{B} \right) \frac{(1 + p)^n - 1}{p(1 + p)^n}$$

with n = design lifetime of the system in years
    p = discount rate.

10.4.2.3 Loss of service

While the system is down for repairs due to a failure, the service normally provided by the system is absent. This may mean that the service has to be temporarily supplied by some alternative means, or it may mean providing no service and forgoing the associated revenue. In either case a cost is caused by the loss of service, which is proportional to the MTTR and to the failure rate. In other words this cost is proportional to the **operational unavailability** of the system.

10.4.2.4 Consequential damages

Finally, there are a number of costs that arise from the fact that a failure has occurred but are not directly related to the system itself. These can be grouped under the heading **consequential damages**, and include such costs as:

- loss of market share due to loss of consumer confidence;
- claims by customers for damages incurred as a result of the loss of service;
- claims by third parties for compensation for personal injury or death resulting from the failure;
- increase in insurance rates.

There are many cases where the consequential damages are dominant; obvious examples are an aeroplane crash or a major fault in a nuclear power station. However, as such catastrophic failures are relatively infrequent, how can the cost be estimated?

One way is to ask what it would cost to insure against the consequences, or at least to compare with existing insurance cover for similar systems. Another way is to build a model that decomposes the cost into a number of independent components, each with its own conditional probability of occurrence (i.e. conditional on the failure having taken place), and then to obtain a distribution for the consequential cost (i.e. using the Monte Carlo method).

10.5 Designing for reliability

10.5.1 Reliability as a design parameter

Reliability is a characteristic of a system—a parameter that describes a certain aspect of the system, in exactly the same sense as noise figure, accuracy, weight and cost (to name a few) are system parameters. As such, **the level of reliability attained by the system should be a conscious result of the design process**. The keywords describing the development of the system reliability within the engineering process are:

> user requirement
> specification
> allocation
> design
> prediction
> analysis
> demonstration.

Not surprisingly, this sequence dovetails into the systems engineering phases first identified in Chapter 2. It also suggests that reliability may be assured through an embedded project properly synchronized with the host project. Outline requirements for such embedded programs are given in various national standards (e.g. MIL-STD-785).

The starting point in the engineering process is always a complete and unambiguous **definition** of what is to be understood by system reliability in the particular case, and the **specification** of a particular value (or set of values for different conditions). In this context three points are noteworthy:

1. The **measure of failure** must be defined in an operational sense; that is, in addition to the aspects discussed in Section 10.2 it is necessary to spell out exactly how the value of the system reliability is to be measured. For example, failure might be measured by operating a system over a 3-month period, collecting data in a prescribed manner, and then extracting the reliability measure from this data according to some rule; or by simulating operation in a defined way. The use of traffic generators to test a communications system is an example of the latter.
2. The **environment** in which the system will operate during the testing must be specified, in particular:

   - climatic conditions;
   - conditions of use (e.g. operating intervals, level of maintenance);
   - manning of the system (e.g. staff levels, degree of experience and training).

3. In specifying a value for the system reliability, the **justification** for choosing this particular value should be included. This is essential if reliability is to be treated seriously in an overall optimization process during system design (see Section 14.2.2).

The allocation process is not peculiar to reliability, but reliability provides a particularly good case for demonstrating a way of going about it. The general problem is that the starting point is a requirement at the system level (i.e. at the top level where the system is represented as a single "element") and then, as the system is partitioned, a requirement has to be placed on each of the resulting elements such that two conditions are fulfilled:

1. The allocated values must result in the prescribed overall value.
2. The allocated values must minimize the overall cost.

The first of these can usually be satisfied relatively easily by developing a reliability block diagram, as discussed in Section 10.2; this results in one equation relating the N element reliabilities. The second condition is much more difficult to fulfil, partly because the cost of providing an element with a particular value of its reliability may not be known, and partly because changing the element will often change other performance parameters, so that reliability cannot be treated strictly in isolation.

As a result of these difficulties, a number of simplified procedures have arisen:

1. *Equal values:* It is assumed that the elements are similar with respect to the cost of providing reliability. Consequently, the same reliability is assigned to each element. The value of this element reliability is then determined by the one equation relating the element reliabilities to the system reliability.
2. *State of the art:* In this procedure past experience (but as recent as possible) is used to determine (or possibly estimate) the failure rates for the elements. These failure rates are multiplied by a constant, which then becomes the only variable to be determined by solving the one available equation, as before. In other words, this

procedure assumes that the most economic solution is to try to change the present, or state-of-the-art, failure rates by the same relative (or percentage) amount.

3. *Cost as complexity:* This method has only limited applicability, in that it assumes the elements to be of the same type (e.g. electronics modules in the form of printed circuitboards). In that case it can be assumed that the complexity of each element is proportional to its cost, and the cost is a parameter usually estimated very early in the design process. The algorithm now takes the failure rate of each element to be proportional to its cost; that is, it equates the most cost-effective solution to one that uses a uniform component reliability level across all elements. The proportionality constant is again determined using the one equation arising from the reliability block diagram.

4. *Minimum cost algorithm:* This algorithm will yield a true optimal solution, given a cost versus reliability function for each element, but it presupposes a system with a reliability block diagram consisting of N elements in series.

10.5.2 Reliability design techniques

10.5.2.1 Overview

Having specified the system reliability and allocated it to the functional elements, the next step is to come up with a design that will achieve this level of reliability. To this end, tools in the form of **rules and procedures** are needed. On the system level some of the most often used ones are:

- developing a consistent set of specifications for the selection of parts, and putting in place mechanisms for controlling that selection process;
- specifying appropriate environmental classes and derating factors;
- use of redundancy and fault-tolerant system architectures;
- applying human engineering to person-machine interfaces in order to reduce the probability of human error;
- fault tree analysis;
- failure mode, effects and criticality analysis (FMECA).

The following sections will briefly discuss the first two of these groups. Redundancy was treated in Section 10.2.4, and human engineering aspects will be the subject of Chapter 13. Fault tree analysis and FMECA will be discussed in Section 10.6.

10.5.2.2 Parts selection and control

Selecting the most cost-effective parts will normally be the prerogative of the equipment designer. The systems engineer should be concerned with **what** the equipment does (i.e. with externally measurable parameters), not **how** it does it, and the equipment specification should be written accordingly.

However, while this is certainly true for equipment that has a well-defined stand-alone role (e.g. equipment making up a home entertainment system), it is often appropriate to specify families or types of parts to be used in equipment designed (or selected) especially for use in the system under consideration. This introduces a **consistency** across the whole system, which rationalizes the provision of spare parts, simplifies the equipment design process and the early cost estimating. It narrows the field of options, and acts as a first line of assurance that the reliability requirements will be met.

The definition of parts types and/or reliability levels within these types is generally contained in the engineering manuals/procedures maintained by all major electronics designers and manufacturers. A very extensive set of definitions can be found in MIL-HDBK-338-2, *Component Reliability*.

In applying such a parts selection program, three points need to be kept in mind:

1. The listing of approved parts (or specification of approved types) must be kept up to date, and it must provide backward compatibility. If it is not up to date, designers will have an added argument for ignoring it (they have a tendency to do this anyway, as they feel that such a listing reduces their freedom as individual designers). If it is not backward compatible in the sense of providing clear cross-referencing between new and old parts or part types, the reliability level of the system may inadvertently decline over its lifetime.
2. Provisions must be made within the engineering management process, usually as part of the review process, for checking compliance with the parts selection program.
3. There must be a defined procedure for requesting and processing exceptions. If the program is too rigid, this will only add another incentive for circumventing it.

10.5.2.3 Environmental classes and derating

Where the choice of parts is left to the equipment designer, this person needs to know the conditions under which the equipment is required to perform. These must be specified in the equipment specifications resulting at the end of the system design phase. For the purpose of reliability design and estimation, a number of **environmental classes** have been defined, as introduced in Section 10.11 and listed in Table 10.12 (see Section 10.8.4). While the descriptions of these classes relate directly to military applications, they can readily be applied to nonmilitary applications, and the advantage of using them is the large amount of data available (e.g. in MIL-HDBK-217) in a form that uses these classes as an index.

Environmental design does not consist of only the selection of appropriate parts or the specification of the appropriate class. For each environmental stress factor (e.g. high or low temperature, shock and vibration) there are a number of techniques that can be applied at both the system and equipment design levels to ensure that reliability requirements are met.

At system level a major technique for responding to environmental stress is the choice of partitioning and the appropriate packaging of the resulting system elements. An example of this is the reliability of a computer system under high temperatures. Instead of using expensive parts or going to redundant equipment configuration, it may be possible to repartition the system so that most of the processing can be done in a central air-conditioned location.

Another technique, somewhat akin to parts selection, which is applied at system level but affects equipment design, is **derating**. It consists of demanding, across all or most of the system, that parts be subjected to only a certain fraction of their rated parameter values. The most common parameter in electronics/electrical engineering is operating voltage, as voltage stress is a major contributor to component failure, but rate of change of current and voltage also are candidates for derating.

It might be expected that temperature would be the most common parameter to be derated. This is in a certain sense true, except that temperature is often taken into account as a design variable rather than subjected to global derating; that is, the temperature versus failure rate function, which usually is relatively well known, is often used as an input to the trade-off process. An example of this is in power electronics, such as large converters or inverters, where the semiconductor elements and their cooling are significant cost factors. Here there is always a trade-off between redundancy (i.e. extra elements in series or parallel), element size and junction temperature in order to reach the required reliability at minimum cost.

10.5.3 Design checklist

The following questions are typical of those that must be asked and answered during the design phase, usually as part of or in preparation for a design review:

1. Are all reliability requirements adequately (i.e. completely, unambiguously) defined?
2. Have critical items been identified?
3. Have storage conditions been taken into account in determining system reliability?
4. Has the design lifetime been considered in choosing the optimum strategy (e.g. redundancy, periodic replacement) for attaining the required reliability?
5. Has reliability for spares and repair parts been considered?
6. Have all derating factors been clearly spelt out in every applicable document?
7. Has the origin of all reliability data used for estimates been properly documented?
8. Does the design account for early failure, useful life and wearout?
9. Are there any requirements for fail-safe system operation, and have these been accounted for in the design?
10. Have alternative designs/strategies been considered (e.g. redundancy versus "high-rel" parts)?

10.6 Fault analysis and failure reporting

10.6.1 Overview

So far in the treatment of system reliability it has been implicit that, while the causes of system failure will be many and diverse, it is possible to define a single criterion for system failure (e.g. that the grade of service decreases below a prescribed limit). This may not always be the case, however, and for large complex systems it will **usually** not be the case. A number of different effects of element failures will occur and be considered to be system failures, and these are called system **failure modes**. This is of course true at any level of detail, right down to parts level. A well-known example is a transistor, which may develop any one of a number of different failure modes: high collector-to-base leakage current, collector-emitter short, open terminals and so on. At higher levels the elements are more complex, and they often have many separate failure modes.

In principle, every parameter results in a failure mode, defined as occurring when the parameter value no longer meets its specification. In practice, however, a few are usually much more common than the rest, and these are then **defined** as the failure modes. Illustrating this again at parts level, consider a resistor. Normally a metal film resistor has two failure modes: open circuit and change of value. However, in rare applications the temperature coefficient is also a critical parameter. If this changes outside its specified value, it could be considered a separate failure mode; but as this is only rarely important, it is not given the status of a failure mode.

10.6.2 Failure modes and effects analysis

Failure modes and effects analysis (FMEA) is a procedure that determines what the effect is at system level (or any level above element level) of a particular failure mode at element level. In principle, FMEA is straightforward; it is a "bottom-up" approach using reliability block diagrams as its basis, allowing the effect of an element failure to be traced. In practice, the art of performing FMEA lies in keeping track of the myriad of different paths from element failure to system failure, and it is to a large extent a "bookkeeping" exercise.

The steps involved in carrying out FMEA are:

1. Develop reliability block diagrams for the system, starting at the lowest element level. There will normally be a set of diagrams for each step upwards in the hierarchy of elements, subsystems and system, with the number of diagrams in the set corresponding to the number of subsystems on the next higher level.
2. Introduce a coding scheme to identify each block, such that the code for a block shows which level it is on and to which block on the next higher level it belongs.
3. Starting at the lowest level, list (or describe) all the failure modes of the elements, and assign an identification number to each one that can be attached to the block code.
4. Using the reliability block diagram, identify and list all the different effects of the failure; these become the failure modes for the next step. The list should be in the form of a pairwise ordering of modes and effects. Each mode results in an effect (including possibly the null effect; i.e. the mode has no influence on reliability), but many modes can result in the same effect; that is, each mode occurs only once in the list, but the same effect can occur many times. Note the very important assumption implicit in this type of ordering, namely that only one failure is present at any one time.
5. Iterate step 4 until the top (i.e system) level is reached. The resulting effects are then the system failure modes.
6. As a result of the coding scheme, every failure mode on the lowest level is linked to a system failure mode by an identified **path**. Now a final **list** can be produced that shows all the element failure modes that result in a particular system failure mode. This list is the result of FMEA.

10.6.3 Fault tree analysis

Fault tree analysis (FTA) is a procedure for obtaining essentially the same result as with FMEA, but in a "top-down" manner. The steps involved are carried out in inverted order:

1. Define the system failure modes.
2. Develop the reliability block diagram for the system in terms of its next lower level subsystem.
3. For each system failure mode, identify what faults on the subsystem level would result in that particular mode. Note, however, that in this case there is no limitation to single faults. On the contrary, all combinations of failure modes on the lower level are allowed, and this can be represented in graphical form using AND and OR logic symbols.
4. Iterate step 3 until the element (i.e. lowest) level is reached. The resulting graphical representation of the logic linking element and system failure modes is the **fault tree**.

FTA is clearly more powerful than FMEA, in that it takes higher order effects into account. However, it is also more difficult to apply and requires considerably more skill and insight, because the practitioner has to **define** the failure modes; they do not arise as effects of lower level failure modes. Therefore a combination of FMEA and FTA is sometimes employed, with FTA being used only to account for steps that have **redundancy** of some sort built in. Clearly, FMEA cannot handle redundancy; failure of a structure with redundancy is always a higher order effect.

10.6.4 Criticality analysis

The two previous procedures, FMEA and FTA, were concerned with developing the logic of failures, the relationships between cause and effect, or what can be considered the **qualitative** aspect of failure. The qualitative aspect consists of introducing values for element failure rates and computing the resulting values for the probability of occurrence of the system failure modes. This probability is called the **criticality** of the particular failure mode. A **critical failure mode** is one whose probability exceeds the specified value.

In this case it is necessary to go back through the logic structure developed using FMEA or FTA, and to identify what element (or subsystem) makes the major contribution to that high probability value. Thus remedial action in the form of subsystem or equipment redesign with respect to reliability is called **criticality analysis**. When combined directly with FMEA in one logic process (often computer-assisted), it is called **failure modes, effects and criticality analysis (FMECA).**

10.6.5 Failure reporting

The reliability methods and considerations so far have been directed mainly at the design process. However, as with every other system parameter, there must be some procedure for verifying that the reliability required by the system specification has indeed been met; that is, the test plan must contain a section on how to establish what level of reliability has been achieved. As with any other parameter, this should be established as early as possible in the system integration process, so that any corrective action required can be undertaken at the least cost. However, by its nature reliability requires observation over a time period; and therefore, instead of testing, it is more a matter of **reporting**.

Equipment and subsystems are operated under defined conditions, often in order to carry out tests on other parameters, and the faults that occur during this operation are reported to a central organizational unit in charge of the reliability program. The report must contain information on what failed, how it failed, and under what conditions it failed. Analysis of these data, and possibly further examination of the failed component are then required to establish why it failed. The collected data are analyzed using statistical methods, and estimates of actual (i.e. achieved) reliability are produced and successively refined as more data become available. These become the basis for deciding when and what remedial action should be undertaken. Such a program—which takes place in the implementation and verification phases—is carried out by the manufacturer and documented in the Reliability, Maintainability and Availability (RMA) Plan for the project and goes under the name of **failure reporting, analysis and corrective action system (FRACAS).**

However, the reliability reporting function does not end with system acceptance by the user. A reporting and data-processing procedure should be in place throughout the lifetime of the system. It should be designed as part of either the RMA engineering effort or the logistic support engineering (see Section 12.1.2) and operated as part of the integrated logistic support (ILS) activity. The benefits of such an ongoing reliability-monitoring program are:

- to discover deficiencies in the system in order to allow corrective action in the form of revision, refurbishment or replacement;
- to discover weaknesses in the design, implementation or verification processes, so that these can be improved for future projects;
- to accumulate statistical failure data and thereby improve the basis for future estimation.

10.7 Case study

10.7.1 Tactical communications system

10.7.1.1 Reliability model

In Section 3.5.2.2 the tactical networks were modeled in terms of normal networks, and properties of this model were developed in Sections 6.5.2 and 9.3.1. Continuing to represent tactical networks by normal networks, a relatively simple model of system reliability can now be developed.

For the purpose of defining system reliability, let the trunk network be composed of three types of elements (see Figure 10.8):

- those parts of a node that, if they fail, cause the whole node to fail (e.g. switch power supply, switch backplane, central processor);

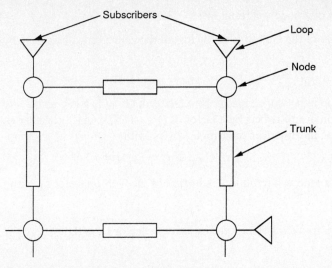

Figure 10.8 Reliability block diagram of the trunk network composed of three element types: nodes, trunks and loops.

- those parts of a node that, if they fail, cause one or more subscribers to become disconnected (e.g. switch TDM loop-group module, MUX CPU and power supply, MUX channel module);
- all the transmission equipment in one link plus those parts of the nodes on either side that, if they fail, cause the link to fail (e.g. switch TDM trench-group module).

These types of element will be called respectively a node, a loop and a trunk, and their failure rates will be denoted by f_n, f_l and f_t respectively.

10.7.1.2 Failure mode analysis

The loss of a node will give the subscribers connected to it a grade of service (GoS) of zero and will reduce the network cross-section by the quantity of trunks attached to the node. The loss of a loop will give the one subscriber connected to the loop a GoS of zero. The loss of a trunk will reduce the network cross-section by one trunk.

When a trunk is lost, not only is the cross-section reduced, but also the average connection length L is increased. This can be taken into account by using a cross-section reduction factor equal to the ratio of the decreased quantity of links to the whole quantity of links, raised to a power greater than unity in order to obtain the reduction in capacity. For the purpose of this study, the power will be taken to be 1.5.

10.7.1.3 Normal network expressions

Let U be the total quantity of subscribers to the network, N the quantity of nodes, N_a the quantity of access nodes and N_l the quantity of links (trunks).

The loss of a node will result in:

- a reduction of the GoS, due to the loss of the connected subscribers, by a factor of r_1, with

$$r_1 = 1 - 1/N$$

- a reduction in network capacity by a factor of $(1 - 1/N)^{1.5}$; and
- a reduction in subscribers by a factor of $(1 - 1/N_a)$; which together give
- a reduction in subscriber capacity C_s, by r_2, with

$$r_2 = (1 - 1/N)^{\frac{1}{2}}$$

The loss of a loop will result in a reduction of the GoS by factor r_3, with

$$r_3 = 1 - 1/U$$

The loss of a trunk will reduce the capacity by factor r_4, with

$$r_4 = (1 - 1/N)^{1.5}$$

Let the mission duration (measured in hours) be denoted by T. Then the probability of an element failing within that time is given by:

$$p(i) = 1 - \exp(-f_i T) \qquad (i = n, 1, t)$$

As the failure of a node influences two of the reduction factors, it is most convenient to redefine the failure probabilities as follows:

$$p(1) = p(2) = p(n)$$
$$p(3) = p(1)$$
$$p(4) = p(t)$$

Then the reduction factors due to failures within T, which will be denoted by $r'(i)$, are given by

$$r'(i) = (1 - p(i))^{e(i)}$$

with

$$e(1) = e(3) = 1$$
$$e(2) = 0.5$$
$$e(4) = 1.5$$

10.7.1.4 System reliability definition

Reliability R(T) is the probability of not having failed after a time T. In the case of a system, failure is usually not an all-or-nothing situation but a degradation of the performance to a point deemed to be unacceptable. As discussed in Section 9.3.1, the global performance parameter is GoS; consequently, it becomes necessary to relate GoS to element failure.

Using the Erlang B model, which was introduced in Section 9.3.1 to relate GoS to the traffic loading q of the network, the value of GoS at the end of the mission is obtained by

$$GoS(T) = r'(1)r'(3) \times E[q/r'(2)r'(4)]$$

where E is the relationship between GoS and q depicted in Figures 9.1 and 9.2.

Assuming the mission duration to be 96 h, the system reliability R will be defined by

$$R = GoS(96)/GoS(0)$$

Thus the system reliability is a function of five parameters:

f_n, f_l, f_t: element failure rates, determined by the failure rates of the constituent equipments;

N: size of the network, measured by the total quantity of nodes;

q: traffic loading of the intact network.

10.8 Summary

1. System **failure** is always due to component failure(s), but system **reliability** is often a complex function of component reliability. Not every component failure leads to a system failure.
2. Reliability is a **probability** and is always **time-dependent**.
3. Failure mechanisms are represented by **reliability models**, which are described by **failure probability density functions**.
4. The reliability of a design is always an **estimate**, based on previous **experience** with actual systems.
5. With the proper interpretation, **software reliability** can be treated on exactly the same basis as hardware reliability, but the models are normally different.
6. To define system reliability it is first necessary to define system failure, and this is usually tied to system **performance**. System reliability models are not concerned with component failure mechanisms but are complex functions of component failure and repair rates, and these functions are closely related to the system **structure**.
7. The benefit of reliability is expressed through the **cost of failure**. This cost is composed of several components, some of which may be highly subjective.
8. Reliability should not be treated as a parameter whose value is simply accepted as whatever it happens to turn out to be. Reliability should be determined by **design** either to satisfy a specified target value or to be used as a parameter in the maximization of cost-effectiveness.
9. In exactly the same way as system behavior may be conveniently described using the concept of states, reliability may be described in terms of **failure modes**.

10.9 Short questions

1. What is a reliability block diagram?
2. What is special about the exponential failure probability density function?
3. Why do systems tend to have a constant failure rate?
4. What is the significance of cut-sets in reliability theory?

10.10 Problems

1. *Communications system*: Consider a small communications system serving four user groups of 20 users each. They are connected to a communications network consisting of a multiplexer (MUX) and a switch for each user group, and a wideband communications channel between adjacent nodes. The MUX has a port common to all users in a group and an interface module for each user. The failure rates for the system components are:

$$
\text{MUX} \begin{cases} \text{common port} & \lambda_m & 2.10^{-4}\,\text{h} \\ \text{user interface} & \lambda_n & 3.10^{-5}\,\text{h} \end{cases}
$$
$$
\begin{aligned} \text{switch} \qquad\quad & \lambda_s & 3.10^{-4}\,\text{h} \\ \text{transceiver} \quad\; & \lambda_c & 10^{-4}\,\text{h} \end{aligned}
$$

and the following assumptions hold:

- All system components have the same mean time to repair (MTTR), namely 3 h.
- The channel between transceivers never fails.
- When a switch fails, all three ports are inoperable.

One important parameter characterizing such a system is its **inherent availability**. However, the formula applicable to a single component,

$$
A = \frac{\text{MTBF}}{\text{MTBF} + \text{MTTR}}
$$

cannot simply be used; it is first necessary to define the inherent system availability. There are several possible definitions, and the one used here is as follows.

Let a **connection** between two users, U_i and U_j, be denoted by X_{ij}, with $i,j = 1, \ldots, 80$ in the present case. Each such connection will have a certain $(\text{MTBF})_{ij}$ and a certain $(\text{MTTR})_{ij}$, and therefore, according to the above formula, a definite value of its availability, A_{ij}. The inherent system availability will now be defined as the average of A_{ij} for all i and j ($i \pm j$). (Note that all quantities are symmetric in i and j, e.g. $X_{ij} = X_{ij}$ and counted as one connection.)

Calculate the inherent system availability for this particular system.

2. *Human reliability*: A new class of nuclear submarines is to be designed. Such submarines must be capable of unsupported operation for a considerable period, say 4 months. All the performance requirements of the submarine as a weapons system have been specified by the Navy; however, in meeting these requirements a number of options and trade-offs remain open to the designer.

An area of major importance, which has to be considered from the very beginning of the analysis phase, is obviously **manning levels** (i.e. how many officers and sailors are needed for the system to fulfill its mission).

(a) Set up a structural approach to the problem of determining the optimal crew size and composition.

This may be in the form of a high-level work breakdown structure (WBS) (say, 6 to 10 work packages), but indicate the necessary relations to other areas of the design process and at what stage decisions are taken.

Viewing the crew as an element of the system, the system performance will depend on crew performance. In particular, some measure of **crew reliability** must enter the determination of system reliability beyond its obvious intuitive meaning.

(b) Give three examples of how human characteristics influence crew reliability.

(c) Discuss the relations between crew reliability and crew size.

10.11 Appendix: Tables for reliability estimation

Tables 10.1 to 10.14 are edited extracts from MIL-HDBK-271C and illustrate the database on which electronic system failure rates are universally estimated.

Table 10.1 Generic failure rate λ_G for bipolar digital devices (TT1 and DTL), for different environments (f/10^6 h)[a]

Circuit complexity	Use environment									
	G_B & S_F	G_F	A_{IT}	A_{IF}	N_S	G_M	A_{UT}	A_{UF}	N_U	M_L
Gates:										
1–20	.0070	.029	.070	.13	.093	.091	.11	.20	.12	.21
21–50	.020	.062	.12	.21	.17	.16	.20	.33	.23	.34
51–100	.032	.094	.18	.29	.24	.23	.28	.45	.34	.47
101–500	.079	.22	.37	.56	.49	.45	.61	.89	.71	.85
501–1000	.13	.34	.56	.82	.73	.67	.92	1.3	1.1	1.2
1001–2000	.29	.78	1.3	1.8	1.7	1.5	2.1	2.9	2.5	2.7
2001–3000	.81	2.1	3.5	5.1	4.5	4.1	5.8	8.1	6.7	7.3
3001–4000	2.2	5.7	9.6	14.	12.	11.	16.	22.	18.	20.
4001–5000	5.9	16.	26.	38.	33.	30.	43.	60.	49.	54.
Bits:										
ROM[b] $\leqslant$ 320	.0083	.022	.036	.053	.048	.043	.060	.085	.070	.078
ROM 321–576	.012	.033	.055	.081	.072	.066	.091	.13	.11	.12
ROM 577–1120	.202	.052	.087	.13	.11	.10	14	.20	.17	.19
ROM 1121–2240	.029	.078	.13	.20	.17	.16	.22	.31	.25	.29
ROM 2241–5000	.045	.12	.20	.30	.27	.24	.33	.48	.39	.45
ROM 5001–11000	.068	.18	.31	.47	.41	.38	.51	.75	.60	.70
ROM 11001–17000	.10	.28	.48	.73	.63	.58	.79	1.1	.92	1.1

Notes: [a] See Tables 10.3 and 10.4 for π_Q and π_L values.

[b] RAM failure rates = $3.5 \times$ ROM failure rates.

Table 10.2 Generic failure rate λ_G for bipolar beam lead, ECL, all linear and all MOS devices, for different environments (f/10^6 h)[a]

	Use environment									
Circuit complexity	G_B & S_F	G_F	A_{IT}	A_{IF}	N_S	G_M	A_{UT}	A_{UF}	N_U	M_L
Gates:										
1–20	.010	.048	.099	.16	.14	.12	.21	.30	.25	.24
21–50	.048	.19	.31	.40	.43	.34	.73	.86	.92	.52
51–100	.076	.31	.48	.59	.68	.54	1.2	1.3	1.5	.78
101–500	.19	.82	1.2	1.4	1.7	1.3	3.1	3.4	3.9	1.7
501–1000	.32	1.4	2.0	2.3	2.8	2.1	5.1	5.5	6.4	2.6
1001–2000	.74	3.1	4.6	5.2	6.4	4.8	12.	13.	15.	6.0
2001–3000	2.0	8.4	13.	14.	17.	13.	33.	35.	41.	16.
3001–4000	5.4	23.	35.	39.	47.	36.	90.	96.	111.	44.
4001–5000	15.	62.	94.	105.	128.	97.	241.	258.	299.	121.
Bits:										
ROM[b] ≤ 320	.021	.087	.13	.15	.18	.14	.33	.36	.42	.17
ROM 321–576	.031	.13	.19	.22	.27	.20	.49	.53	.62	.26
ROM 577–1120	.048	.20	.31	.35	.42	.32	.78	.84	.98	.41
ROM 1121–2240	.072	.30	.45	.52	.63	.48	1.2	1.3	1.5	.61
ROM 2241–5000	.11	.46	.70	.80	.96	.74	1.8	1.9	2.2	.94
ROM 5001–11000	.17	.70	1.1	1.2	1.5	1.1	2.7	2.9	3.4	1.5
ROM 11001–17000	.25	1.1	1.6	1.9	2.2	1.7	4.1	4.5	5.2	2.2
Linear ≤ 32 transistors	.011	.052	.12	.20	.16	.15	.22	.35	.27	.33
Linear 33–100 transistors	.023	.11	.24	.41	.35	.31	.48	.73	.60	.66

Notes: [a] See Tables 10.3 and 10.4 for π_Q and π_L values.
[b] RAM failure rates = 3.5 × ROM failure rates.

Table 10.3 Quality factors π_Q for use with Tables 10.1 and 10.2

Quality level[a]	π_Q[b]
S	0.5
B	1
B-1	2.5
B-2	1
C	8
C-1	45
D	75
D-1	150

Notes: [a] See Table 10.13 for descriptions of quality levels.
[b] π_Q values shown here are different from those in Table 10.13.

Table 10.4 Learning factors π_L for use with Tables 10.1 and 10.2

$\pi_L = 10$ under any of the following conditions:

1. new device in initial production;
2. major changes in design or process have occurred;
3. there has been an extended interruption in production or a change in line personnel (radical expansion).

The factor of 10 can be expected to apply until conditions and controls have stabilized. This period can extend for as much as 6 months of continuous production.

$\pi_L = 1$ under all production conditions not stated in 1, 2 or 3 above.

Table 10.5 Generic failure rate λ_G for discrete semiconductors, for different environments (f/10^6 h)[a]

Part type	G_B & S_F	G_F	A_{IT}	A_{IF}	N_S	G_M	A_{UT}	A_{UF}	N_U	M_L
						Use environment				
Transistors										
Si NPN	.017	.11	.28	.59	.26	.59	.60	1.2	.84	.96
Si PNP	.025	.17	.46	.96	.41	.96	.96	1.9	1.4	1.5
Ge PNP	.025	.25	.75	1.6	.84	1.6	.78[b]	1.6[b]	2.1[b]	2.5
Ge NPN	.072	.66	2.0	4.3	2.2	4.3	3.3[b]	6.6[b]	5.4[b]	6.6
FET	.046	.31	.78	1.6	.70	1.6	1.7	3.4	2.3	2.6
Unijunction	.15	1.0	2.7	5.6	2.4	5.6	6.3	13.	9.0	9.0
Diodes										
Si, gen. purpose	.0051	.036	.098	.20	.090	.20	.24	.48	.33	.33
Ge. gen. purpose	.0066	.078	.25	.51	.30	.51	.44	.87[b]	.75[b]	.81
Zener & avalanche	.016	.096	.24	.51	.22	.51	.54	1.1	.72	.84
Thyristor	.023	.16	.43	.90	.40	.90	1.0	2.0	1.4	1.4
Si microwave det.	.19	2.2	6.0	12.	3.9	12.	7.5	25.	17.	46.
Ge microwave det.	.41	5.6[b]	18.[b]	35.[b]	na	35.[b]	na	na	na	na
Si microwave mix.	.25	3.0	8.0	16.	5.1	16.	17.	34.	23.	64.
Ge microwave mix.	.72	10.[b]	31.[b]	61.[b]	na	61.[b]	na	na	na	na
Varactor, step recovery, tunnel	.24	1.5	3.9	8.1	3.5	8.1	8.6	17.	12.	13.
LED	.034	.14	.25	.49	.45	.35	.91	1.8	1.4	.88
Single isolator	.051	.21	.38	.74	.68	.53	1.4	2.7	2.1	1.30

Notes: [a] See Table 10.6 for π_Q values.
    [b] This value is valid only for electrical stress, $s \leq 0.3$.
    na Do not use in these environments, since temperature normally encountered combined with normal power dissipation is above the device ratings.

Table 10.6 Quality factors π_Q for use with Table 10.5

Part type	JANTXV	JANTX	JAN	Non-MIL hermetic	Plastic
Microwave diodes	0.3	0.6	1.0	2.0	—
All other types	0.1	0.2	1.0	5.0	10.

Table 10.7 Generic failure rate λ_G for resistors for different environments (f/10^6 h)[a]

Construction	Style	MIL-R-Spec.	G_B & S_F	G_F	A_{IT}	A_{IF}	N_S	G_M	A_{UT}	A_{UF}	N_U	M_L
Resistors, fixed							*Use environment*					
Composition	RCR	39008	.00051	.0032	.0037	.0075	.0046	.0066	.014	.027	.021	.02
Composition	RC	11	.0025	.016	.018	.038	.023	.033	.069	.13	.10	.099
Film	RLR	39017	.0012	.0031	.0043	.0088	.0033	.0062	.016	.032	.021	.028
Film	RL	22684	.0061	.015	.022	.044	.017	.031	.079	.016	.10	.14
Film	RN	55182	.0014	.0033	.0049	.01	.0037	.007	.018	.036	.024	.032
Film	RN	10509	.0073	.017	.025	.05	.019	.035	.09	.18	.12	.16
Film, power	RD	11804	.012	.026	.055	.11	.026	.078	.15	.29	.19	.46
Film, network	RZ	83401	.026	.072	.17	.34	.14	.24	.80	1.6	1.2	1.1
Wirewound, accurate	RBR	39005	.0085	.019	.058	.12	.020	.078	.22	.44	.17	.39
	RB	93	.043	.094	.29	.58	.10	.39	1.1	2.2	.85	1.9
Wirewound, power	RWR	39007	.014	.044	.073	.15	.037	.091	.19	.37	.25	.54
	RW	26	.072	.22	.36	.73	.19	.45	.94	1.9	1.3	2.7
Wirewound, ch. mount	RER	39009	.0079	.021	.045	.090	.024	.056	.11	.22	.16	.34
	RE	18546	.040	.010	.22	.45	.12	.28	.56	1.1	.79	1.7
Resistors, variable												
Wirewound, trimmer	RTR	39015	.014	.034	.078	.16	.077	.15	.19	.37	.24	1.1
	RT	27208	.072	.17	.39	.79	.39	.74	.93	1.9	1.2	5.5
Wirewound, prec.	RR	12934	.84	2.1	5.5	11.	4.6	11.	14.	29.	18.	132.
	RA	19	.31	.84	2.4	4.8	2.1	9.5	na	na	na	na
Wirewound, semi-prec.	RK	39002	.31	.84	2.4	4.8	2.1	9.5	na	na	na	na
	RP	22	.31	.78	2.0	3.9	1.7	7.8	na	na	na	na
Non-wirewound, trimmer	RJR	39035	.02	.067	.12	.23	.95	.23	.27	.54	.34	1.8
	RJ	22097	.10	.33	.58	1.2	4.8	1.2	1.4	2.7	1.7	9.2
Composition	RV	94	.12	.49	1.1	2.1	.88	4.1	6.6	13.	6.8	18.
Non-wirewound, prec.	RQ	39023	.086	.35	.65	1.3	.58	1.3	2.8	5.7	2.6	10.
	RVC	23285	.096	.34	.6	1.2	.51	1.2	2.2	4.4	2.0	9.6

Notes: [a] See Table 10.9 for π_Q values
na Not normally used in this environment.

Table 10.8 Generic failure rate λ_G for capacitors, for different environments $(f/10^6 \text{ h})^a$

Dielectric	Style	MIL-R-Spec.	G_B & S_F	G_F	A_{IT}	A_{IF}	N_S	G_M	A_{UT}	A_{UF}	N_U	M_L
Capacitors, fixed												
Paper	CP	25	.011	.022	.057	.11	.047	.057	.16	.31	.15	.29
Paper	CA	12889	.012	.031	.087	.17	.083	.087	.47	.90	.53	.44
Paper/plastic	CZR	11693	.0047	.0098	.025	.05	.021	.025	.072	.14	.064	.13
Paper/plastic	CPV	14157	.0021	.0042	.0088	.018	.0088	.0088	.025	.05	.023	.044
Paper/plastic	CQR	19978	.0021	.0042	.0088	.018	.0088	.0088	.025	.05	.023	.044
Paper/plastic	CHR	39022	.0028	.006	.012	.024	.012	.012	.035	.069	.032	.06
Paper/plastic	CH	18312	.02	.042	.084	.17	.087	.084	.24	.49	.22	.42
Plastic	CFR	55514	.0041	.0086	.022	.043	.018	.03	.067	.13	.075	.13
Plastic	CRH	83421	.0023	.0048	.0096	.019	.010	.0096	.028	.055	.025	.048
MICA	CMR	39001	.0005	.0022	.0059	.012	.0043	.0084	.044	.088	.042	.042
MICA	CM	5	.003	.013	.035	.071	.026	.050	.27	.53	.25	.25
MICA	CB	10950	.09	.19	.42	.85	.3	.60	1.9	3.8	1.4	3.0
Glass	CYR	23269	.0003	.0014	.0037	.0075	.0066	.0053	.027	.054	.028	.026
Glass	CY	11272	.001	.0043	.011	.022	.020	.016	.082	.16	.084	.079
Ceramic	CKR	39014	.0036	.0076	.033	.066	.0098	.016	.068	.14	.032	.12
Ceramic	CK	11015	.011	.023	.099	.20	.029	.047	.20	.41	.096	.35
Ceramic	CCR	20	.0008	.0032	.008	.016	.0058	.011	.058	.12	.070	.057
TA, sol.	CSR	39003	.012	.026	.078	.16	.035	.052	.15	.29	.14	.26
TA, non-sol.	CLR	39006	.0061	.014	.082	.16	.049	.069	.14	.28	.15	.23
TA, non-sol.	CL	3965	.018	.043	.24	.49	.15	.21	.42	.83	.46	.69
Al oxide	CU	39018	.074	.23	1.2	2.3	.96	1.6	4.8	9.7	5.3	5.5
Al dry	CE	62	.090	.36	1.9	3.7	1.7	2.6	10.	21.	12.	8.7
Capacitors, variable												
Ceramic	CV	81	.32	1.6	2.5	4.8	4.2	3.5	24.	48.	19.	31.
Piston	PC	14409	.099	.54	1.2	2.3	1.7	1.5	9.2	18.	22.	7.5
Air, trimmer	CT	92	.4	3.0	4.8	9.4	8.0	6.8	49.	98.	37.	60.
Vacuum	CG	23183	1.2	6.2	15.	29.	15.	21.	140.	270.	94.	na

Notes: [a] See Table 10.9 for π_Q values.

na Not normally used in this environment.

Table 10.9 Quality factors π_Q for use with Tables 10.7 and 10.8

Failure rate level	$\pi_Q{}^a$
L	1.5
M	1.0
P	.3
R	.1
S	.03

Notes: [a] For non-ER parts (styles with only two letters in Tables 10.3 and 10.13) $\pi_Q = 1$ providing that parts are procured in accordance with the part specification. If procured as commercial (non-MIL) quality, use $\pi_Q = 3$. For ER parts (styles with three letters), use the π_Q value for the "letter" failure rate level procured.

Table 10.10 Generic failure rate λ_G for inductive, electromechanical and miscellaneous parts, for different environments (f/10^6 h)[a]

Part type	G_B & S_F	G_F	A_{IT}	A_{IF}	N_S	G_M	A_{UT}	A_{UF}	N_U	M_L
Inductive										
Low power pulse transformer	.003	.0048	.041	.082	.017	.047	.069	.14	.065	.12
Audio transformer	.006	.0096	.082	.16	.034	.094	.14	.28	.13	.24
High power pulse & power transformer filter	.019	.053	.31	.60	.13	.35	.46	.92	.98	.86
RF transformer	.024	.038	.33	.64	.14	.38	.56	1.1	.52	.96
RF coils, fixed	.0016	.004	.021	.042	.0096	.048	.039	.078	.038	.12
RF coils, variable	.0032	.008	.042	.084	.019	.096	.078	.16	.077	.24
Motors	na	15.	19.	19.	24.	19.	41.	41.	49.	na
Relays										
General purpose	.13	.30	.65	1.3	.89	.81	2.8	5.6	2.9	16.
Contractor, high current	.44	1.0	2.2	4.5	3.0	2.8	9.6	19.	10.	56.
Latching	.10	.24	.52	1.0	.71	.65	2.2	4.5	2.3	13.
Reed	.11	.26	.55	1.1	.75	.69	2.4	4.8	2.5	14.
Thermal bimetal	.29	.69	1.5	3.0	2.0	1.9	6.4	13.	6.7	37.
Meter movement	.90	2.1	4.6	9.2	6.3	5.8	20.	40.	21.	na
Switches										
Toggle & push button	.035	.011	.18	.35	.15	.61	1.8	3.5	.84	24.
Sensitive	.15	.44	.74	1.5	.59	2.5	7.4	15.	3.4	100.
Rotary	.22	.67	1.1	2.2	.89	3.8	11.	22.	5.1	150.
Connectors (per pair)										
Circular, rack & panel	.0062	.029	.12	.24	.053	.12	.17	.34	.23	.18
Printed wiring board	.0031	.028	.060	.12	.036	.060	.090	.18	.11	.090
Coaxial	.0084	.032	.13	.26	.060	.10	.18	.36	.24	.20
PC wiring boards										
Two-sided	.0012	.0024	.005	.01	.0048	.0048	.012	.024	.012	.024
Multilayer	.15	.30	.63	1.3	.60	.60	1.5	3.0	1.5	3.0
Connections[b]	See MIL-HDBK-217C									
Tubes	See MIL-HDBK-217C									
Lasers	See MIL-HDBK-217C									

Notes: [a] See Table 10.11 for π_Q values.
[b] Usually negligible compared to system failure rate.
na Not normally used in these environments.

Table 10.11 Quality factors π_Q for use with Table 10.10

	Quality level	
Part type	*MIL-spec.*	*Non-MIL*
Inductive	1	3
Motors	1	1
Relays	1	3
Switches, toggle & sensitive	1	20
Switches, rotary	1	50
Connectors	1	3
PC wiring boards	1	—
Others	—	1

Table 10.12 Environmental symbol identification and description

Environment	π_E Symbol	Nominal environmental conditions
Ground, benign	G_B	Nearly zero environmental stress with optimum engineering operation and maintenance.
Space, flight	S_F	Earth orbital. Approaches ground, benign conditions without access for maintenance. Vehicle neither under powered flight nor in atmospheric re-entry.
Ground, fixed	G_F	Conditions less than ideal to include installation in permanent racks with adequate cooling air, maintenance by military personnel and possible installation in unheated buildings.
Ground, mobile	G_M	Conditions more severe than those for G_F, mostly for vibration and shock. Cooling air supply may also be more limited, and maintenance less uniform.
Naval, sheltered	N_S	Surface ship conditions similar to G_F but subject to occasional high shock and vibration.
Naval, unsheltered	N_U	Nominal surface shipborne conditions but with repetitive high levels of shock and vibration.
Airborne, inhabited transport	A_{IT}	Typical conditions in transport or bomber compartments occupied by aircrew without environmental extremes of pressure, temperature, shock and vibration, and installed on long mission aircraft such as transports and bombers.
Airborne, inhabited fighter	A_{IF}	Same as A_{IT} but installed on high performance aircraft such as fighters and intercepters.
Airborne, uninhabited transport	A_{UT}	Bomb bay, equipment bay, tail, or wing installations where extreme pressure, vibration, and temperature cycling may be aggravated by contamination from oil, hydraulic fluid and engine exhaust. Installed on long mission aircraft such as transports and bombers.
Airborne, uninhabited fighter	A_{UF}	Same as A_{UT} but installed on high performance aircraft such as fighters and interceptors.
Missile launch	M_L	Severe conditions of noise, vibration, and other environments related to missile launch, and space vehicle boost into orbit, vehicle re-entry and landing by parachute. Conditions may also apply to installation near main rocket engines during launch operations.

Table 10.13 Quality factors π_Q

Quality level	Description	π_Q
S	Procured in full accordance with MIL-M-38510, Class S requirements.	1
B	Procured in full accordance with MIL-M-38510, Class B requirements.	2
B-1	Procured to screening requirements of MIL-STD-883, Method 5004, Class B, and in accordance with the electrical requirements of MIL-M-38510 slash sheet or vendor or contractor electrical parameters. The device must be qualified to requirements of MIL-STD-883, Method 5005, Class B. No waivers are allowed.	5
B-2	Procured to vendor's equivalent of screening requirements of MIL-STD-883, Method 5004, Class B, and in accordance with vendor's electrical parameters. Vendor waives certain requirements of MIL-STD-883, Method 5004, Class B.	10
C	Procured in full accordance with MIL-M-38510, Class C requirements.	16
C-1	Procured to screening requirements of MIL-STD-883, Method 5004, Class C and in accordance with the electrical requirements of MIL-M-38510 slash sheet or vendor or contractor electrical specification. The device must be qualified to requirements of MIL-STD-883, Method 5005, Class C. No waivers are allowed.	90
D	Commercial (or non-MIL standard) part, hermetically sealed, with no screening beyond the manufacturer's regular quality assurance practices.	150
D-1	Commercial (or non-MIL standard) part, packaged or sealed with organic materials (e.g. epoxy, silicone or phenolic).	300

Table 10.14 Parts with multi-level quality specifications

Part type	Quality designators
Microelectronics	A, B, B-1, B-2, C, C-1, D, D-1
Discrete semiconductors	JANTXV, JANTX, JAN
Capacitors, established reliability (ER)	L, M, P, R, S
Resistors, established reliability (ER)	M, P, R, S

Table 10.15 Differences between hardware and software reliability features

Hardware	Software
Failure can be caused by deficiencies in design, implementation, operation and maintenance	Failure is primarily due to design errors, with production (i.e. copying), use and maintenance (excluding corrections) having negligible effect.
Failures can be due to wear or other energy-related phenomena. Sometimes warning is available before failure occurs.	There is no wearout phenomenon. Software failures occur without warning.
Reliability is time-related, with failures occurring as a function of operating (or storage) time.	Reliability is not time-related. Failures occur when a program step or path that is in error is executed.
Reliability is related to environmental factors.	The external environment does not affect reliability, except insofar as it may affect program inputs.
Reliability can be predicted in theory from a knowledge of design and usage factors.	Reliability cannot be predicted from any physical bases, since it entirely depends on human factors in design.
Failures can occur to the components of a system in a pattern that is, to some extent, predictable from the stresses on the components, and other factors.	Failures are not usually predictable from the analysis of separate statements. Errors are likely to exist randomly throughout the program, and any statement may be in error.
Hardware uses standard components as basic building blocks.	There are no standard parts in software, although there are standardized logic structures.

Source: Adapted from MIL-HDBK-338.

10.12 References

Arsenault, J.E. & Roberts, J.A. (eds), *Reliability and Maintainability of Electronic Systems*, Computer Science Press.

Billinton, R. & Allan, R.N., *Reliability Evaluation of Engineering Systems*, London: Pitman Publishing, 1983.

Dhillon, B.S. & Singh, C., *Engineering Reliability*, New York: Wiley, 1981.

MIL-HDBK-217, *Reliability Prediction of Electronic Equipment,* Washington DC: US Department of Defense.

MIL-HDBK-338, *Electronic Reliability Design Handbook,* Washington DC: US Department of Defense, 1984.

MIL-STD-810, *Environmental Test Methods,* Washington DC: US Department of Defense.

O'Connor, P.D.T., *Practical Reliability Engineering,* New York: Wiley, 1985.

Von Alven, W.H. (ed.), *Reliability Engineering*, ARINC Research Corporation, Englewood Cliffs, N.J.: Prentice Hall, 1964.

CHAPTER 11

Maintainability

11.1 Introduction

11.1.1 Definition of maintainability

The **maintainability** of a system is a characteristic deriving from its design and installation, and is identified with the ease, efficacy, safety and cost of maintenance actions taken to retain or restore its performance. The range of issues implicit in this description makes it all but impossible to achieve a quantitative definition that is both useful and universal in its application. Possibly the most widely accepted (as per MIL-STD-721B) defines maintainability as *the probability that a system will be retained in or restored to a specifed condition in a given period, when maintenance is performed in accordance with prescribed procedures and resources.* However, this definition is not used extensively in the same way as the definition of reliability is used.

The objective of maintainability is to design and develop systems and equipment that can be maintained in the least time, at the least cost, and with a minimum expenditure of support resources, without adversely affecting the item performance or safety characteristics.

The realization of the objective of maintainability requires involvement in the **total design process**. This includes participation in system design, equipment design, production, installation and the activation of a system. Traditionally, maintainability has been mainly concerned with keeping equipment operational—with how to combat the effects of component failure by suitable measures in the equipment design phase. Such measures include modularization and integrated test facilities as well as easy accessibility, the clear and complete labeling of all components and test points, and structured easy-to-follow maintenance manuals. However, with the generally increasing importance of systems, more and more attention is being given to the systems aspects of maintainability.

Firstly, there is the impact of equipment maintainability on system **lifecycle costs**. A particular value of such a quantity as system availability can be achieved by a wide variety of combinations of individual equipment availabilities and system structures, but the costs will differ. Specifying the optimum solution therefore makes it necessary to allocate maintainability targets as part of equipment specifications.

Secondly, a system is not maintained by a multiplicity of independent groups or entities, one for each piece of equipment. Maintenance can only be carried out in a cost-effective manner if there exists a well-designed **maintenance organization**, whose structure is adapted to the structure and boundary conditions of that particular system. This maintenance organization is a subsystem in its own right, with its own elements and relations, and all that has been said about systems and their design in this textbook holds for the maintenance organization.

A further aspect of maintainability is emerging as an increasingly important factor in the lifecycle cost of a system—one that has nothing to do with maintenance in the sense of replacing faulty or worn components or returning equipment. It is caused by the progressively more rapid change in technology generally, which, when combined with the long timespans necessary for the realization of large systems, tends to shorten the useful lifetime of the systems. In extreme cases it may even make a system obsolete by the time it is put into service.

It is clear that this aspect of maintainability has a severe effect on the cost-effectiveness, and it therefore becomes imperative to extend the scope of maintainability to include the ease and economy with which system effectiveness can be maintained in the face of **changing boundary conditions**. In other words, maintaining a system is extended to making modifications and extensions as is required to meet the demands of changing circumstances.

This aspect of maintainability has become most urgent in the case of data-processing equipment. While the cost of the hardware, in terms of processing and memory capacity per dollar, is continually sinking, the cost of producing software has remained largely constant. Thus, relatively speaking, the software costs have been skyrocketing, and the only way to keep cost-effectiveness from dropping is to increase the effective lifetime of the software. This is done by periodically introducing new versions of the programs, and the organization of this activity, including the updating of the documentation, has become a major maintainability task.

The primary impetus for maintainability development is attributable to the US Department of Defense and to military programs. It should be noted, however, that, during the same period that the US government was advancing the concept of maintainability, general industry was taking measures to counter similar problems of rising costs in support of their products. The basic principles applied by general industry were identical to those set forth in military specifications and standards, although they varied in formulation and application to suit the needs of individual companies. The concept was first introduced by the US military services in 1954, and the first formalized specification of a maintainability program appeared as MIL-STD-470, *Maintainability Program Requirements* and its accompanying handbook, MIL-HDBK-472.

It follows from the definition of maintainability that it is determined by activities that take place during the creation of a system (i.e. during the five phases of the systems engineering process). From the above introductory discussion, however, it should also be clear that maintainability is closely linked to maintenance and integrated logistic support (ILS). The boundaries between these areas are not always clearly defined and may also vary between organizations. The approach taken in this textbook is as follows:

In order to recognize explicitly both the fact that a system must contain the means of sustaining itself as well as the important role played by humans as integral components of the system, it is convenient to consider any system as consisting of four subsystems:

> functional subsystem
> operating subsystem
> repair subsystem, and
> support subsystem.

The functional subsystem consists of the equipment (hardware and software) necessary to transform the inputs into outputs (sometimes called the prime mission equipment). The operating subsystem encompasses the human involvement in that transformation process, and is characterized by such concepts as organization, procedures and skills. These two subsystems together produce the service that satisfies the need expressed in the project definition, and while the former has traditionally received the main part of the attention and effort during the design phase, they both contribute to the demand for maintenance and support—that is, they influence the maintainability.

The repair subsystem is that part concerned with the direct maintenance of the functional subsystem and consists of both personnel and equipment (e.g. test equipment), whereas the support subsystem is concerned with providing those support functions that allow the human involvement to proceed on a sustained basis. With regard to the repair subsystem, the support subsystem typically provides the spare parts, consumables and facilities required for an ongoing operation; for both the operating and support subsystems the repair subsystem provides personnel management, training, and documentation.

The relationships between the subsystems are illustrated in Figure 11.1.

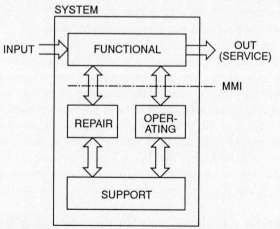

Figure 11.1 The four main subsystems.

11.2 Types of maintenance

11.2.1 Corrective maintenance

Corrective maintenance consists of all the activities performed *to restore an item to satisfactory condition after a malfunction has caused degradation of the item below the specified performance*. The major tasks associated with corrective maintenance are:

Preparation	Gathering tools; obtaining, setting up and calibrating maintenance aids; warming up equipment and so on.
Localization	Determining the location of a failure to the extent possible without using accessory support equipment.
Isolation	Determining the location of a failure by using accessory support equipment.
Disassembly	Equipment disassembly to the extent necessary to gain access to the item that is to be replaced.
Interchange	Removing the defective item and installing the replacement.
Reassembly	Closing and reassembly of the equipment after the replacement has been made.
Alignment	Performing any alignment, minimum tests and/or adjustment made necessary by the repair action.
Verification checkout	Performing the minimum checks or tests required to verify that the equipment has been restored to satisfactory performance.

This definition and list of tasks reflect the fact that the major part of corrective maintenance effort is spent at equipment level, but this should not obscure the importance of system level activities that are crucial to corrective maintenance, particularly **fault location**.

When a fault is detected at system level, it first has to be traced to a single piece of equipment, and in a large system this can be a time-consuming and costly task. However, good system design practice (and this is particularly true of systems with inherent data-processing facilities) will always strive to include a suite of test and fault-locating facilities and routines that facilitate fault location. Indeed, the design of these facilities or subsystems can constitute a considerable part of system design. In addition there needs to be a system-level corrective-maintenance manual that supports the fault location procedures and allows the fault to be pinpointed at equipment level using information gained at system level.

11.2.2 Preventive maintenance

Preventive maintenance is performed *to retain a system in satisfactory operational condition* by inspecting the system, detecting and preventing incipient failures, overhaul, lubrication, calibration and so on. Preventive maintenance is also called **scheduled maintenance**, to emphasize that it is designed into the system and planned as part of the normal system operation. It is a positive action rather than a reaction to external or random influences, as is the case with corrective maintenance.

Preventive maintenance is appropriate wherever system operation causes a gradual degradation of system elements and/or system performance, and also wherever elements have a well-defined lifetime. By inspection, or simply by monitoring the number of operating hours, it may be possible to carry out maintenance without causing any disruption to system operation. Typical examples include the periodic painting of external surfaces, replacing brake linings, changing motor oil and cleaning air filters.

Preventive maintenance is not appropriate when the cause of system degradation is the random failure of system elements. As long as the failure rate of a component does not increase, there is no reason to replace the component. This is the normal case for electronic equipment; as long as it works properly, leave it alone!

11.2.3 Adaptive maintenance

Adaptive maintenance is performed *to retain a satisfactory level of cost-effectiveness and performance under changing external conditions* (e.g. operational requirements, new technology). It consists of modifications and extensions, each of which may occur through all the phases of systems engineering.

The main point about adaptive maintenance in the current context of systems design is that it is not something unexpected that just happens to become necessary at some time in the life of the system. The expected degree of adaptive maintenance and the form that it is most likely to take are estimated in the analysis phase, and the capability of the system to be modified accordingly is designed into it.

Two typical measures are:

- **modularization**, whereby those parts of the system most likely to require future modification are either formed into separate modules or all grouped into one module;
- the provision of **interfaces** or "hooks", to which future additions can be connected with a minimum of effort.

Checking that a developing design is taking the requirements for adaptive maintenance adequately into account should be an item on every design review.

11.2.4 Maintenance indices

1. *Mean time between maintenance (MTBM)* includes both preventive and corrective maintenance requirements. Relative to corrective maintenance, MTBM considers primary (i.e. random) failures, secondary (i.e. dependent) failures, quality and manufacturing defects, operator- and maintenance-induced failures, and others.
2. *Mean time between replacements (MTBR)* is the mean time between equipment or item replacements for both preventive and corrective maintenance. MTBR forms the basis for spare/repair-parts determination. Note that a maintenance action (represented by the MTBM factor) does not always result in an item replacement.

3. *Mean time to repair (MTTR)* is the (arithmetic) mean active corrective-maintenance time.
4. *Mean maintenance time (M)* includes both corrective and preventive maintenance.
5. *Maintenance downtime (MDT)* is the total time during which an equipment item is not in condition to perform its intended function. MDT includes logistics time and waiting or administrative time.
6. *Logistics or supply time* is that portion of nonactive maintenance time during which maintenance is delayed solely because a needed item is not immediately available.
7. *Waiting or administrative time* is the portion of nonactive maintenance time that is not included in logistics or supply time.

11.2.5 Connection between reliability and maintainability

Both maintenance and reliability (or the lack of it) have as their common underlying cause the deterioration of all system elements, as discussed in Chapter 4. Consequently, there must be a strong connection between them.

This can be approached by first realizing that there are two important **timescales**:

- the **design lifetime** (DLT) of an object (e.g. system, equipment, component), which is the time for which it is expected to be able to be in operation without major refurbishment;
- the mean time between failure (MTBF) of the object or the typical (or average) MTBF of the components making up the object.

There are now two important cases:

MTBF $\gg$ DLT
MTBF $\ll$ DLT

An example of an object that falls into the first case is a bridge. Clearly, a bridge is not expected to fail (e.g. fall down) within its design lifetime. Another example is a deep space probe; it is designed to survive its mission time with a high probability. In both of these examples **no** corrective maintenance is expected—a general characteristic of objects in this class. On the other hand, preventive maintenance may vary widely, from quite high for a steel bridge such as the Sydney Harbour Bridge to nil for a deep space probe.

Electronic systems normally fall into the second case, for the following reason. The MTBF of individual components (e.g. resistors, capacitors, ICs) is generally considerably higher (e.g. 10^7h) than the design lifetime (e.g. 10^5h) of the system. However, as there is always a very large number (e.g. 10^4) of components in a system, component failure will be a common occurrence in the operation of an electronic system. Corrective maintenance thus becomes an important factor in determining system cost-effectiveness, whereas preventive maintenance is often negligible.

The above also explains why electronic systems (or any system that conforms to the second case) have a constant failure rate. After a period of operation the system will contain a steady state mixture of old and new components, and no matter what the failure rate characteristics of the individual components are, the system will see failures occurring at random and with a rate that equals the double average over time and components, which is a constant.

In this steady state reliability is no longer such an appropriate system parameter; much more important is the probability of finding the system in its operating state. This is called the **availability**, and it will clearly be a function of both reliability, in the form of the MTBF, and maintainability, in the form of a measure of the time for which the system is inoperable due to maintenance. Depending on which maintenance indices are used, different definitions of availability can be obtained:

1. *Inherent availability* is the probability that a system or equipment, when used under stated conditions, without consideration for any scheduled or preventive action, in an ideal support environment (i.e. available tools, spares, manpower, data and so on) will operate satisfactorily at a given point in time. It excludes ready time, preventive maintenance downtime, logistics time, and waiting or administrative downtime. It may be expressed as

$$A_i = \frac{MTBM}{MTBM + MTTR}$$

MTBF (mean time between failures) is a measure of the reliability of the equipment or system.

2. *Achieved availability* is the probability that a system or equipment, when used under stated conditions in an ideal support environment (i.e. available tools, spares, manpower, data and so on) will operate satisfactorily at a given point in time. It excludes logistics time and waiting or administrative downtime. It includes active preventive- and corrective-maintenance downtime. It may be expressed as

$$A_a = \frac{MTBM}{MTBM + M}$$

3. *Operational availability* is the probability that a system or equipment, when used under stated conditions in an actual operational environment, will operate satisfactorily at a given point in time. It includes ready time, logistics time, and waiting or administrative downtime. It may be expressed as

$$A_o = \frac{MTBM + ready\ time}{(MTBM + ready\ time) + MDT}$$

11.3 Maintainability program

11.3.1 General

The maintainability program is a schedule of all activities pertinent to achieving the maintainability objectives. These activities are carried out in parallel with other systems engineering activities, and consequently they can be correlated with the phases of systems engineering.

11.3.2 Definition phase

The maintainability-related effort in the definition phase consists mainly of **identifying** the major maintenance requirements and **defining** to what extent they are to be included in the project. It is quite common for the client to have a strong input at this point, and to integrate the maintenance requirements into an already existing organization. If feasibility studies are carried out at this stage, maintainability should be included as a significant factor.

11.3.3 Analysis phase

During the analysis phase the requirements identified during the definition phase are translated into a project-related approach to achieving the optimum system maintainability (as defined e.g. by maximum cost-effectiveness). The result of this activity is included in the system definition and consists essentially of two parts: the maintainability program plan and the technical support information.

The maintainability program plan contains both management and engineering aspects and should cover all or most of the following items:

- definition of the program objectives;
- how the goals will be achieved;
- methods of reducing maintenance skills requirements;
- methods of improving maintenance and diagnostic routines;
- techniques for optimizing the frequency and extent of preventive maintenance;
- methods of minimizing specialized tools, test equipment and test programs;
- design and constructing techniques that will maximize system MTBF and minimize maintenance action and repair time;
- prediction of MTBF, MTTR and MDT;
- demonstration test plans;
- an agenda of items to be covered in design reviews;
- how failure analysis and corrective action reports will be used to improve and upgrade the system design (i.e. the learning or maturing curve);
- an overall logistics support plan, covering instruction books, manuals, handbooks, engineering documentation, data management and so on.

The **technical support information** will contain such items as packaging concepts, standardization requirements, applicable standards and/or legislation, and existing system or equipment characteristics.

11.3.4 Design phase

In this phase the maintainability requirements are **described** and **specified** in quantitative terms. Also, the items in the program plan are further developed and integrated with the system requirements so as to result in **specific requirements** on the individual items of equipment, forming a part of the equipment specifications. An important task in this phase is to effect an overall optimization by finding the best trade-off between the various system elements (which are **not** restricted to hardware or equipment).

The major impact of maintainability on design is no doubt in the area of equipment design. However, maintainability requirements should play an all-important role in systems design as well. The following are a few points to consider:

1. *Transportability and accessibility:* The weight and size of the various pieces of equipment must be related to the access to the site of installation. This may require special requirements to be placed on the equipment design, or special access routes to be constructed, or a particular sequence to be developed for this subsystem or piece of equipment.

2. *Liability and other legal aspects:* Where several manufacturers are involved in a project, a subdivision of the maintenance organization may be necessary in order not to invalidate warranties and performance guarantees. This implies coordination and certain formal procedures in carrying out the maintenance.

3. *Self-monitoring and fault diagnosis:* Just as equipment has monitoring circuits and error detection and locating routines built into it, systems have the same type of features, just suitably scaled up. Instead of monitoring circuits (e.g. a voltage min/ max monitoring circuit), systems have monitoring equipment (e.g. a monitoring receiver to test a radar transmitter); and instead of routines (e.g. firmware), systems have special parts of the operating procedures that exercise the system functions (e.g. using friendly intruders to test an early warning system).

   Such procedures should in general contain four groups of instructions:

   - how to exercise or excite the system;
   - what data are to be gathered;
   - how, where and when the data are to be analyzed;
   - what to do with the results of the analysis (usually, initiate some form of corrective maintenance).

   The design of such self-monitoring and fault-diagnostic features can be a very significant part of the system design effort.

4. *Training:* It is all too easy to think of systems as consisting mainly of equipment and to underestimate completely the importance of the operating personnel (e.g. Three Mile Island). Maintaining their skills is vital to overall system performance, and the system design must include facilities to carry out the corresponding training in a reliable and cost-effective manner.

   It must be recognized both that people tend to forget and that there will be a turnover in personnel, and training facilities must take both factors into account. An example of such facilities can be found in large power-system control stations, where the operator's console may be run in a training mode. In this mode, fault conditions can be simulated (e.g. from a supervisor's console) and the operator can respond, but all without actually influencing the power system.

11.3.5 Implementation phase

During this phase the maintainability program plan is further implemented. However, the system engineer will generally not be significantly involved in this phase. It is up to the equipment manufacturer to meet the maintainability requirements laid down in the equipment specifications.

11.3.6 Verification phase

The demonstration test plan is carried out, demonstrating that the maintainability objectives have indeed been met. The system engineer may or may not be involved in this phase, depending on the extent to which the client's operating personnel are able to verify compliance.

11.4 Maintenance planning and design

11.4.1 Maintenance concept

The **maintenance concept** is *a narrative statement or illustration that provides a high-level definition of the means of maintaining an equipment item or system*. The statement relates the tasks that should be performed at each level of maintenance, the test equipment and tools that should be used in the maintenance of each item, and the required skill levels of the maintenance personnel who perform the identified tasks.

The maintenance concept forms part of the maintainability program plan insofar as several of the items in that plan are heavily influenced by the maintenance concept. However, as the word "concept" implies, it forms the foundation on which the plan builds, and it is developed in the very early stages of the system engineering process. The concept may be roughly developed during the definition phase and then refined in the early stages of the analysis phase. Indeed, the project definition may make explicit reference to the maintenance concept in delineating areas of responsibility. Normally, the finished concept will form part of the system specification.

The maintenance concept defines criteria covering repair philosophies, maintenance support levels, personnel factors, maintenance time constraints and so on. It is truly a system-oriented function, because maintenance necessarily involves all the primary elements of the system to be maintained, and because the maintenance functions themselves become elements in the overall system. Some of the major points to be considered in a maintenance concept are maintenance levels, repair policy and system operational requirements.

11.4.1.1 Maintenance levels

Maintenance will usually not be carried out at a single level. While the definition and number of levels will vary from system to system, the following comments are generally valid.

The lowest level is the **organizational level**, so called because maintenance on this level is carried out by personnel belonging to the user organization, particularly the operating personnel. The maintenance functions on this level are limited to the simplest and most often occurring functions (e.g. replacing lightbulbs, switching over to standby equipment) and should not require any test or diagnostic equipment other than whatever is provided as an intrinsic part of the equipment. Another function often required at this level is analysis and information gathering in accordance with prescribed routines (e.g. logic diagrams, decision trees) and the passing on of this information to the next level.

The lowest level is sometimes also called **onsite maintenance**. This term can have a slightly different meaning, in that it may imply that the item is too bulky to transport and can be maintained only on site. However, in the electronics field the two terms should not cause any confusion.

The next level is the **intermediate level** or **central maintenance facility**. Such facilities are equipped and staffed to repair modules, conduct performance tests on subsystems and perform module level modifications on facilities and systems within their area of responsibility. They will be equipped with specialized test equipment and tools, spare parts, and specialized diagnostic hardware/software required to repair the equipment removed as a result of lower level maintenance activities. Such facilities may be provided by the user organization or by the equipment vendor or manufacturer.

The final level is so-called **depot maintenance**, which provides the maintenance program with the capability for completely overhauling and rebuilding equipment as well as for performaning highly complex maintenance actions that are beyond the resources of the field maintenance organization. The depot will often be synonymous with or closely related to the factory where the equipment is produced.

11.4.1.2 Repair policy

A number of possible repair policies are applicable to a system design. A system may be designed to be nonmaintainable, partially repairable or fully repairable.

A **nonmaintainable** item, usually modular in construction, is not repaired when it malfunctions but is replaced by a like spare. If this policy is selected for the system, design criteria promote a positive self-test capability and the encapsulation (or hermetic sealing) of all item components. In this case there is no requirement for internal accessibility, plug-in subassemblies, test points or additional maintainability design criteria; nor is there a requirement for ancillary test equipment, specific personnel skills or elaborate maintenance support facilities.

A **partially repairable** system implies that some or all of the system elements (i.e. equipment items) can be repaired by replacing subassemblies. As an example, in the case of the combustion control system the detector may be either nonmaintainable or repairable (replaceable sensor element), whereas the processing electronics will almost certainly be repairable.

Finally, a **fully repairable** system promotes the level of maintenance all the way down to component level. This policy places the greatest requirements on support in terms of personnel and training, tools and test equipment, spare/repair parts, documentation and facilities.

11.4.1.3 System operational requirements

The system operational requirements may be said to form the boundary conditions for the maintenance concept. They include:

- a definition of system **deployment requirements** (e.g. quantity of operational installations, number of systems per installation, location of operational installations and distances from support facilities, operational environments);

- an identification of **effectiveness requirements** or system **quantitative factors**, which may be expressed in terms of single factors or as a combination of availability, operational readiness, dependability, MTBF, MTTR, maintenance manhours per operating hour and so on;
- an identification of system **constraints** covering specific operational limits (e.g. availability or educational level of personnel).

11.4.2 Documentation and feedback

The general rules to be followed in producing documentation are assumed to be familiar. Here only a few rules specific to maintainability will be mentioned:

1. *Purpose:* It is necessary to distinguish between documentation intended for training purposes and documentation used during the maintenance operation. The latter type must either be in a form that is convenient and suitable for the technician to carry along and use during the maintenance operation (e.g. ringbinder, foldout) or be attached to the equipment. Documentation attached to equipment is particularly useful for the lowest level of maintenance.
2. *Correlation with physical equipment:* Maintenance instruction must refer specifically to the equipment or item in question. If different models of the same general type of equipment exist, each model must be treated separately. In particular, this requirement implies that some mechanism must be available that allows the documentation to be updated each time modifications are carried out.
3. *Correlation with maintenance level:* Each level of maintenance should have its own documentation. It is senseless to supply the operating personnel with a detailed manual suitable for depot maintenance and to expect them to try to pick out the few items that are applicable to the organizational level.
4. *Quantity:* The documentation must be supplied in a quantity that is appropriate to the size of the organization that is going to use it. This means not only the number of people but also their physical separation (i.e. the extent to which they can share the same set of documentation).

An extremely important part of any maintainability program is the feedback from the field, which starts as soon as the first pieces of equipment go into operation. This feedback, usually in the form of **maintenance reports** and **operating logs**, serves to:

- verify the design levels of availability;
- pinpoint equipment and/or components that have a high failure rate;
- verify the operational procedures;
- optimize spare parts holdings.

However, good feedback requires a considerable amount of planning, and the following are a few points to consider:

1. What types of reports are to be produced?
2. When are they to be produced—in regular time intervals (e.g. hourly or daily logs), or on the occurrence of some event (e.g. event logs, repair reports)?
3. What information are they to contain, and in what form (e.g. narrative, tables)?
4. Are preprinted forms to be used?

5. How are the forms to be routed, and how will compliance be checked?
6. What is to be done with the information? (Remember, nothing is more demoralizing than producing data that nobody looks at and that leads to no tangible result.) How is it to be processed, and who is to receive it?

It requires a lot of thought to design the right forms and set down the procedures that will produce significant and useful data. It requires even more work, however, to think through the whole process of these data flowing through an organization, being processed in different ways and resulting in a variety of different actions, from equipment redesign and modifications through incoming component inspection to revised maintenance procedures—and how to **force** the data to flow, and the appropriate actions to be taken in the face of human laziness!

11.4.3 Example: SCADA system

As an example of what is required in an industrial system today, Table 11.1 presents the maintainability requirements for a SCADA system, as formulated by the Queensland Electricity Commission.

Table 11.1 Maintainability requirements for a SCADA system

The system shall be designed, manufactured, installed and commissioned with maintainability characteristics that will ensure that the availability goal will be achieved and that the Principal has all the prescribed procedures and resources to maintain the system during the lifetime of the system. These characteristics shall include, but shall not be limited to, the following:

(a) Equipment self-test diagnostics and troubleshooting procedures shall be provided to localize any failure or malfunction to the lowest field replaceable module. Online periodic diagnostics shall be provided to maintain a confidence level in the functioning of various elements of the system. On-demand diagnostics shall be provided to gain a level of confidence in the functioning of various items and to isolate types of malfunctions. A comprehensive set of offline diagnostics shall be provided to diagnose faults and to check the operational state of the repaired items.
(b) The system shall have a comprehensive set of alarm indicators and messages to assist maintenance staff in identifying faulty units or components.
(c) Readily accessible test and/or break points shall be provided to facilitate fault isolation and signal injection. The placement of components on cards shall allow access to test probes and connectors.
(d) Suitable grips or handles shall be provided to facilitate the safe removal and installation of heavy or bulky units.
(e) Facilities shall be provided to preclude the interchange of units or components of a same or similar form that are not in fact interchangeable and to preclude the improper mounting of units or components.
(f) Facilities shall be provided to ensure the correct identification, orientation and alignment of cables and connectors.
(g) Sensitive adjustment points shall be located or guarded so that adjustments will not be disturbed inadvertently.
(h) Internal controls shall not be located close to dangerous voltages or other hazards. If such locations cannot be avoided, the controls shall be appropriately shielded and labeled. All groundable parts shall be positively identified.
(i) Preventive maintenance shall be possible without affecting the online system functions. The preventive maintenance program shall minimize "wearout" failures.

(j) The maintenance procedures shall be as simple as possible and shall be based on exchangeable items, standard instruments and standard tools. An inventory system shall be provided by the Contractor to ensure that spare parts are available.

(k) The equipment shall be organized in a modular fashion so that troubleshooting can be based on an orderly selection and execution of predefined diagnostic procedures. A planned approach to troubleshooting shall be provided to simplify the problem of training, to minimize the necessary qualifications of maintenance personnel to a competent technical level, to eliminate the need for special test equipment and to ensure that the mean time-to-repair parameters will be achieved.

(l) Facilities shall be provided physically independent of main RTU, to simulate inputs and outputs, test faulty I/O modules and train personnel.

(m) Facilities and procedures shall be provided so that minimum disturbance to online system functions is caused during the testing of parts of the control system.

11.5 Summary

1. **Maintainability** is a characteristic of a system, just as reliability and performance are. However, there is no exact (i.e. mathematical) definition of maintainability; it is only reflected in the **overall cost of maintaining** a system.
2. A system can generally be considered to consist of four subsystems: **functional, operating, repair** and **support**. Maintainability always involves all three and is inseparable from the activity of **maintenance**.
3. There are three types of maintenance: **preventive, corrective** and **adaptive**. The latter increases in importance as system lifetime increases relative to the inverse of the rate of change of the operating environment.
4. The **maintainability program** (or plan) is a description of all the tasks that are to be undertaken within the engineering process in order to reach the desired level of maintainability.
5. The **maintenance plan** is a description of how maintenance is to be carried out once the system has been created. Developing the maintenance plan is part of the maintainability program.
6. Because **repair** is an activity, designing the repair subsystem involves designing an organization.

11.6 Short questions

1. Define maintainability.
2. What is the difference between maintainability and maintenance?
3. Name two important activities that are part of the maintainability program.
4. What is adaptive maintenance?
5. What is the maintenance concept? In what phase of the engineering process is it developed?
6. What is the difference between maintenance policy and maintenance concept?
7. What is understood by "maintenance levels"?

11.7 Problems

1. *Airport maintenance:* In the early part of the analysis phase of a project to design a new airport, you have been given the task of developing the maintenance concept. The first thing you must do is to write down (for your own use) a small work breakdown structure (WBS) for this task. It may look something like:
   1. Determine the main elements of airport maintenance in general.
   2. Determine the characteristics of these elements for this particular airport.
   3. Determine the main factors that influence the cost-effectiveness of these elements and the possible choices available for their realization.
   4. Design a lifecycle cost model of the maintenance activities; make preliminary choices.
   5. Define the model and integrate it (to the extent necessary) into the airport lifecycle cost model.
   6. Define the maintenance concept based on the optimum cost-efficiency of the maintenance activities.

   Addressing points 1 and 3 only:
   (a) Show how you determine the main elements in a top-down fashion. Elements on the same level should have roughly the same importance.
   (b) Be sure that the title of each element is specific and descriptive.
   (c) The factors to be determined under point 3 can also be developed in a general top-down fashion, but only to a limited degree before correlation with the elements has to take place. Show this process clearly.
   (d) Present the final result, namely the factors and choices, in a clear and structured form.

2. *Maintained production system:* A system consists of two subsystems: a production subsystem that can produce a yearly output Q, and a maintenance subsystem. The initial cost of the production subsystem is V, measured in units of Q; the yearly cost of the maintenance subsystem is W, also measured in units of Q.

   The output value Q assumes that the production subsystem is operating all year without interruption. In reality the production will be interrupted by maintenance, and the mean downtime each time maintenance is required is MDT $= 1/\mu$, where μ is the maintenance rate. From a preliminary study of different maintenance options it appears that μ depends on W in the following way:

   $$\mu = 0.05\sqrt{W} \quad [\text{h}^{-1}]$$

   The mean time between maintenance (MTBM) is of course a property of the production subsystem alone, and experience with similar systems indicates that the rate $\lambda = 1/\text{MTBM}$ can be expressed as

   $$\lambda = 0.001(1 + 3/\sqrt{V}) \quad [\text{h}^{-1}]$$

   The system is being designed for an operating life of 20 years; and as a best guess it is assumed that the discount rate p can be taken as constant over the system's lifetime, with a value of p $= 15$ percent.

(a) Develop an economic model of this system. The output of the model should be a measure of the system's profitability or viability.

(b) Determine the values of V and W that optimize the output of the model. For a numerical method, use V = {0.1,0.5,2.0} and W = {0.02,0.1,0.5}.

(c) If, in addition to the costs V and W defined above, there were a yearly operating cost (e.g. personnel cost), how large could this cost be without making the system unprofitable?

(d) Give an example of a system for which this model would be applicable.

3. *Power generation repair policy:* The continuous power requirements for an isolated mining venture are met with a power station comprising two identical generators. The power station is operated and maintained by an independent authority under an agreement with its sole customer: the mining company. The agreement establishes pricing policies, together with incentives for maintaining power services without interruption.

Each generator has a capacity C of 5 MW. The total load supplied by the two-generator set varies on a daily basis, lying with **linearly decreasing probability** between limits of C and 1.8C. This diurnal pattern (including the peak load) remains unchanged year round.

The failure rate λ for each generator under operating conditions is known and is constant. No improvement in the associated mean time to failure (MTTF) of 2000 h can be achieved by the operating authority. However, repair policies (and their related costs) are its responsibility, and have to be set in accordance with the following provisions of the agreement:

- Interruptions to the **variable** load (which can be shed by the authority at any time) affect the tariff paid by the customer for all power supplied.
- Interruptions to the **base** load (which in turn stop all mining activities) involve penalty payments by the operating authority to the mining company.

Extensive studies by the operating authority indicate that a policy of immediate repair on failure, subject to the availability of the entire maintenance team, involves costs that vary with the mean time to repair (MTTR) according to the relationship

$$\text{expected annual repair costs} = \$4 \times 10^5 \left(3 + \frac{100}{\text{MTTR}}\right)$$

where MMTR = mean time to repair one generator, in hours.

(The same studies show that an alternative policy of parallel repair of failed generator units would increase repair costs by approximately 80 percent, due to the need for a second maintenance team.)

Under the agreement between the operating authority and the mining company, the tariff rate for power actually supplied is set by the formula

$$\text{adjusted rate (r)} = r_{max} (1 - 2t_1/T)$$

where r_{max} = rate payable in the absence of any interruption over a 12 month period

= 20 cents/kWh

t_1 = total time during the year when the demand cannot be fully met

T = total time in a year, in same units as t_1.

Annual penalties Q payable by the operator for interruptions to the base load are determined under the formula

$$Q = \$150,000t_2$$

where t_2 = time when the base load is not met, in hours.

(In the following questions it can be assumed that mathematical models based on constant repair rates are acceptable.)

(a) Define the system states that can arise within the power station, with respect to the operational status of the two generators.
(b) Determine the steady state probability of the system being in each of these states, expressed solely in terms of the **failure** and **repair** rates λ and μ.
(c) Define the **criterion** by which the operating authority should set its repair policies.
(d) List all factors relevant to that criterion, based on the **information given above**, and develop the equations for each of these factors.
(e) Determine the actual MTTR that the authority should seek to achieve. (Hint: Make suitable simplifying assumptions in establishing expected or most likely costs.)

11.8 References

Blanchard, B.S., Lowery, E.E., *Maintainability–Principles and Practices*, New York: McGraw-Hill, 1969.

Cunningham, C.E. and Cox, W., *Applied Maintainability Engineering*, New York: Wiley, 1972.

MIL-HDBK-472, *Maintainability Prediction*, Washington DC: US Department of Defense, 1972.

MIL-STD-470, *Maintainability Program Requirements*, Washington DC: US Department of Defense, 1966.

MIL-STD-721C, "Definitions of effectiveness terms for reliability, maintainability, human factors, and safety". Washington DC: US Department of Defense, 1981.

Muth, E.J., "An optimal decision rule for repair vs replacement", *IEEE Trans. Reliability*, vol. R-26, August 1977, pp. 179-81.

Sasaki, M. et al., "System availability and optimum spare units", *IEEE Trans. Reliability*, vol. R-26, August 1977, pp. 182-7.

Chapter 12

Integrated logistic support

12.1 Introduction

12.1.1 Definition

Integrated logistic support (ILS) is a composite of all the support considerations necessary to assure the effective and economic support of a system for its lifecycle. It is an integral part of all other aspects of system acquisition. ILS is characterized by harmony and coherence among all the logistic elements.

There are two associated but separate aspects of ILS:

- the work carried out during the system engineering process, which addresses the ILS factors affecting engineering requirements and how these factors will be applied in meeting the requirements—called **logistic engineering**;
- the work carried out after implementation in order actually to support the system— called **logistic support**.

The logistic engineering activities are defined and scheduled in an **Integrated Logistic Support Plan (ILSP)**, which is normally produced in the analysis phase; the logistic support activities are defined in documents that are created under the ILSP.

12.1.2 Logistic engineering

The logistic engineering tasks are performed in order to achieve a supportable and cost-effective system and result in the establishment of the optimal logistic requirements for the deployment and operational phases of the system lifecycle. They include the following:

1. *Maintenance concept development* sets out basic boundary conditions, such as the number of maintenance levels, organizational responsibility for each level and relevant national objectives.

2. *Logistic support analysis (LSA)* consists of:
   - *maintenance engineering analysis (MEA)*, which facilitates: the systematic and complete development of maintenance requirements; the sorting and combining of logistics data; the determination of the quantity of maintenance equipment, personnel and spares; inputs to system effectiveness and lifecycle cost analyses in terms of required factors; and the identification of system calibration and measuring standard requirements;
   - *repair level analysis (RLA)*, which determines the most economic means and the allocation of maintenance between the various levels of maintenance defined in the maintenance concept;
   - *repairable item assessment*, which sorts all items of hardware into two groups at each level: repairable and nonrepairable—the latter then becoming repair parts at that level;
   - *breakdown spares assessment*, which determines the levels of spares or repair parts to be stocked;
   - *technical support analysis*, which identifies the types and quantities of support and test equipment.

3. *Preventive maintenance design* determines the most cost-effective manner and level of planned (i.e. scheduled) maintenance, and documents the results in a **Technical Maintenance Plan (TMP)**.

4. *Repair subsystem design* is the design of all facilities (fixed and/or mobile) needed to carry out corrective maintenance in accordance with the maintenance concept and with due consideration to the results of the technical support analysis.

5. *Technical database development* produces a database that is capable of cross-referencing maintenance documentation, manuals, drawings, associated support equipment, functional performance and parts identification. It produces all handbooks, manuals and computer software documentation needed for the operation and maintenance of the system.

6. *Supply support development* identifies all supply support items and provides for their supply to sites, the accounting for them and their maintenance. Further information that may be provided includes:
   - the **spares** to be procured, by original manufacturer's part number, quantity, supply source and price;
   - details of the **inventory management system** to be utilized;
   - details of the **facilities and materials-handling equipment** required for the storage of the spares holdings, including antistatic storage, handling and packaging devices for PCBs and so on.

7. *Training support* includes the identification and specification of training requirements, the design of training courses, the development of all training facilities, and the initial training of instructors and operators.

12.1.3 Logistic support

The main logistic support activities, which take place after the system has been put into service, are:

1. *Supply support*: this includes:
   purchasing spare parts and consumables
   storage and warehousing
   inventory control
   handling and shipping to site.
   These supply support activities are defined in a **Supply Support Plan**, which in addition includes the following details:
   - the **spares** to be procured, by original manufacturer's part number, quantity, supply source and price;
   - details of the **inventory management system** to be utilized;
   - details of the **facilities and materials-handling equipment** required for the storage of the spares holdings, including antistatic storage, handling and packaging devices for PCBs, and so on.

2. *Training:* At periodic intervals classes will have to be taken through appropriate training programs, to:
   train replacement personnel
   retain/refresh existing personnel.
   Also, course content may have to be revised from time to time in order to keep up with the evolution of the system.

3. *Documentation maintenance:* All the documentation needed to operate and support the system has to be revised to accord with approved changes to the system; amendment lists have to be prepared and distributed; and new versions have to be printed. Managing the maintenance of an adequate documentation library for a large and complex system is a full-time task for one person. The documentation will typically include:
   operating manuals
   maintenance manuals
   servicing schedules
   system description manuals
   interface manuals
   programming manuals
   program source listings
   program documentation
   installation manuals
   support equipment manuals.

4. *Fault reporting and analysis:* Fault-reporting procedures are overlaid on maintenance activities. They ensure that all information available about faults at the time of their occurrence and throughout the restoration/repair process is directed back to a central location where it is analyzed.

12.1.4 Relation to RMA

From the foregoing definition of ILS activities it is clear that there is a close relationship between ILS and the specialist engineering areas of **reliability, maintainability and availability (RMA)**, as they were treated in Chapters 10 and 11. It may be said that ILS activities are determined, both quantitatively and qualitatively, by the results of RMA activities. The latter are part of the prime mission equipment design and determine those characteristics of the equipment that lead to a need for logistic support. The level of reliability will determine the need for repair and spare parts; the level of maintainability will determine the need for special support items (e.g. special tools, instruments, training). The RMA activities end when the system has been created; the logistic support activities then have to cope with the problems of keeping the system running. In order to facilitate this, the logistic engineering will design and provide certain support facilities and subsystems.

As a result of this relationship it is necessary to have close coordination between RMA, logistic engineering and regular systems engineering activities. This coordination is an important part of the **engineering specialty integration** described in Section 2.4.1.

12.2 Integrated Logistic Support Plan

12.2.1 Overview and Table of Contents

The **Integrated Logistics Support Plan (ILSP)** forms the basis for all ILS activities. However, in addition to defining and scheduling ILS activities, it also contains what may be termed "boundary conditions" on these activities in the form of a number of concepts and strategies. While the ILS activities are always chosen from the set of activities described as logistic engineering in Section 12.1.2, the concepts and strategies have to be formulated separately for each project. They usually have to be formulated in conjunction with the user. This work starts as part of the project definition and is then refined and finalized in the early part of the analysis phase.

The contents of an ILSP will typically be as follows:

Part A   INTRODUCTION
      1. Scope
      2. Applicable documents
      3. Definitions and abbreviations
      4. System description
      5. Project management

Part B   CONCEPTS AND STRATEGY
      6. Operations concept
      7. Test and evaluation concept
      8. Maintenance concept
      9. Acquisition strategy

Part C   ILS ELEMENTS
   10. Logistic support analysis
   11. Preventive maintenance design
   12. Repair subsystem design
   13. Supply support development
   14. Training support

Part D   ILS MANAGEMENT
   15. Organization
   16. Configuration management
   17. ILS schedule and milestones

The introduction serves mainly to make the ILSP a self-contained document. However, the system description should concentrate on those aspects of the system that are of particular importance to ILS (e.g. the effects of system failure or outage) and should include relevant background information (e.g. availability of trained manpower, geographic characteristics that influence the supply situation, availability of local repair facilities). It is not necessary to go into any detail on system performance.

The other three parts of the ILSP will be discussed briefly in the following three sections.

12.2.2 Concepts and strategy

The operations concept is concerned with two broad areas of system operations. On the one hand, it defines how the system is to be used or deployed; that is, it sets out under what circumstances the system will have to be in its various operating modes, how it must respond to external circumstances (e.g. emergencies), and how it is integrated into any overall scheme of operations. Also, this is the place to document any assumptions about the operational environment (e.g. operating frequency, total operating hours per year).

On the other hand, the operations concept addresses how the system is to be staffed: permanently or on demand, with employees or contract personnel, using existing personnel categories or training system-specific specialists, number of shifts and working hours, union rules and so on. The staffing concept is often heavily influenced by factors outside the direct influence of the systems engineer, and these factors need to be identified at an early stage.

The acquisition strategy also is often determined either by existing policy or by nontechnical factors. As it will normally affect both the engineering management in general (e.g. through the work breakdown structure, WBS) and the ILS engineering in particular, it must be available early in the project. Issues treated in the acquisition strategy include the use of a prime contractor versus multiple contractors and a systems integrator, single or phased contract(s) for each contractor, and to what stage competition will be maintained (i.e. two or more contractors working in parallel).

12.2.3 ILS elements

The ILS elements are essentially those listed in Section 12.1.2. However, this section of the ILSP must not simply be a general description of these activities, such as a textbook gives; that would be absolutely worthless. It must be a detailed definition of what each element is going to consist of **in this particular case**, the extent and level of detail to which each task will be carried out, what documentation will be produced, and the interrelations and dependencies that exist with one element and vis-à-vis other elements in this particular project.

12.2.4 ILS management

The ILS elements form an easily identifiable subset of the project activities, and as such they will normally be carried out by a separate team under the leadership of an ILS manager. The organization of the ILS team addresses such issues as the formation of specialist groups, the duration for which these groups need to exist, the use of outside specialists, and the locations and facilities required.

There is one aspect of ILS that distinguishes it from the other areas of systems engineering and that has a significant influence on the ILS management, namely the fact that some ILS activities are carried out in a more-or-less continuous fashion from logistic engineering to logistic support, as these were defined in Section 12.1.1. For example, spare parts supply is initiated in the implementation phase and carried on from there throughout the lifetime of the system, and the transition from the engineering phase (contractor responsibility) to the operating phase (user responsibility) places considerable demands on the management function.

12.3 Summary

1. **Integrated logistic support (ILS)** is concerned with the effective and economic support of a system for its lifecycle. As such it is very closely related to **reliability, maintainability and availability (RMA)**.
2. ILS activities fall into two groups:
   engineering tasks (logistic engineering)
   support tasks (logistic support).
   The first take place while the system is being created, the second after the finished system has been put into operation.
3. **Logistic engineering** includes developing/carrying out:
   maintenance concept
   logistic support analysis
   preventive mintenance design
   repair subsystem design
   technical database
   supply support
   training support.
4. **Logistic support** includes:
   supply support
   ongoing training
   documentation maintenance
   fault reporting and analysis.

5. The main document within ILS is the **Integrated Logistic Support Plan (ILSP)**. It defines and schedules the ILS activities and sets down the concepts and strategies to be followed/applied in carrying out the activities.
6. The ILSP is finalized in the early part of the analysis phase and covers both engineering and support.
7. The main parts of the ILSP are:
    introduction
    concepts and strategy
    ILS elements
    ILS management.
Note the inclusion of management; ILS is closely related to specialty engineering integration.

12.4 Short questions

1. What is the relationship between RMA and ILS?
2. Define and describe the two major groups of ILS activities.
3. How does ILS relate to the systems engineering process?
4. What is the purpose of the ILSP?
5. What is included in an ILSP?
6. Explain why ILS has such a major management component, and how this ties in with systems engineering management.
7. What is logistic support analysis (LSA)? How does it relate to systems analysis?

12.5 Problems

Support of a ship: A ship constitutes a special type of system in that, when at sea it is an almost closed system, but when in port it can interact strongly with other systems. This must have a strong influence on the way maintenance is carried out and how it is organized; that is, the maintenance concept for a ship must have certain features common to all ships.
(a) Set up the framework (i.e. headings) for such a maintenance concept, say, main section headings and the first level of subheadings.
(b) Choose one subheading that will be strongly dependent on the fact that the system is a ship, and develop it in more detail. Discuss possible options and how choices may be made between them (or how an optimum mix may be found).

12.6 References

Blanchard, B.S., *Logistics Engineering and Management,* 2nd edn, Prentice Hall, 1981.

Integrated Logistic Support, DEMPS Project Report No. 84-2, Ottawa: Canadian Department of National Defence, 1984.

System Engineering: Management Guide, Defense Systems Management College, Fort Belvoir, VA: US Department of Defense, 1990.

Chapter 13

Human engineering

13.1 Introduction

In the narrowest sense, **human engineering (HE)** is concerned with the interface between man and machines—the **man-machine interface (MMI)**. In particular it is the physical part of this interface, called **ergonomics**, that has been furthest developed and most widely applied. It forms an important part of equipment design, and design criteria as well as a large amount of detailed information, resulting from years of practical experience, are available (e.g. in MIL-STD-1472C).

However, although this aspect of HE is obviously important for systems, because systems are built up of equipment, it is a wider definition of HE that is of most interest in systems engineering. Instead of viewing the human as a **user** of the system (or equipment), the human is an **integral part** of the system, and the system performance depends on the combined performance of human and equipment. HE can then be defined as *the total set of activities concerned with optimizing the combined performance of humans and equipment*—a definition that broadens the scope of HE way beyond ergonomics, into behavioral science and the way the brain processes information.

To discuss human performance in quantitative terms it is necessary to be able to measure this performance, and by far the two most important parameters are **accuracy** and **speed**. Speed is perhaps the most obvious, and this has for a long time been reflected in the use of time studies. However, even in quite simple operations there has to be some inspection or quality control of the product. Units that do not pass inspection are at least not counted and at worst result in an additional penalty reflecting the losses incurred through loss of material or the cost of reworking. Thus, in most cases both speed and accuracy are involved in measuring performance.

In electrical and electronic systems speed can often be a secondary consideration; there is nothing to be gained by forcing operators to work or react above a very modest speed. However, accuracy is almost always important, and in some cases it is absolutely crucial (e.g. power system operation, aviation control). The central theme in the HE of electrical systems is therefore: What factors influence the human performance within a system, and how must the system design be carried out in order to optimize this performance—in particular, to minimize the error rate?

Of course, the final goal of systems engineering is to optimize the overall system performance (or cost-effectiveness), and there probably will be trade-offs between human performance and equipment performance. For example, providing an operator with processed rather than raw data may improve the operator's performance, but the failure rate of the complex equipment needed to do the processing may result in a poorer overall performance. It is the essential characteristic of systems engineering that it takes into account **all** the factors that influence system performance, which means considering such diverse disciplines as theoretical physics and psychology. Systems engineering provides a method for handling and processing this vast and diversified amount of data in a systematic fashion. One aspect of this method is the breakdown into groups whose elements are closely connected, and HE must be understood as one such group. It is in this sense that it is possible to speak of **human performance** as separate from system performance.

It also follows from the above that HE is concerned with **all** the man-machine interfaces within a system and includes operation, maintenance, and support activities. It is a part of the design of all the three corresponding subsystems, and is managed as part of the engineering specialty integration (see Section 2.4.1). HE can be regarded as an **embedded** project which is executed according to the normal systems engineering methodology; that is, in a phased manner and with concurrent consideration of its application to all parts of the system.

Human performance is determined by three main factors:

- the **human**;
- the **activity** to be performed;
- the **conditions** under which the activity is to be performed.

These three factors will be considered separately in the following sections; but again, it must be clear that they are interrelated and that the boundary is not always sharp.

Also the extent to which the systems engineer can influence these factors may vary widely depending on the type of system. In designing a radar system the designer has full control over all three. The operator can be selected and trained to any degree necessary; the activity is narrowly defined in operating manuals; and the conditions can be controlled to a very large degree—in military applications even down to the operator's food and exercise. On the opposite side of the scale is the automobile designer, who has very little control over any of the factors. The driver can vary from a reckless inexperienced youth through a mature well-balanced driver in prime condition to a sleepy drunk driver. The activity is predefined; only minor variations (e.g. manual or automatic gearshift) are possible. The conditions vary from hot dusty desert roads with a washboard surface to icy snow-covered surfaces at $-50°C$. Nevertheless, in the design of the specific man-machine interface there still remains considerable scope for good design (e.g. easy-to-read instruments, comfortable seating, good view, easy steering).

13.2 The human as a system component

13.2.1 A simple model

Humans may, as far as their performance in a system is concerned, be looked upon as consisting of three elements: **sensors**, a **processing** capability and **actuators**, as shown in Figure 13.1.

Overall performance can be described in terms of the **response** (i.e. the action of the actuators as a function of a given stimulation of the sensors). However, the number of possible different stimuli is so astronomical that it is out of the question to record the corresponding responses. The best that can be done at present is to note some general characteristics of the three elements in Figure 13.1.

However, the characterization of the elements must recognise the fact that there is a considerable spread in the numerical values of any one characteristic quantity obtained from different persons. This can, on the one hand, be taken into account by using **distributions** rather than single numerical values. Assuming normal distributions, reference is then made to a 95 percent confidence interval, corresponding to the mean value $\pm 2s$, where s is the standard deviation. On the other hand, the statistical variability can be reduced by **parametrizing** the characteristics; values will be considerably smaller than for the global population. Typical parameters are age, sex and educational level.

One aspect of human performance that must not be forgotten is the human capacity to **learn**; that is, the processing element is highly **adaptive**, and its performance depends on previous experience. This aspect has some very important implications for systems engineering:

1. Obviously, functionally oriented training will improve the performance of corresponding tasks. This is the process of acquiring **skills**.
2. Overall system performance will improve during an initial period after a system is put into operation as a result of the improvement in human performance as its human operators adapt to the system environment. A knowledge of this so-called **learning curve** allows the performance of a system to be judged after only a short time in service.
3. The learning process, in the sense that it is habit-forming, can have negative consequences; in particular, it can make the system unable to cope with the **unexpected**. This is a very serious problem in such positions as power system operators or high-readiness military-system operators and must be countered by continuous retraining and emergency state simulation. However, it also manifests itself in many other, less dramatic ways, such as overlooking a change in a familiar picture. The mind calls up the familiar picture or situation rather than analyzing the actual sensory input.

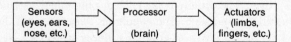

Figure 13.1   The human as a system component.

13.2.2 The sensors

13.2.2.1 Visual sensory characteristics

Vision is the most important means of sensory input. The **quality** of the visual input (i.e. the signal-to-noise ratio) is determined by three factors:

- the **strength** of the light entering the eye;
- the **contrast** between the background and the information-carrying elements and/or between different elements;
- the type of **coding** used (e.g. text or pictures, black-and-white or color).

The **quantity** of the visual input (i.e. the information content of a single picture as well as the rate of change of this content) also is determined by three factors:

- the **coding** used;
- the **number of elements** present;
- how the temporal **change in data** is presented.

An example of the latter aspect is the following. A page of text is displayed on a VDU, relaying information to an operator on which action is required. If a new page is displayed, it may be that only a very small portion of the information, say only a single value, has changed; but if this fact is not conveyed to the operator (e.g. by flashing or a different color), the operator has to read the whole page to ascertain this fact, and the information input to the operator is unnecessarily large. The result may be an overloading of the visual sensory input channel and consequent erroneous response.

The **luminance** of a surface is the amount of light emitted per unit surface, measured in stilb, with 1 sb = 1/60 part of the luminance of a black body at the temperature of solidifying platinum (2046.7 K). Conversely, the unit of **intensity** of a source is measured in candela, where 1 cd = 1/60 that of 1 cm^2 of a black body at the above temperature. Hence

$$1 \text{ sb} = 1 \text{ cd/cm}^2$$

For reflecting surfaces that luminance is determined by the **illumination level**, measured in lux, with lx equal to the illumination level at 1 m distance from a point source with an intensity of 1 cd, and by the **reflectivity** of the surface. If the luminance is denoted by L, the illumination by E and the reflectivity by r, with $0 \leq r \leq 1$, then

$$L = rE/(4\pi \times 10^4)$$

To simplify the numerical factor, the apostilb is defined as

$$1 \text{ asb} = 1 \text{ sb}/(\pi \times 10^4)$$

The **contrast** between an element (e.g. letter, figure) and its background is given by L_e/L_b, where L_b is the luminance of the background and L_e that of the element. Thus, contrast is a function of the viewed object, not of the illumination level.

The **visual angle** under which an object or element is seen is, for small elements, the linear size of the element divided by the distance between the eye and the element, as shown in Figure 13.2.

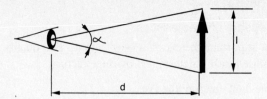

Figure 13.2   Visual angle α which for small angles equals l/d.

The smallest detail that the eye is capable of resolving, expressed as a visual angle, is referred to as **minimum visual acuity**. It is a function of contrast, background luminance and age. Age shows little influence up to about 40 years, after which minimum visual acuity starts to increase significantly. A typical value for minimum visual acuity is 1 minute of arc. However, it is recommended never to go below 2 minutes for any detail and to let any figure whose shape has to be recognized subtend at least 12 minutes of arc.

For **characters**, the minimum height should be

$$H = \frac{\text{viewing distance}}{200}$$

which results in the values given in Table 13.1.

Table 13.1   Character height H as a function of viewing distance (both in mm)

Viewing distance	Character height
< 500	2.5
500– 900	4.5
900–1800	9
1800–3600	18
3600–6000	30

For best results the width of characters should be $\frac{2}{3}H$ and their thickness about $\frac{1}{6}H$. The distance between **lines** should be at least $\frac{1}{2}H$.

It is not only a question of being able to see an information element; it is also a question of the ease with which a person can go on viewing small elements over an extended period (i.e. the question of eye strain). This has led to the formulation of recommended illumination levels for certain **types of work**, as shown in Table 13.2.

Table 13.2   Recommended illumination levels for various types of work

Type of work	Illumination level (lx)
General office work or workshop	400– 800
Drafting	750–1500
Small parts assembly	1000–2000
Very fine work	2000–4000

However, it should be noted that, for illumination levels above 1000 lx, problems of glare and unwanted reflections become much more noticeable; so special care has to be taken in these circumstances.

A particular case that merits special mention is where information is presented on **CRT screens** (i.e. VDUs), for several reasons:

1. This type of information presentation has increased explosively in the last 10 years, particularly with the advent of computers with dialog man-machine interfaces.
2. The length of time spent concentrating on the display is longer than for most other types of displays. Thus the question of fatigue and long-term effects has attained great prominence.
3. The problem of unwanted light reflection is particularly important.

To explain the latter more fully: CRT screens are a case where both the background luminance L_b and the charater luminance L_e are composed of both a light component produced by the VDU and a reflected luminance L_r. This latter component can be expressed by

$$L_r = qE + rL_o$$

where q is the scattering coefficient, describing the normal scattering of the ambient illumination E from a rough surface, and r is the reflection coefficient describing the direct reflection of a light source of luminance L_o.

If specular reflections are disregarded (i.e. $L_o = 0$), character contrast drops slowly as the illumination increases. With artificial lighting the loss of contrast is not critical if character luminance can be adjusted accordingly. However, a considerable reduction in contrast that cannot be compensated for may occur if primary light sources are reflected in the screen surface. In a typical case this problem becomes noticeable as the specular component of L_r surpasses 5%; and as it increases up to 50% the contrast ratio falls from its optimum value of 5 to below 3, resulting in a 20% decrease in visual performance.

Only in exceptional cases (e.g. radar screens) can the problem of specular reflection be solved by reducing the general illumination level. In most other cases the operator has to view alternately both the VDU screen and some document, and the visual performance in reading the latter improves with increasing illumination. A compromise has to be found, and tests show the optimum value of E to be around 400 lx. However, what can be done is to reduce the reflection coefficient r (and automatically also, to some degree, q), by means of a special antireflective coating on the CRT screen or by putting a filter in front of it.

Although contrast and illumination level are primary factors determined by the eye's characteristics, more mechanical factors (e.g. field of vision, speed of eye movement and of focusing) also must be taken into account.

The **field of vision**, characterized by the apex angle of the corresponding cone when both eyes and head are held still, can be divided into three zones:

- the field of distinct vision, where objects are sharply focused, about $1°$;
- the middle field, where objects are not seen clearly but where movement and strong contrast are sensed, about $40°$;
- the outer field, bounded by forehead, nose and cheeks, where objects are not noticed unless they move (i.e. peripheral vision), up to about $70°$.

To move the field of distinct vision (e.g. when reading), the eye makes a series of jumps, with a time between jumps of about 100–300 ms and a time of 30 ms for the movement itself. Going from one line to the next takes about 120 ms.

Accommodation is the change of focus when shifting the field of distinct vision from a near to a far object or vice versa. **Adaptation** is the change in sensitivity that occurs as a function of background illumination. The times needed for these changes vary widely; accommodation and adaptation to high light intensity can take place in as little as 50 ms, whereas adaptation to low light levels (i.e. night vision) can take half an hour.

Finally, **color** may be used as a means of visual coding. Common choices of colors for indicator lights and annunciators are shown in Table 13.3.

Table 13.3 Use of color in indicator lights and annunciators

Colour	*Indication*
Red	Fault
Flashing red	Emergency conditions
Yellow	Caution, marginal conditions (e.g. redundant system on maintenance)
Green	Operating normally, "online"
White	Alternative functions with no "right" or "wrong" implications (e.g. system 1, system 2)
Blue	Advisory

13.2.2.2 Auditory sensory characteristics

The auditory input channel has of course always been very important as a receiver for speech and music. However, as a man–machine interface in electrical systems it has, up till now, been of much less importance than the visual channel (except in entertainment electronics), as it has been difficult for electronic equipment to generate speech. This situation is changing rapidly, and it may not be long before verbal communication becomes the dominant type of man–machine interface.

Audio signals are used to convey short simple messages that usually call for **immediate action**. Some examples are the alarm clock, telephone bell, doorbell, foghorn, starter's gun and sirens of various types. Some audible signals convey more complex messages (e.g. the pinging of a sonar), and Morse code may convey arbitrarily complex messages, although in the latter case the actual sensory signal is simply an on-or-off signal of constant pitch.

The ear can detect a very wide range of **sound levels**. Some common sounds and their relative sound-pressure levels are shown in Figure 13.3.

The sensitivity of the ear is **frequency**-dependent, and audio signals should lie in the high-sensitivity range of 300–3000 Hz. This frequency dependence is shown in Figure 13.4, which is taken from an ISO-Norm. Note that the reference sound pressure (i.e. O dB) of 2.10^{-4} μb corresponds to an energy density of only 10^{-12} W/m^2, and that normal conversation represents about 10^{-6} W/m^2.

Because the absorption of sound increases with increasing frequency, frequencies below 1 kHz should be used for long distances; and if the signal must bend round obstacles or pass through partitions, the frequency should be below 500 kHz.

As with visual signals, the question of **contrast** is very important. Most environments will have some background noise, and a signal must distinguish itself from this by having a different frequency or an additional coding (e.g. an amplitude and/or frequency modulation). Conversely, unwanted and distracting signals can be masked by a suitable "random" signal (e.g. a hum of distant voices, soft nondescript

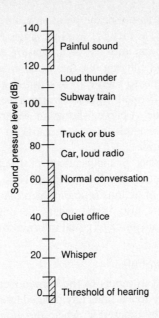

Figure 13.3 Relative sound-pressure levels of various common signals.

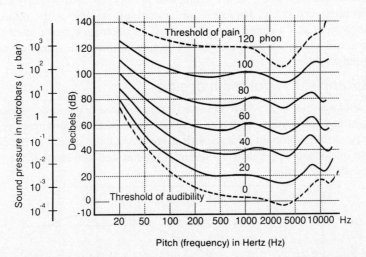

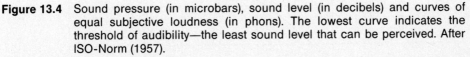

Figure 13.4 Sound pressure (in microbars), sound level (in decibels) and curves of equal subjective loudness (in phons). The lowest curve indicates the threshold of audibility—the least sound level that can be perceived. After ISO-Norm (1957).

music). Very low background-noise levels do not necessarily provide a good working environment, because then any incidental noise (e.g. door closing, pencil falling, coughing) provides a distinct distraction.

In describing and measuring noise, the phon values displayed in Figure 13.4 are of somewhat reduced usefulness, as they are valid for single frequencies. Instead, so-called **weighted sound levels** have come into use as a measure of loudness. Weighting sound level is essentially a process of filtering out the sound energy in the lowest and highest frequencies, at which, according to Figure 13.4, sensitivity is less. Hence sound pressure has less significance in these frequency ranges. Various weighting curves are in use, but the most common one is the A-curve, shown in Figure 13.5. Many psychological studies have shown that noise levels measured in dB(A) give a reliable assessment of subjective disturbance from noise.

The auditory input channel is independent of head orientation and personnel location, and also independent of illumination level. Thus auditory signals are useful as **alarms** to initiate a sequence of actions (one of which may be to obtain more detailed, visual information). They are also useful when the operator's attention has to be directed towards one particular source of visual information, such as a VDU screen or a process (e.g. rolling mill).

13.2.2.3 Other sensory channels

Other senses that can be used as inputs to the human elements of a system are touch, taste and smell. Of these, only touch finds any premeditated use. The position of toggle switches may be ascertained by touch (e.g. where the illumination is low, such as on a battlefield at night), and the same is also possible with properly designed rotary switches and other similar devices. Furthermore, touch serves as a feedback signal in many cases, the most common example being a keyboard. The movement of the key and the resistance at the end of that movement signal that the key has been properly operated.

Smell is used very rarely. However, one important example is the odorization of natural gas to serve as a warning in case of a leak. In electrical systems smell may serve as an unplanned alarm in the case of overheating or fire. The smell of ozone can sometimes indicate arcing.

13.2.3 Example: control room design

An example of a situation in which all these factors must be taken into account is given in the design of control rooms, such as may be found in the area control center of a power utility. A large amount of data with a greatly varying degree of detail is to be presented to one or more operators; so several VDUs as well as a large mimic panel (for system overview) are utilized.

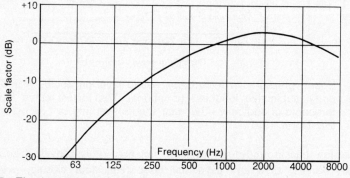

Figure 13.5 The A-weighting curve for sound level, in dB(A).

From Table 13.1 it follows that characters on a VDU at a distance of 1.8 m should be at least 9 mm, or that symbols on a mimic board at a distance of 5 m must be at least 30 mm. The size of invariable symbols and texts can be about two-thirds of this, since in time the operating staff will know those from memory. To enable the operator to monitor and control the system with ease, the most important monitoring aids must lie in the operator's primary field of vision, and the control devices must lie within easy reach, as shown in Figures 13.6 and 13.7.

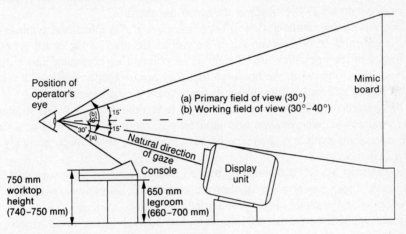

Figure 13.6 Side view of a control room layout.

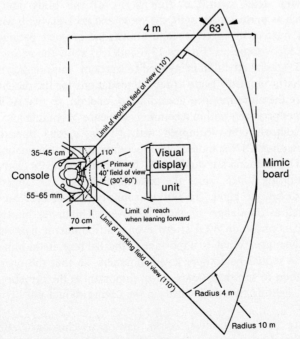

Figure 13.7 Top view of a control room layout.

The mimic board presents to the operator an overall picture of the layout and state of the system, and also gives indication of alarms. It is not intended that the board should provide the operator with detailed information, since this is available from display units.

It is therefore important that the operator should be able to perceive changes on the mimic board when gazing at the display units; so the board should be of such a size that it does not extend outside the operator's working field of view. As can be seen from Figures 13.6 and 13.7, the board should therefore subtend an angle not greater than 110° horizontally or 25° vertically at the eye of the operator.

There is a second consideration to borne in mind. The graphics and symbols on the mimic board must be of a certain size to be visible. However, if the board is not set at right angles to the operator's line of sight at all points, foreshortening may reduce the apparent size of objects on the board so that they are no longer visible with certainty. For this reason a curved, rather than a straight, board is recommended.

The operators are likely to spend most of their time studying VDU displays. To ensure accuracy of interpretation and limit fatigue, these are positioned at a distance where the normal symbol displayed subtends an angle of 16 minutes, normally 1.8 m from a 23 inch screen.

The height of a VDU is dictated by the fact that the operator's primary field of view in the vertical plan lies between the horizontal and 30° below horizontal. The VDUs are positioned in the center of this primary field, as shown in Figure 13.6.

Color can be used in the VDU presentation to highlight selected aspects of the diagrams or text. Up to eight shades can be selected for graphics, characters and background.

Anthropometry deals with the dimensions of the body and with human characteristics such as muscular strength, the speed and accuracy with which objects can be grasped, and movement space. Precise knowledge of such parameters results in design guidelines for workplaces. Figures 13.6 and 13.7 show the necessary dimensions for a control desk, special attention being paid to legroom, working height and reach.

In addition, anthropometry leads to recommendations for the design and layout of controls. Keys are therefore grouped according to function, so that as little movement as possible becomes necessary within a common sequence of commands.

In modern control rooms equipped with VDUs, special importance must be attached to the lighting. DIN standard 5035 recommends an illumination of 500 lx for control rooms. This illumination guide must be considered in the context of modern control rooms utilizing VDUs, since the advantages gained (i.e. slightly reduced eye fatigue, and a somewhat higher resolution capacity of the eye) are countered by disadvantages such as the danger of glare, fadeover of luminous information displays and reduction of contrast on VDUs. An even distribution of light is therefore not necessarily an advantage. What is important is the lighting at the mimic boards and workplaces. These should be equipped with dimmers, so that the operating staff can vary the illumination to suit their needs. Most important is the prevention of reflection on video screens; lighting fixtures located on the ceiling should not be reflected on the video screens.

In addition to being decorative, colors have a psychological effect on people. Generally speaking, dark colors have a depressing effect and bright colors a pleasant soothing effect. Colors also have an illusory effect and may be used to create an impression of warmth and distance. For control rooms the decor should be warm, soothing or neutral, to ensure that the operating staff feel at ease; brown, yellow and green shades are mainly used. These colors also match the wood veneer frequently used on display panels and the wooden desks. Pure luminous colors are not to be employed in the room design, these being reserved for color coding.

Various types of floor covering may be used. Where possible, carpeting is preferable to a plastic floor covering, since it improves acoustics and the room's ambience. A textured floor covering is preferable to a plain one, and antistatic properties are an advantage.

The efficiency of people's physical and mental capacities is strongly influenced by room temperature, among other things. Consequently, the specified degree of comfort and industrial hygiene requirements for air-conditioning (as found in the relevant literature) must be followed for control rooms. It is assumed that the work involved spans "mental activity" to "light manual work in a sitting position":

1. Optimum room temperature and humidity conditions are:

   temperature (°C): 18   20–21   24
   humidity (%):   70   50   30
   (In summer the temperature can be higher by 2–3 °C.

2. The difference between the room and the outside temperature preferably should not exceed 4 °C when airconditioning is used. In hot countries this is impossible, but the following guide limits should be considered with respect to control-room staff efficiency when defining airconditioning requirements.

Control room temperature	*Reaction*
25 °C	Discomfort
26 °C	Perspiration
28 °C	Diminished efficiency

3. The air recirculation rate should not exceed 0.2 m/s.
4. The air renewal rate is based on the manning of the room. Control rooms are usually sparsely manned; so a fresh air requirement of 10–30 m^3/h per person suffices. However, the air in the room must be renewed at least once every hour.

In addition to these important reference values, oxygen content, air purity and static electricity conditions are relevant parameters that must be considered.

13.2.4 Processing capability

Signal processing is carried out mainly in the brain, and to a much smaller extent in a distributed fashion in certain sections of the nervous system. The processing can be subdivided into three stages: perception, intellection and movement control. **Perception** is recognizing an object for what it is; **intellection** is reasoning, problem-solving and decision-making; and **movement control** is the correct translation of the result of the intellection (i.e. desire or will) into a physical output (e.g. movement of arm, hands and fingers to turn a knob).

Errors can occur in all three stages. One object can be mistaken for another (e.g. mistake a coiled snake for an old piece of rope); a wrong decision can be made (e.g. try to accelerate in front of a car crossing one's path); and a wrong movement can be made (e.g. depress the wrong key on a typewriter). A basic aim of human factors engineering is to recognize the sources of such errors and to eliminate or reduce their effect as far as possible. One source of error is **overloading**. Processing sensory data takes time, some typical values being:

	(ms)
Sensory receptor	1– 38
Neural transmission to brain	2–200
Cognitive processing delays	70–300
Neural transmission to muscle	10– 20
Muscle activation time	30– 70
Total	113–528

Thus, the reaction time is typically 300 ms, and this does not include any problem-solving or decision-making. If signals or information reach an operator at intervals approaching this time, a marked decrease in performance can be expected. This was demonstrated by an experiment in which typists had a buzzer and pushbutton installed next to their typewriter. At random times during their regular typing work the buzzer would sound, and they were then required to respond by pushing the button. Rather than measuring the response time, the response reliability (i.e. the percentage of signals that were acknowledged) was measured. The result is shown in Figure 13.8. As might be expected, performance decreased with increasing frequency, and in this case the frequency at which the decrease became noticeable was about three signals per minute.

The frequency at which a decrease in performance becomes noticeable does of course depend on many factors, such as the complexity of the task to be performed, the environment (i.e. the signal-to-noise ratio) and the training of the operator. Take as an example the case where the signal is in the form of a lamp lighting up, there are a number of lamps of different colors, and for each color a different button has to be pressed in acknowledgement. The reaction time is roughly 150 ms plus 100 ms for each color; that is, for four colors the reaction time is about 550 ms. Obviously, the reaction time alone is going to limit useful signal frequency.

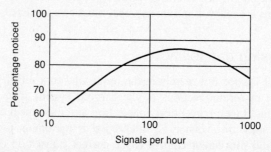

Figure 13.8 Relation between the frequency of signals requiring reaction and observed performance.

The reaction time and information-processing ability of the operator must be taken into account in the design of man-machine interfaces, particularly in determining the amount of data to be presented and the update time. (Note that only the variable part of a display is considered in this context.) The state of a network or a process control system normally requires many pieces of data (e.g. analog values, status information), and the reaction time of the operator may vary from 10 s for a simple alarm up to 2 minutes for a more complicated system disturbance. It follows that it would be senseless to go to great lengths and expense to reduce the update time below about 2 s.

Next to signal frequency, the **coding** of information is of primary importance when it comes to reducing error rate. For example, if several possible signals are presented by means of pilot lamps arranged in a row, there is a certain probability that a signal will be incorrectly recognized. If color is used as a coding factor in addition to position in the row, the error probability will be decreased.

As the set of signals (e.g. the number of bits) becomes more complex, more complex coding is needed. In other words, the effectiveness of a single coding factor **saturates**. Some typical values are given in Table 13.4.

Table 13.4 Saturation values of coding factors

Coding method	Maximum number	Effectiveness
Geometric shapes	5	Fair
Pictorial shapes	10	Good
Area	3	Fair
Brightness	2	Poor
Frequency	2	Poor

At the other end of the complexity scale there is the information characterizing the state of a large network, such as the network of a large power-supply authority. To perform their functions the operating personnel must be informed in detail of all process conditions and states. Most important for the personnel is information on the process, not simply data.

When selecting the information to be displayed, only information to which the operating personnel are able to react should be shown. Furthermore, digital numerical values should be displayed only where analog quantities do not suffice. The impression is often gained that, due to ignorance of the actual information requirements of the personnel, far too much information is displayed. During a state of emergency, for example, the personnel are practically inundated with data, with nobody taking responsibility for selecting from this flood the information that is important to define the actual situation existing in the power system.

The methods used today to display information are not necessarily user-oriented. It is in this sector in particular that the potential offered by the digital computer and by CRT displays is not fully utilized. Display of the power system in different colors according to its switching state is a first step in this direction. It enables the personnel to see at a glance which parts of the system are without voltage after a disturbance, and which parts of a subordinate distribution system are galvanically coupled and not linked via transformers to a higher voltage level. A presentation of the power system in different colors is much easier to take in at a glance than a simple topology display showing which circuit breakers are open.

A further possibility is automatic hierarchical picture selection. Here the operator is not only made aware by means of an alarm of limits that have been exceeded or disturbances that have occurred, but is also guided by an automatic sequence of pictures (using blow-ups or focusing) to the point where the trouble lies.

Analysis of system disturbances has shown that they are almost always accompanied by drastic changes in the load flow pattern in the power system. The control system's primary task is therefore to inform the operator of these changes. An exact quantitative display is not necessary; a qualitative analog representation of the load flow condition offers far more transparency to the operator.

The operating point of a generator may, in addition to being shown digitally in terms of the active and reactive power, be represented graphically on a P/Q diagram. Use of this method enables the operator's fear of running the generator with unconventional excitation to be eradicated.

Of course, coding does not only enter into the perception or input side of the processing performance; it can also have a marked effect on the correctness of the decision or output. One example is the grouping of controls that belong together (i.e. are in the same functional class). Another aspect is the automatic association that the brain makes between a command and its expected effect. If a hoist or lift is controlled by two pushbuttons, the buttons should be arranged vertically, with the "up" button on top; any other arrangement would increase the error rate. Or consider the indicator lights on a car; if they are operated with the right hand, one would expect pushing the lever up to cause the indicator on the left side to blink.

13.2.5 Mechanical characteristics

The final element in the simple model of the human as a system component, as presented in Figure 13.1, consists of the various actuators that allow the human to influence the system. These are mainly mechanical in nature and are characterized by such parameters as reach, strength and size.

Figure 13.7 gives an example of the area that can be conveniently **reached** by the hands of a seated person. However, it must be emphasized that in this figure, and in many others, only the area or volume that is accessible on a continuous basis is being considered. Much larger areas or volumes become practicable for operations that take place only occasionally.

The **forces** that a human can be expected to exert also depend very much on whether the corresponding motion is carried out only occasionally or frequently. An example of the former is the operation of a manually operated disconnector or isolating switch, where a force of 50–200 N may be required; an example of the latter is a keyboard, where the force required may be about 1 N. A pushbutton or toggle switch should be designed for an operating force of about 3–6 N, and a rotary switch should require a torque of about 1 Nm.

However, it is not only a question of being able to reach or being able to exert the required force; it is also a question of performing over an extended period (i.e. a question of physical fatigue). In this connection the most important aspect for operators of electronic systems is probably their **seating**, as most operators perform their work seated.

13.3 The activity

It is obvious that some activities are more difficult to perform than others and therefore more prone to errors. For example, typing is more difficult than lifting crates onto a conveyor belt. However, what is not always so obvious is that generally there is considerable latitude in designing the activity. What is predetermined is the **effect** of the activity on the system; by interposing one or more stages of transformation, the activity can be designed to suit the characteristics of the human. An important example is afforded by the entry of data via a terminal in dialog form. Clear wording of the prompting text, logical sequence, convenient format for multivalued data, useful default values, simple control codes and error handling, and so on, all help to simplify the activity and improve performance.

To understand the possibilities available in designing an activity, it is necessary to have a clear picture of the nature of an **activity**. An activity is *a functionally defined sequence of tasks*, where a **task** is *a single distinct act of perception, intellection or movement control*. Performance with respect to a single task is determined by factors discussed in Section 11.2. In meeting the functional requirement on the activity, some choice in the combination of tasks and in the sequence of their execution is often available.

With respect to the choice of **sequence**, two aspects should be kept in mind:

1. An activity may be easier to perform for one particular choice of sequence, or it may even be mandatory to perform tasks in a particular sequence. In such cases extra devices or constructs may be employed to ensure that the correct sequence is employed (e.g. interlocks, sequentially displayed text).
2. A certain sequence may be logical or may be expected due to previous experience. This is particularly the case if the activity is similar to another, often-performed activity. Such natural or expected sequences should be respected wherever possible.

Of great importance in designing an activity are the problems of **boredom** and **fatigue**. There must be a correlation between attentiveness or reaction time and signal frequency and resultant exertion on the other hand. As an example of the former, Figure 13.8 shows how performance may decrease when the signal frequency becomes too low.

13.4 The conditions

The conditions under which an activity is performed fall into two classes: the physical conditions (e.g. lighting, temperature, humidity, noise), and what may be termed the psychological conditions.

The **physical** conditions should obviously be adapted to the physical characteristics of the human body, as discussed in Section 11.2, and can be more or less readily expressed in quantitative terms.

The **psychological** conditions are much more subtle and correspondingly difficult to quantify. They include such factors as salary, social prestige, fulfillment of personal ambitions, security, privacy and what is generally known as motivation. Any serious treatment of these factors must rely heavily on the behavioral sciences and is somewhat outside the scope of this book. However, it is important for the systems engineer to be aware of the **relations between people**; they are just as real as the relations between man and machine, and in some cases essential. A lot of the information flow within a system may actually take place through so-called "informal" verbal communication channels; so properly designed facilities (e.g. canteens, coffee bars, rest areas) can go far towards promoting a desirable level of interaction.

One point that deserves mention when it comes to designing human activities within a system is the question of **feedback**. It is very important that personnel can see some tangible result of their work and, in particular, that this result gives a quantitative indication of the quality of the work. There is nothing more demoralizing than to feel that one is working for no good reason and that it does not seem to matter if one exerts oneself or not. If the process does not intrinsically provide a visible result, the systems engineer must provide feedback loops specially for this purpose.

13.5 Human engineering activities

13.5.1 Overview

As with all the other specialist engineering activities that together make up a large part of systems engineering, HE is applied throughout the whole engineering process. Early in the analysis phase some of those functions in which human effort plays a part are identified. For example, one function that is almost always present and that always involves human effort is maintenance; another is the integrated logistic support (ILS). Until the end of the design phase, HE aspects of the system are continuously being considered, expanded, refined and developed into completed system elements, just as is the case with hardware or software elements.

HE activities on the system level aim to achieve the integration of personnel into the system functions by:

- effective staffing at the man-machine interface;
- efficient use of human capabilities during system operation and maintenance;
- cost-effective demands on personnel resources, skills and training.

To achieve this the HE effort must reflect the effort of the systems engineering methodology in general, and must include active participation in the following three areas of system development:

analysis
design and development
test and evaluation.

The main types of activities that either are wholly concerned with HE or have a strong HE component are:

task definition
operational sequence definition
task analysis
personnel requirements definition
error analysis
safety analysis.

13.5.2 Task definition

Early in the analysis phase, when a start is made on defining a system to meet the objectives of the project definition, it becomes necessary to formulate the degree of human participation in the operation and support of the system. This must be done in order to define the main functional elements, compare various types of systems, develop high-level models and so on. Following the top-down aproach, the successive stages of this are identification of the areas or functional elements where human activity is involved (e.g. operation, maintenance), then description of the types of activities involved (e.g. driving a vehicle, entering data via a keyboard, carrying out first-level repair), and finally preparation of structured (i.e. systematically numbered) definitions of the tasks. These definitions will be developed and become more detailed and more completely described as the design phase progresses.

13.5.3 Operational sequence definition

As was emphasized earlier, systems engineering differs from equipment engineering in that, in addition to the HE aspects of the user interface, systems have humans as elements, and the performance of a system depends on the performance of those human elements. This performance has both static and dynamic aspects. The latter arise from the fact that the processes involved in the operation of the system, in order for it to provide the required service, consist of sequences of tasks interspersed with operations of the non-human elements of the system.

For example, a supervisory subsystem may alert an operator to the fact that something is wrong with a process; the operator responds by interrogating the subsystem; the subsystem searches its database and presents detailed information on the problem; the operator analyzes the data and issues commands to the process via the subsystem; and so on. Each step in this sequence of events involves an interchange of information and/or the execution of a prescribed action, but it also requires this activity to take place within a certain time in order for the system performance to meet its specification.

There are different ways of describing such operational sequences, but a common one is by means of **operational sequence diagrams**. They are in the same class as PERT diagrams in project management or data flow diagrams in software design. They use graphic symbols to depict the different types of elements within a sequence such as manual operations, automatic operations, human decision processes and information processes (i.e. acquisition, transmission, presentation in the direction to the operator, entry, transmission, execution in the direction from the operator). Each element is given an allocated and/or maximum execution time, resulting in an execution time for each sequence. If there are many interlinked sequences, it is possible to carry out a critical path analysis and so on, just as for project management tasks.

13.5.4 Task analysis

Task analysis is concerned with obtaining a quantitative characterization of a task in terms of such parameters as duration, intensity, required skill level, frequency of occurrence and error rate. This is achieved by decomposing the task, as it is defined in its task definition, into elementary tasks that are known from other systems and for which data are available. A typical example of such an elementary task is to read a line of text on a VDU screen that instructs the operator to hit a particular key on a keyboard, and then actually to carry out that action.

In addition to the characteristics of the task itself, the **effects** of the task must be considered. How critical is the task? How accurately must it be carried out? What happens if it is not (or incorrectly) carried out? These considerations will have a strong influence on training requirements.

Finally, task analysis leads to requirements on the man-machine interface in the form of what data are required at various stages in the execution of the task, how quickly the data must be available, for how long the data must be available, and so on. It follows from this that the HE activities must be progressed to at least the completion of the task analysis before the detailed specification of the man-machine interfaces can be commenced.

13.5.5 Personnel requirements definition

Using the data obtained in the course of the previous three activities, the personnel required to operate and maintain the system can be defined. This is, however, far more complex than just summing over the requirements for each task. Some of the factors that require consideration are the following:

- the physical locations where the tasks are to be carried out; (is it reasonable for several tasks to be carried out by the same person, or by persons based at one location?);
- the temporal relationship between the tasks; (can they be carried out at different times, or is there an expected degree of overlap?);
- the range and depth of skills required; (is it practical to have multiskilled personnel? Can skills be identified for which demand is great enough to warrant a separate group of dedicated specialists?).

13.5.6 Error analysis

After the first design of a human element of a system and its associated man-machine interfaces has been carried out, it is necessary to look at the reliability of that element. What (classes of) human error can occur, how likely are they to occur, and what will be their effects on the performance of the system?

The answers to the first two questions are to be found in areas of behavioral science and can be closely related to such "peripheral" issues as motivation, reward and working conditions. Consequently, an electrical engineer will normally have to seek specialist assistance to answer these questions. The last question can be attacked in exactly the same manner as applies in the case of hardware or software errors, by means of a failure modes and effects analysis (FMEA) methodology using fault trees and so on (see Section 10.6.2).

13.5.7 Safety analysis

Finally, there is a requirement to assess the safety of the system, identify hazards to personnel (and equipment), estimate the probability of their occurrence, and determine what should be done to reduce the risk to an acceptable level. It is quite common to approach the safety issue from two different directions: on the one hand, by putting safety requirements on the **design** of an object; on the other hand, by developing safety procedures in the **use** of the object, or in carrying out certain activities.

The former take the form of safety factors, examples being that a steam generator (or boiler) must be designed to withstand six times its rated pressure, or that the electrical insulation of a 230 V a.c. circuit must withstand a test voltage of 2.5 kV. Examples of the latter are requirements to wear safety helmets, to have both a visible break and a visible earth connection to a high-voltage circuit before attempting to work on it, and to secure loads on vehicles.

In most cases system safety can be achieved only by applying requirements of both kinds, and the greatest threat to safety (aside from carelessness) lies in a lack of coordination and communication between the equipment designer and the applications or system designer. The equipment designer intends the equipment to be used in a particular way or under particular circumstances; but if this is not communicated to the system designer, the latter can come up with operating or maintenance procedures that just happen to create a safety hazard for that equipment.

The formal safety analysis procedure is closely related to failure analysis (in reliability engineering) and error analysis. Both diagrammatic techniques (e.g. fault trees, safety state diagrams) and probabilistic evaluation techniques (using distributions) can be developed. The safety-related activities are carried out in accordance with a system Safety Program Plan, which is developed early in the analysis phase and refined throughout the analysis and design phases.

13.6 Summary

1. **Human engineering (HE)** encompasses the activities concerned with optimizing the combined performance of human and equipment within the context of a system.
2. The HE problem is tackled by describing human performance as a function of the **environment** in which the activities are carried out, and optimizing the combined system in an iterative process.
3. The environment is determined not only by **equipment** design but also by such factors as **satisfaction** and **motivation**.

4. The two most often used parameters in characterizing human performance are **speed** and **accuracy**, with the latter usually dominating in electrical and electronic systems.
5. **Human performance** is determined by three main factors:
    the human
    the activity
    the conditions.
6. The human as a system component may be described in terms of a model consisting of three elements: **sensors**, a **processor** and **actuators**. The sensors (e.g. eyes, ears) and actuators (e.g. mouth, hands) are much easier to characterize than the processor (i.e. brain).
7. Besides their physiological characteristics, the performance of all sensors depends on the **signal-to-noise ratio**.
8. The performance of the processing function depends to a large extent on the **coding** of the information. The manner in which information is presented is a central issue in HE.
9. The main **HE activities** are:
    task definition
    operational sequence definition
    task analysis
    personnel requirements definition
    error analysis
    safety analysis.

13.7 Short questions

1. Define HE.
2. Which are the two parameters most commonly used to characterize human performance?
3. Name the three main factors that determine human performance.
4. Describe a single model of the human as a system component.
5. What factors influence the quality of an optical signal?
6. What is a typical figure for minimum visual acuity?
7. Give a typical value for the time it takes to react to a simple binary signal.
8. How can the conditions under which an activity is performed be classified?
9. Name four of the most important HE activities.

13.8 Problems

Telephone keypad: A normal telephone keypad consists of twelve keys: the number 0 to 9 and two function keys (e.g. * and #). In addition to ordinary dialing, the following functions are to be realized:

- two digital abbreviated phone numbers, to be programmed by the user;
- single-keystroke redialing of the last-used number;
- redirecting to another extension (two separate cases);

- automatic redialing of a busy number as soon as it becomes available;
- disconnecting the telephone for a fixed time interval.

Using this information:
  (a) Decide how to realize these functions with the available keys in as simple and logical a manner as possible.
  (b) Write a corresponding user manual.

13.9 References

Bailey, R.W., *Human Performance Engineering*, Englewood Cliffs, NJ: Prentice Hall, 1982.

Brown Boveri Review, Special Issue on Power System Control, 1/2, 1983.

Grandjean, E, *Fitting the Task to the Man,* Taylor & Francis, 1980.

IEE Proceedings, Special Issue on Design, vol. 130, part A, no. 4, June 1983.

Kerin, J.T., "Lighting for VDUs", *The Energy Journal,* April 1983.

McCormick, E.J., *Human Factors in Engineering and Design,* New York: McGraw-Hill, 1976.

MIL-H-46855B, *Human Engineering Requirements for Military Systems, Equipment, and Facilities,* Washington DC: US Department of Defense, 1979.

MIL-STD-1472A, *Human Engineering Design Criteria for Military Systems, Equipment and Facilities,* Washington DC: US Department of Defense, 1983.

Osborne, D.J., *Ergonomics at Work,* New York: Wiley, 1982.

Chapter 14

Cost-effectiveness

14.1 Introduction

14.1.1 Universal applicability of cost-effectiveness

It is appropriate to end this book with a chapter on **cost-effectiveness (CE)**, because it is the most **universally applicable** of all the measures that can be applied to the engineering process. No matter what the system, there is always a compromise between cost and effectiveness, and achieving the optimal balance is the ultimate aim of systems engineering. In some systems the balance may be heavily in favor of effectiveness, as is the case where human lives are at stake. Even then, however, the highest possible effectiveness or safety is not wanted as this would imply essentially infinite cost, and the systems would never be completed and go into service.

Not only is CE universally applicable to the systems engineering process; it is also a measure that links engineering with other spheres of human activity and concern (e.g. economics, ecology, politics). Given an increasing awareness of the limits imposed by the finite nature of the world, CE becomes **the** parameter by which any project is scrutinized and compared to other projects competing for the same resources.

Finally, CE is universal in the sense that it encompasses all the engineering specialities (e.g. reliability engineering, maintainability engineering, human engineering) that have been introduced and applied throughout this text. Any change made to one or more of these in the engineering design process will affect the system CE. Thus, only after an understanding has been gained of all the activities that shape systems engineering is it possible to define CE and to appreciate what is required in order to obtain a truly optimized design.

14.1.2 Definition of cost-effectiveness

Cost-effectiveness *(CE)* is *the ratio of system effectiveness (SE) to lifecycle cost (LCC)*; that is,

$$CE = SE/LCC$$

In this text, in contradistinction to the definition used in many other texts, *SE* will be expressed in dollars; so *CE* will be a dimensionless parameter.

It follows from this definition that a discussion and development of CE could be separated into one of SE and one of LCC, and this is the approach that will be adopted. However, it must always be remembered that any decision taken within the engineering process will have a simultaneous effect on both SE and LCC, and that to gauge the true effect of such a decision the resultant change in CE must be calculated.

The definition therefore leads to a (conceptually) simple **universal criterion** governing all engineering decisions:

> A decision is **good** if it results in an increase in cost-effectiveness; and in a choice between several options the **best choice** is the one that results in the greatest increase in cost-effectiveness.

Note that, while this criterion is appropriate for engineering decisions and for decisions between alternatives competing for the same resources, it may not always be entirely suitable for investment decisions. Consider the case of a system design for which it is estimated that SE is 1 million dollars and LCC is 0.5 million. The "profit" over the system lifetime will therefore be 0.5 million, and CE equals 2. A design change is being contemplated that would increase LCC to 1 million but increase SE to only 1.8 million. This would mean a decrease in CE from 2 to 1.8; so according to the criterion it would not be a good decision. However, the "profit" would increase from 0.5 million to 0.8 million; therefore, if there were no better opportunity for investing the additional 0.5 million, it might still be an advantageous decision.

This case also illustrates another issue, namely the difference between cost-effectiveness and **cost-benefit** as a measure of a system's "worth" or "goodness". Cost-benefit is the difference between revenue and cost, or, as will be discussed in the next section, between value and cost. In the present case, the cost-benefit before the change was 0.5 million, after the change 0.8 million. Whether cost-effectiveness or cost-benefit is the most appropriate measure will depend on the particular project and on what aspect of the project one is considering.

14.2 System effectiveness

14.2.1 Definition of system effectiveness

System effectiveness (SE) is a measure of how well a system performs the functions that it was designed to perform, or how well it meets the requirements of the system specification. It is often defined as *the probability that the system can successfully meet an operational demand within a given time when operated under specific conditions.*

This definition implies a number of important aspects of SE:

1. Operating time may be critical, and in general SE is a function of time.
2. Maintenance is not excluded; "specified conditions" will in most cases include both preventive and corrective maintenance
3. "Operational demand" implies that there are two separate classes of system failures:

   - The system is in an inoperable condition when needed.
   - The system fails to perform satisfactorily throughout the required operating period.

4. The inclusion of both "operational demand" and "specified conditions" shows that failure (i.e. failure to meet operational demand) and the conditions under which the system is used (i.e. operational stresses) are related.

From this it follows that, while SE is obviously influenced by system design and implementation, it is equally influenced by the way the system is used and maintained, by the logistic system that supports the operation, and by operational administration, through personnel policy, rules governing equipment use, fiscal control and many other policy decisions.

The definition given above, and expressed in terms of a probability, is particularly useful for systems that are required to operate for a prescribed, relatively short period (i.e. systems fulfilling a mission, as is the case for most military requirements). For most other systems the period of operation is the lifetime of the system, and this is usually very long compared with the timescales for other events affecting the system (e.g. component failure, personnel retraining, preventive maintenance).

As a result the system settles into a macroscopic steady state (see Section 11.2), which is characterized by an average performance or, perhaps more to the point, by an average deviation from "perfect" performance, as defined in Chapter 9. However, as the performance of a system is usually a complex multidimensional parameter, measuring it in terms of a probability is not very appropriate. The proper approach is to determine the decrease in the **value** of the system as a function of the decrease in performance. The definition of SE can then be formulated as *the value of the system over its design lifetime.*

14.2.2 Factors affecting the value of a system

Obviously, every system has some value; otherwise its development would not even be contemplated. Equally obviously, this value must in some way depend on how well the system performs; if it did not perform at all, its value would be zero. The problem arises in trying to step from a qualitative statement, like "High reliability is very desirable", to a quantitative statement, like "An increase in the reliability from 0.92 to 0.94 is worth $60,000". It is then found that the value of a system, particularly its dependence on various performance parameters, is often a highly subjective matter. Nevertheless, it is a problem that must be solved, because without assigning some value to system performance there is no basis for taking rational engineering decisions. The following example is used to illustrate how an engineer might go about solving the problem for a simple case.

Consider a system that provides something, either goods or a service, the value of which is determined by its market price. The present price is known, the future price is estimated based on some model of the market and the competition, and the dependence of price on volume is determined similarly. The value of the system, or SE, is then the difference between the unit price and the cost of the raw materials per unit multiplied by the volume, suitably calculated over the design lifetime with respect to a fixed point in time, usually the present. (The cost of raw materials may be zero in the case of a service.) At this point SE is a function of one variable only, namely volume.

The next step is to take into account the operational availability of the system (i.e. the fraction of the scheduled operating time for which the system is actually able to operate). A decrease in availability from its ideal value of unity causes a decrease in SE, usually by a proportionate amount, but possibly by somewhat less if advantage can be taken of a short-term undersupply by increasing the unit price. SE is now a function of two variables: volume and availability.

However, as discussed in Section 12.4.2.4, a consequential loss of value may arise from system failure(s). As a result it may not be equally damaging to have a few long interruptions as to have many short ones, even though availability is the same. In other words, reliability is also important, and SE is now a function of three variables: volume, availability and reliability.

So far, only factors that affect output quantity have been considered. However, most products and services have to meet some set of requirements; if they do not, the unit price obtained will be less than if they do, and possibly zero (i.e. rejection). So the value will also become a function of the several parameters that characterize the product. This function is determined by the market (i.e. the users). For example, in a telephone system the quality of the transmission, as characterized by an intelligibility test or a suitably defined signal/noise ratio, influences the value of the service (i.e. what people are willing to pay). However, in this case the value is far from being a linear function of intelligibility; below a certain level of intelligibility the value is zero, whereas above a somewhat higher level the value remains almost constant. A value function such as in this example can only be determined by conducting tests with users; it is not an absolute quantity, but depends on the **perception** of a representative user group.

Such nonlinear behavior is quite common, and the limiting case of a step function corresponds to a fixed value specification of the parameter. This value must be achieved for the system to be accepted, but any better performance is not accorded any higher merit. It is of course the simplest way to specify a parameter, but it is often somewhat unrealistic and can mask real optimization opportunities.

14.3 Lifecycle cost

14.3.1 Basic concepts

In any project **the systems engineering process must focus at all times on the service to be provided by the system**, and only in a secondary or derived sense on the system itself. This principle is central to systems engineering. It has been reflected in various issues discussed throughout this text and should by now be firmly entrenched in the reader's mind.

Normally, the user places requirements on the service, and the user's judgement of the success (or otherwise) of the project is determined by the extent to which these requirements for service are met by the system. Consequently, the relevant cost for CE purposes must be the cost of providing the service, either per unit time or over the design lifetime of the system. This consists not only of the nonrecurring (i.e. one-off) cost of acquiring the system, but also of the recurring (i.e. per unit time) costs of operating and maintaining the system (also called the **cost of ownership**). This total cost is the **lifecycle cost (LCC)**.

However, it is not sensible or even very useful to just add up all the estimated costs over the lifetime of the system. Because of the cost of capital (i.e. interest) and inflation, costs incurred at different times have a **different relative value**, and to compare them they must be discounted.

Let the **discount rate** (the effective cost of capital, i.e. the commercial interest rate which depends on the **risk** associated with the project, plus any commissions and charges) be denoted by p, measured in units of per unit time; and let C be a cost item incurred at time t. Note that the value of C is the actual dollar value of the cost at the time it is incurred (as cash flow); it is the amount that appears in the books and enters into a cash flow calculation. There is no question of adjusting it for inflation or anything like that; such effects are included in the discount rate. Then the **present value** (i.e. at t = 0) of C, denoted by C', is given by

$$C' = C/(1 + p)^t$$

assuming that the interest is **compounded** (i.e. calculated and added to, or subtracted from, the capital) every unit of time. The time t is constrained to integer values but may be either negative or positive.

Of particular interest is the case of a **series of** n **equal payments**, where one payment (or cost item, e.g. maintenance cost) C is made at the end of each accounting period (i.e. at t = 1, 2, 3, ... , n). Then

$$C' = C \left(\frac{(1 + p)^n - 1}{p(1 + p^n)} \right)$$

However, very often the cash amounts are not equal, even if they are for the same amount of work or the same item each time; it has to be assumed that they will increase in time due to inflation. Let the **inflation rate** be denoted by q, again per unit time. Then

$$C(t) = C(0) (1 + q)^t$$

and the problem can be transformed back to one of a series of equal payments by introducing the **effective discount rate**,

$$p' = p - q$$

Using the effective discount rate is sometimes described as calculating in **constant dollars** (i.e. dollars with the same value or **purchasing power** as they have at time t = 0).

While the above relations represent the most common way of combining costs that occur at different times, as must be done in a LCC calculation, there exist a different point of view and a different way of carrying out the calculation. This approach, although it leads to the same result, is in some ways more directly related to real life and normal accounting practices.

Consider a system with a fixed design lifetime T and a cost C that is incurred at a time $0 \le t \le T$. Then the cost K of providing the capital to fund expense C (i.e. the accumulated interest) over the lifetime of the system, referred to the end of the lifetime (i.e. $t = T$), is given by

$$K = pC \sum_{i=0}^{m-1} (1 + p)^i$$
$$= C[(1 + p)^m - 1]$$

where $m = T - t$. This cost is carried as a separate item, and together with C it forms the **future value** of C, denoted C*. (This of course becomes C', or $C/(1 + p)^t$, for negative values of t.) The basic principle is the following: Interest is attracted during any period by the **net** capital borrowings (or funds in use) during the period, and is deemed payable at the end of the period. Unremitted interest at the end of a period adds to the net borrowings for the next period.

Similarly, the value of K for a series of n equal payments C made at the end of each unit of time is given by

$$K = C \left(\frac{(1 - p)^n - (1 + np)}{p} \right)$$

which again of course adds up with the n principal payments to the compound amount formula

$$C* = C \left(\frac{(1 + p)^n - 1}{p} \right)$$

For the purpose of optimizing the CE of a system, there is no intrinsic advantage in using future value rather than present value. However, expressing the cost of capital K as a separate cost item does have three advantages:

1. This method of itemizing costs corresponds to that used in normal accounting; the periodic value of K is an item that will appear in the books each period and that will affect the cash flow in each period.
2. The algorithm for computing the periodic value of K is very simple; it is just the total value of funds expended up to the beginning of the period, multiplied by p.
3. Other costs associated with providing capital (e.g. fees for available but unused credit) are easily and consistently accounted for and included in a LCC model.

A small example will illustrate the two points of view and their equivalence. To emphasize the concept of **cash flow**, the example is a project that has both costs and revenues. The project has a lifetime of 5 years, and the costs and revenues occurring in each year are shown in Table 14.1. Adding these items gives a net cash flow for each year. In the first 2 years, when investments are heavy and revenues have not yet started to flow, the cash flow is negative. This situation turns around in year 3, and the last 2 years show a strong positive cash flow. For simplicity, it is assumed that all items occur at the end of the year.

In the first approach to a **discounted cash flow** analysis (same as LCC, but with both revenues and costs), the yearly cash flows are discounted back to the beginning of year 1 (or end of year 0) using the present value factors, $1/(1 + p)^i$, as shown in Table 14.2. The result is a net present value (NPV), as shown in Table 14.3.

Table 14.1 Costs, revenues and resultant cash flows for a small project of 5 years' duration (values in 10^4)

Item	1	2	3	4	5
			Year, i		
Management	10	20	20	10	5
Equipment	20	50	50	—	—
Materials	40	25	50	50	50
Labor	—	30	60	60	60
Maintenance	—	5	10	10	10
Total out	70	130	190	130	125
Product sales	—	50	200	250	220
Equipment resale	—	—	—	—	30
Total in	—	50	200	250	250
Cash flow	−70	−80	10	120	125

Table 14.2 Present value factors, $1/(1 + p)^i$, for four discount rates p

p (%) \ i	1	2	3	4	5
5	0.952	0.907	0.864	0.823	0.784
10	0.909	0.826	0.751	0.683	0.621
15	0.870	0.756	0.658	0.572	0.497
20	0.833	0.694	0.579	0.482	0.402

Table 14.3 The cash flows from Table 14.1 discounted by the factors from Table 14.2, resulting in net present value (NPV) (values in 10^4)

i \ p(%)	5	10	15	20
1	−66.64	−63.63	−60.90	−58.31
2	−72.56	−66.08	−60.48	−55.52
3	8.64	7.51	6.58	5.79
4	98.76	81.96	68.64	57.84
5	98.00	77.63	62.13	50.25
NPV	66.20	37.39	15.97	0.05

The first thing to note is that the NPV decreases with increasing discount rate. This is true of any project for which the cash flow increases, on average, throughout the project. Secondly, if the cash flow is negative in the first part of the project, as is true of any project requiring an initial investment, there exists some discount rate for which the NPV becomes zero. This is the **internal rate of return (IRR)**.

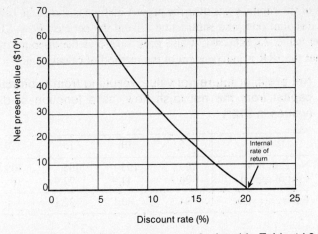

Figure 14.1   Net present value of the project, as calculated in Table 14.3, showing the dependence on the discount rate.

The IRR constitutes the most useful single characterization of the financial viability of a project. It represents the **break-even** discount rate that will just allow repayment of the initial investment. If the actual discount rate (i.e. interest rate plus any other related financial charges) is less than the IRR, a profit will result. However, if the discount rate is higher than the IRR, the NPV will be negative, and the project will result in a loss. In the above example the IRR equals 20%. This is shown clearly in Figure 14.1.

The third thing to note is that Table 14.1 does not include the cost of providing the capital, although that cost (i.e. interest and fees) must constitute an actual cash item.

In the second approach to a discounted cash flow, using future value rather than present value, the first thing to do is to determine the **capital demand** arising from the cash flow in Table 14.1, as shown in Table 14.4. This demand then leads to **capital costs**, which depend on the discount rate, and they also are included in Table 14.4.

Table 14.4 Yearly capital demand, the resulting total capital bound in the project, and the cost of that capital (values in 10^4). Note that the cash flow outcome in one period only attracts interest in the succeeding period.

$p(\%)$	i	2	3	4	5
Net capital demand		70	80	-10	-120
5	Capital	70	153.5	151.2	38.8
	Interest	3.5	7.7	7.6	1.9
10	Capital	70	157.0	162.7	59.0
	Interest	7.0	15.7	16.3	5.9
15	Capital	70	160.5	174.6	80.8
	Interest	10.5	24.1	26.2	12.1
20	Capital	70	164.0	186.8	104.2
	Interest	14	32.8	37.4	20.8

The sum of the capital costs over the project lifetime is shown in Table 14.5 for each value of the discount rate, and subtracting it from the net cash flow, $1,050,000, yields the future net value. As expected, it also goes to zero as the discount rate reaches 20%, indicating that the IRR is independent of the method of calculation.

Table 14.5 Net profit, or future net value, resulting from subtracting the cost of capital from the net cash flow, as a function of discount rate p (values in 10^4)

$p(\%)$	5	10	15	20
Net cash flow	105	105	105	105
Cost of capital	20.7	44.9	72.9	105
Future net value	84.3	60.1	32.1	0

14.3.2 Cost items

The LCC is made up of a number of **cost items**. A cost item is the smallest cost that is calculated or estimated as a separate entity. The number of cost items used (or in other words, the average size of a cost item relative to the total LCC) depends on the point in the engineering process at which the calculation is carried out. The set of cost items is developed in parallel with the development of a work breakdown structure (WBS), and it is indeed usual to tie a cost item to a work package at a certain level of the WBS. The level is chosen so that responsibility for a cost item can be assigned to a single person.

However, while a task is analyzed by decomposing it into activities chosen from a predefined set, the cost of executing a task is calculated by decomposing it into **cost types** (sometimes also called **cost categories**) chosen from a predefined set. This set is in itself developed in a structured or hierarchical fashion as the engineering process develops. At the highest level there are only three cost types:

labor costs
material costs
capital costs.

A cost item is identified by one element from each of two index sets: the set of tasks and the set of cost types. In addition, there must be an indication of when each cost item is incurred. Consequently, a cost item is identified by three index values; in other words, the "space" of cost items is three-dimensional.

In developing the set of cost items the most difficult part is to develop the WBS, as this WBS must encompass all the work associated with developing, designing, manufacturing, installing, commissioning, operating and maintaining the system over its whole lifetime. It is not enough to consider only the project activities (and costs), as was the case in Section 2.5.2.

The following breakdown has been found to be both applicable to a wide range of systems and simple to use in the sense of making it easy to assign costs to an account:

1. ENGINEERING
   11 Research and development
   12 Analysis
   13 Design

2 IMPLEMENTATION
   21 Acquisition
   22 Integration and test
   23 Installation
   24 Commissioning
3 LOGISTIC SUPPORT
   31 Repair subsystem
   32 Documentation
   33 Training
   34 Maintenance
4 OPERATION
   41 Management
   42 Quality assurance
   43 Operation
5 DECOMMISSIONING
   51 Equipment disposal
   52 Redundancy payments
   53 Site amelioration

Some typical costs that fall within each of these main items are as follows:

11 Research and development: the costs of any investigations and studies carried out specifically to support or create the technology needed for the system, an allocated share of the costs of more general R&D programs and license fees for the use of technology.

12 Analysis: the costs of economic feasibility studies, environmental impact studies, market investigations, visits to inspect existing systems, system analysis and developing the system specification.

13 Design: the costs of all activities connected with producing the complete set of system element specifications, such as: modeling, simulation, optimization and mockups; developing databases; producing drawings, parts lists, engineering plans and test requirements; and developing the specifications themselves.

21 Acquisition: the costs associated with acquiring all the system elements, such as setting up manufacturing facilities, unit manufacturing costs, purchasing, incoming inspection, transport, storage and handling.

22 Integration and test: the costs associated with designing and setting up test facilities, rental of test equipment, interface verification, subsystem tests, modifications resulting from unsatisfactory test results, system acceptance tests and test documentation.

23 Installation: the costs of site establishment, site works, auxiliary equipment/facilities (e.g. air-conditioning, sewerage, power, lighting, conduits), cabling (if not prefabricated as a part), site inspections, development of installation instruction, and documentation maintenance.

24 Commissioning: the costs associated with putting the system into service, such as onsite training, cut-over, initial maintenance and spares provisioning, initial supervision and start-up costs (e.g. filling fuel tanks).

31 Repair subsystem: the costs of designing, constructing and supplying all repair facilities, including the facilities for managing the repair work (computer systems), any special jigs, fixtures or test equipment, and any costs resulting from setting up the assured supply of spare parts.

32 Documentation: the costs associated with developing, producing and maintaining all documentation, such as operating and maintenance manuals, spare parts lists and cabling schedules.

33 Training: the costs of developing training courses, writing training manuals, conducting training, assessing training needs and providing training facilities—also the costs of attending the training courses.

34 Maintenance: the costs resulting from actually carrying out all three types of maintenance, including the costs of materials (i.e. consumables and spare parts) and labor on all maintenance levels, as well as monitoring and fault-reporting systems.

41 Management: the costs associated with project management and the management of system operations, such as the development and maintenance of all plans, contract management, personnel management and the management of system operations.

42 Quality assurance: the costs of setting up and carrying out quality assurance, such as supplier qualification, inspections and verifications, test equipment calibration and the traceability of standards, and all types of quality assurance audits.

43 Operation: all costs associated with the human element of the system (e.g. salaries, social costs, amenities, transportation), as well as energy costs, taxes, licenses, rents and leasing costs.

51 Equipment disposal: costs associated with dismantling and disposing of eqiupment. (Of course, such costs can be negative, i.e. represent revenue.)

52 Redundancy payments: all payments due to personnel on termination of operations, including relocation costs, "golden handshakes", and retraining.

53 Site amelioration: cost of restoring site to its original condition, possibly including detoxification.

14.3.3 Example: Microwave Landing System

14.3.3.1 Introduction

The Microwave Landing System (MLS) is an air-derived radio navigation system providing approach and landing guidance to aircraft by means of time division multiplexed microwave transmissions from gound-based elements to aircraft in its coverage sector. It is described by Henry W. Redlien and Robert J. Kelly.

Over a 10-year span from 1974, three LCC studies of the MLS were undertaken by the US Federal Aviation Administration (FAA). The first, in 1974, was concerned with the potential economic discriminants among four systems under competitive development at that time within the USA. The second was prepared in support of submissions to the International Civil Aviation Organization (ICAO), culminating in the choice of the Time Reference Scanning Beam (TRSB) signal format in 1978 as a new international standard. The third, released in 1982, provided a detailed LCC forecast for the planned introduction of MLS ground equipments over a 20-year period. It served as the cost segment of a cost-benefit analysis supporting policy decisions.

In Section 14.3.3.2 (Table 14.6) some edited extracts from the document circulated in 1974 are reproduced to illustrate the preceding sections in a more specific way. The complete document sets out the **requirements** for data; it is not furnished with the estimates themselves.

In the **Introduction** the background and purpose of the exercise are clearly stated. The argument for including airborne equipment in the assessment should be noted, bearing in mind the nature of the decision facing the FAA at that time and the fact that the airborne equipment would be purchased substantially by nongovernment users.

In the **General Plan** the emphasis placed on supporting data should be noted. Such a requirement imposes a heavy commitment on contractors. The strategy of considering several representative systems (both civil and military) is again consistent with the specific purpose of the LCC study (the actual configurations are not relevant here). The cost factors are defined under four categories: acquisition, installation, operation/ maintenance and logistic support. (Note that certain government costs (e.g. program management, flight testing) and previously expended funds for R&D work are not included, although they could be added by FAA personnel to the cost-benefits analysis.) The basic assumptions for costing are also detailed precisely, with information on relevant factors such as system lifetime, lot sizes, upgrade intervals and discount rate. All these are included in an attempt to achieve standardization—that is, a comparison of "apples with apples"! This pattern is continued in the next section of the requirement.

In the **Detailed Plan**, several segments of which are reproduced, the inclusions expected by the FAA under each of the major cost categories are specified. Reference to the "standard cost system" indicates the acceptability of cost data supplied in the same manner as is normally required for tendering to the US government (e.g. using standard forms). The level of supporting justification for each approved costing technique should be noted. Along with the standard conditions, the use of multilevel cost breakdown worksheets (with over 150 cost items per configuration) is an attempt to structure the submissions for evaluation purposes. While the **amount** of information required on acquisition costs may be the main point of the initial extracts from the Detailed Plan, the **character** and detail of the information sought under the final two major categories are more important in the context of systems engineering and preceding chapters (especially Chapters 10–12). A small segment of the cost breakdown section dealing specifically with these final categories is reproduced also.

Overall, the document shows how an LCC program has to become specific to the system and be "thought through" if the resulting costs are to serve their intended purpose. In particular, the limitations of an ill-considered "skirmish" with LCC should be noted!

14.3.3.2 FAA document extracts

Table 14.6 Microwave Landing System lifecycle cost data requirements, prepared by the Federal Aviation Administration in 1974

TABLE OF CONTENTS

INTRODUCTION
GENERAL PLAN FOR LIFECYCLE COSTING
  I SELECTION FOR SYSTEM REPORTING
  II BREAKDOWN OF SYSTEM COSTS
  III COMMON BASE FOR ECONOMIC ANALYSIS
  IV MODEL GUIDELINES
  V DISCOUNT RATE

DETAILED PLAN FOR LIFECYCLE COSTING

INTRODUCTION

The objective of the Microwave Landing System (MLS) Program is to develop a new approach and landing system that will satisfy the entire range of civil and military operational requirements on a fully compatible basis with modular configurations to permit satisfaction of individual user needs on a cost-effective basis. This system will be an integral part of FAA's upgraded third- and fourth-generation Air Traffic Control System and provide the necessary coverage and accuracy in the terminal airspace where safety, efficiency, and effective air traffic management must prevail.

At the present time, the MLS Program is in the second phase of a three-phase program where a major decision, that of selection of the scanning beam or Doppler scan technique, is to be made. In this phase (feasibility demonstration) of the program, there are technical performance tests being conducted to demonstrate the potential of the four different systems to meet overall performance requirements and thus, to discriminate between techniques on a technical basis. However, since the selection of a technique and signal format will impact upon the users, both from a technical and economical standpoint, there must be economic factors introduced into the decision-making process. Therefore, system lifecycle costs will be used to provide the necessary economic factors for assistance in the Government's decision on system technique and signal format. This method of costing was chosen to reflect the Government's concern for the cost factors that affect all users of the airways.

When all the elements of cost have been submitted by each contractor, they will provide the necessary economic inputs to the assessment process for the MLS; provide input to the current FAA cost-benefit study; and serve as a reference in the FAA budget cycle. Since the major elements of lifecycle cost will be reported directly or be provided in a form where the cost can be directly derived, it should be of benefit to the users and operators of the Air Traffic Control (ATC) system. Particular value may accrue to the operating services of FAA where the budget process requires planning to a five-year projection. Lifecycle factors of system purchase, maintenance and operation, training and logistics support can all be taken from the data provided. Likewise, the air carrier, business aircraft operators, and the general aviation users can use the data for planning of future avionics equipment.

It should be recognized that the economic estimates of the different systems must be logical, reasonable and be able to withstand the critical cost review of the aviation community. For these reasons, and because the economic factors will be a significant factor in the crucial decisions of system technique and signal format, every effort must be made to provide usable cost estimates.

The purpose of the cost analysis is to provide insight into the economic merits of the systems proposed by the contractors and to identify engineering approaches which show promise of offering a lower cost to the eventual users of the system. Since each contractor is charged with the responsibility to consider system trade-offs that will result in the most economical system for his technique, it now becomes the responsibility of the Government to weigh the cost advantages of the four competing techniques. The output of this weighing process will then provide additional input to the techniques assessment process and ultimately assist in the selection of the most effective system.

Specific guidelines and requirements for cost estimating by the contractors are outlined in the Detailed Plan for Lifecycle Costing. The schedule for cost estimating has been structured to generally follow that of the cost schedule for Phase III of the MLS Development Program. In order to permit validation of contractor provided cost data, specific guidelines are included on the major factors affecting estimates for future systems.

GENERAL PLAN FOR LIFECYCLE COSTING (LCC)

Lifecycle cost procedures have been established to take advantage of the present Statement of Work (Phase II) Program where provisions have been made for cost considerations in the planning process. The SOW requires the contractors to provide cost estimates for prototype equipment as well as a limited production option plan. Incorporated throughout the Statement of Work are the various elements that require engineering analysis by the contractor and form the basis of the detailed LCC plan. It is the purpose of this procedure to derive the initial and sustaining costs of the proposed systems from the on-going program in an orderly and logical manner. It must be emphasized that the reporting of cost in this document does not in any way alleviate the contractors' responsibility to conduct an economic trade-off analysis for their system which is to be reported in Section 2.6 of the Revised MLS Development Program Plan. The end product of this analysis, plus experience gained in the Feasibility Demonstration Phase and the use of product engineering expertise will form the basis for cost derivation in the document. Cost data will not be considered alone by the Government and must be fully supported in accordance with the requirements set forth in the detailed plan.

I SELECTION FOR SYSTEM REPORTING

In order to cover a broad spectrum of user requirements and yet keep the cost reporting, substantiation, and subsequent analysis to a functional minimum, there have been three primary system configurations selected for the lifecycle cost considerations. These configurations are the civil "D" and "K" corresponding to categories of service I and III, respectively, and the military "G" (transportable) system. In addition, since some contractors have proposed unique low cost configurations for the general aviation user, there are provisions for costing one optional system.

The military "G" transportable configuration was selected for lifecycle costing due to its widespread application in the three Military Services (USAF, USN/USMC and USN) and because it also represents the greatest departure from the civil configurations. However, the Military Services have not yet achieved full commonality in their requirements for the "G" configuration and it is therefore necessary to narrow in on the system for one of the Services in order to keep the contractor's lifecycle costing effort to a reasonable level. Accordingly, the Army "G" has been selected for this activity since it has potentially the greatest number of ground and airborne installations. The Army "G" cost estimates should be representative of an eventual common military "G" configuration with appropriate cost adjustments by the Government to reflect the particular installation philosophy (split site or collocated) of the individual Services. It should also be noted that the civil configuration estimates will be usable by the Military Services in deriving cost estimates for the MLS configurations to be installed at fixed main operating bases.

II BREAKDOWN OF SYSTEM COSTS

The cost over the life of the systems has been divided into four major categories:

- a. Acquisition
- b. Installation
- c. Operation and Maintenance
- d. Logistic Support

Within each category there are major elements and, where appropriate, subelements and units. In general, the elements are separated into ground and airborne systems with the major subsystems further broken down to the unit level. Detailed instructions for the cost estimates are provided in the detailed plan and are broken down to the unit level. The breakdown of the military "G" configuration is being limited to acquisition (system hardware) and installation costs only. Due to the wide variation in maintenance and logistic support procedures between the Military Services, it was not considered practical at this point in time to estimate these cost categories.

III COMMON BASE FOR ECONOMIC ANALYSIS

In order to maintain a relevant comparison between competing techniques in estimating over the 25 year life of the MLS, there must be a common baseline provided for both ground and airborne equipments. This will ensure that all contractors estimate to the same number of systems, and eliminate large disparities in marketing projections and subsequent costs in later years.

For both the ground and airborne systems, an estimate has been made for the number of systems to be employed over the full life range. In addition, estimates for the Military "G" are provided. The first 5 year lot buy for ground systems is based upon the FAA 10-year National Aviation System Plan and includes the limited production option quantities. Estimates for the ground and airborne systems are included in Enclosure 2 and are outlined in 5-year increments. The contractors should assume that the total is composed of equal yearly buys over the 5-year period.

IV MODEL GUIDELINES

The FAA intends to construct an economic model that will accept the inputs from the lifecycle costs provided by each contractor. Total lifecycle costs will be examined after an evaluation of the contractor's supporting rationale has been concluded. Sensitivity of the model to the element, unit, and subsystem level cost will be analyzed prior to its actual use in order to determine the most critical cost factors. Exercise of the model will then focus on these critical factors in order to identify any significant cost advantages of the four competing techniques.

Since the MLS Development Program is now scheduled to conclude in 1977, and the system has been planned to be operation until at least the year 2000, a 25-year study life has been established. This is entirely consistent with the life of electronic systems and will meet the design requirements set down by RTCA SC-117. Lot purchases are assumed to begin in 1977 and concluded in 1998 based on the 5-year purchase increments. This purchase procedure is considered to be reasonable, at least within FAA, where large numbers of systems, of a complex nature, can be programmed for over this timespan. While the competitive nature of the civil avionics industry results in a rather dynamic situation with constant revision and updating of equipment to meet the operational needs of the aviation community, a 5-year purchase plan appears reasonable when considering major system changes of this nature.

Although the FAA has responsibility for implementing and maintaining the ground system, it is necessary to consider airborne costs as well in order to arrive at the most cost-effective overall system. Optimizing only the ground system costs might impose an unnecessary cost burden on the aviation user. Therefore, a system approach for the MLS lifecycle costing has been selected and will be carried out in the assessment and selection of the optimum technique and signal format. This is in keeping with the Statement of Work which requires the contractors to develop a complete MLS system (ground and airborne) with supporting plans for the maintenance, logistic support and all other economic factors associated with their proposed system.

V DISCOUNT RATE

A discount rate of 10 percent has been selected to conform to Government guidelines and to reflect a realistic profit margin for the avionics industry. This discount rate will be used to transform the projected costs of the ground and airborne equipment over the life of the system into present worth at constant 1974 dollars. The discount rate will be applied by the Government during the cost model exercise. The contractors shall use 1974 dollars as the basis for their cost estimates and shall not include inflation factors.

DETAILED PLAN FOR LIFECYCLE COSTING

I INTRODUCTION

The definition of cost elements is intended to clarify the cost data requirements for lifecycle cost purposes by defining the content of cost elements in detail, identifying the sources for derivation of cost data from the Statement of Work, and by incorporating standard conditions which affect cost computations. The information provided by these definitions when used in conjunction with the projection of microwave landing system requirements through the year 2002 should provide the basis by which contractors can develop the required cost data for both the ground and airborne systems.

II FORMAT

The format follows that of the cost breakdown work schedule and corresponding paragraph numbers are used. The cost breakdown work schedule contains cross-reference for the data elements between the Statement-of-Work, cost schedule reference, and the nomenclature of the document. The knowledge acquired during fabrication, installation, test, and operation of the feasibility demonstration hardware, the prototype hardware design, and the development of the limited production option plan, as identified by the specified references, should be applied in the derivation of cost data. The definitions apply to configuration "K" with Category III airborne equipment. Configurations "D" and "G" and associated airborne equipment are treated by exceptions that are explained in the body of the work. Generally, these exceptions are simplifications which made certain cost elements inapplicable. Configuration "G" should be treated as a military system.

Certain standard conditions have been established to assure that all contractors derive costs on the same basis. These standard conditions, such as aircraft interface, airport layout, soil condition, logistic paths, maintenance, and training information, as included in the relevant cost element sections.

III DETAILS

A. Acquisition [*Extract*]

1. Ground equipment

This section of the schedule should contain the cost per unit of the ground hardware including factory test cost. Equipment should be costed as though built to FAA specifications except for the military "G" where MIL SPECs are required. Cost should include parts, labor, G/A, overhead, capitalization for production and packaging, handling, and transportation costs for a one-way trip of 1000 miles. Profit should not be included. (The Government will apply standard markups to this basic cost in order to more accurately reflect the actual cost to the user.)

This section of the cost schedule should also contain the supporting rationale used to derive ground system hardware costs.

The "Supplemental Instructions to Offerors for the Preparation of Cost or Pricing Data", dated October 25, 1973, from ALG-340 to each contractor, contains guidelines for the derivation of material, labor and overhead costs. A clarification of the supporting rationale is as follows:

Ground Systems hardware costs may be derived from a standard cost system or parametrics at the contractor's option. If a standard cost system is used, the following information should be provided:

Labor

- The incremental buildup from leveled time to proposed time (e.g. leveled time + PF&D = standard time, standard time + rework factor = proposed time).
- The source of leveled time (e.g. direct observation time study).

Material
- Identify attrition factors (e.g., scrap, waste, spoilage).
- Identify any material cost curves.
- Identify economic purchase lot considerations (e.g. price breaks).

For overhead and G&A, the date of the last audit and the name of the auditor activity should be identified.

If learning curves are used to extend cost data, the following information should be identified:

- The curve hypothesis used (e.g. Wright's cumulative average, Crawford's unit curve).
- The technique used for determining the historical curve from data points (e.g. least square methods).
- The description of the Y-axis (e.g. direct manufacturing labor hours).
- The source of curved data (e.g. identify similar equipment).
- The slope of the curve (expressed as a percent).
- The unit at which the curve intersects with the labor standard (value of the unit at which the employees attain standard performance).
- The derivation of unit one (e.g. direct mathematical relationship to the labor standard or other base point).

If parametrics are used, the level at which parametric cost relationships are developed should be detained to at least subsystem level. The identifying information for the parametric cost relationships should be supplied in a cost rationale at the most detailed level for which they were applied. For example, if a parametric relationship is utilized to derive transmitter costs, the necessary information should be supplied in that section of the cost schedule.

Where parameters are used, the following information should be provided:

- Identify the dependent and independent variables in a cost estimating relationship (CER).
- Identify data points used in developing the CER.
- Identify the source of these data points (e.g. equipment nomenclature).

2. Initial spares

A. Ground

This section of the cost schedule should contain the cost of initial spares necessary to support a unit ground system and the supporting rationale used to derive these costs. Costs may be derived on a lot basis. However, costs should then be allocated on a $ per unit system per year basis for entry into the cost breakdown work sheets. Costs may be derived from hardware acquisition cost computations. Costs should include packaging, handling and transportation charges for a one-way trip of 1000 miles. Costs should be consistent with hardware costs. Required quantities should be consistent with the maintenance philosophy and reliability estimates.

A similar rationale should be employed for initial spares of airborne equipment.

(1) Initial spares are defined as those replacement modules (throwaways), repairable modules and assemblies, and repair parts necessary to establish the recommended stocking levels at the central depot facility and airport maintenance facilities consistent with the number of units.

(2) For the purpose of computing initial spares, the number of units for any time period should be the increment accrued during the initial time period as indicated in the projection of ground system requirements.

(3) Initial spares should be allocated to system hardware categories of angle equipment or DME. No further differentiation is necessary.

3. Special Test and Handling Equipment

A. Ground

This section of the cost schedule should contain the acquisition cost of the special test and handling equipment necessary to support a unit ground system and the supporting rationale used to derive these costs. Costs may be derived on a lot basis. However, costs should then be allocated on a $ per unit system basis for entry into the cost breakdown schedule. Costs should include parts, labor and packaging, handling, and transportation charges for a one-way trip of 1000 miles. Equipment should be associated with the system hardware category for which use it is primarily intended. Quantities of units required should be consistent with the number of ground systems installed in the time period, and with the maintenance philosophy and reliability estimates.

A similar rationale should be employed for special test and handling equipment for the airborne system.

4. Initial Training

A. Ground

This section of the cost schedule should contain the man-hours of training required to train the personnel required to support a unit ground system. The supporting rationale should contain the basis for this estimate. For the purpose of cost estimation, it should be assumed that all personnel to be trained have an appropriate background and that only skills and information specific to the peculiarities and details of the microwave landing system need be taught.

Categories of training should be defined in accordance with the use of personnel as envisaged in the particular maintenance philosophy.

Differentiation of training requirements might be required for:

(1) Function (electromechanical technician, electronic technician).

(2) Level of maintenance (central facility, on-site). For each category so defined, provide the total number of class and laboratory hours required to complete training per system.

A similar supporting rationale should be developed for initial training of personnel supporting a unit airborne system.

5. Initial Documentation

A. Ground

This section of the cost schedule should contain initial documentation costs on a $ per unit system basis and the supporting rationale by which these costs were derived. Costs should include all costs associated with the production of:

(1) A complete set of operating and maintenance manuals (reproducible).

(2) A complete set of drawings and schematics (reproducible).

(3) A set of installation drawings.

(4) A spares list for central facility and on-site stocking.

A similar rationale should be provided for the cost of initial documentation for the airborne system.

B Installation [*Extract*]

1. Ground

This section of the cost schedule should contain the cost of site preparation and facility installation on a $ per unit system basis. The supporting rationale should be consistent with the scenario provided in the subsequent section, the experience gained by the contractor during the feasibility phase and prototype and limited production option estimates.

A. Site preparation

This cost category is intended to include those costs associated with excavation and foundations for shelters or facilities. A semi-prepared condition is assumed so that initial clearing and grading or the establishment of access roads is not required. Trenching for cabling should be allocated to the associated subsystem if buried cabling is to be used. (The contractor should assume a $2.00/ft/trenching and backfill cost (excluding cable cost) in estimating the cost category.)

For the purpose of cost estimating, the following typical soil conditions are to be assumed:

(1) There is no significant ground water.
(2) Rocks requiring blasting are not present.
(3) The soil is primarily sandy clay with a compressive strength of 4 tons per square foot after initial consolidation.

For the purpose of uniform cost computation of installation costs, the following standard airport conditions are defined:

Configuration "D" and optional configuration
- For cost purposes, the configuration "D" airport should be considered to contain a 7000-foot runway. Location of MLS ground equipment should be idealized without considering possible obstructions.
- A cable run of 6000 feet to the control tower from the front azimuth site should be assumed.

Configuration K
- For cost purposes, the configuration "K" airport should be considered to contain a 14,000-foot runway. Location of MLS ground equipment should be idealized without considering possible obstructions. A cable run of 9500 feet to the control tower from azimuth site should be assumed.

Configuration "G" (military)
- The military "G" costs should be based on a 4000 foot runway

B. Facilities

The cost of installing shelters including test areas, storage and work areas, and materials not specifically listed as test or handling equipment should be provided.

In addition, the cost of the electronic equipment installation and checkout should be included. The contractor shall assume that the overall installation effort is a "turnkey" operation using contractor personnel. Flight inspection costs shall not be included in the cost estimates; these will be estimated by the Government.

C. Operation and Maintenance (Extract)

Maintenance costs will include all activity necessary to keep the equipment operating within the design goals of the reliability and maintainability plans. The man-hours required for preventive and unscheduled maintenance will be generated by factors of the mean-time-to-repair (MTTR), mean-time-between-failure (MTBF) and the experience gained in maintaining the equipment in Phase II. Traceability can then be made through the reliability and maintainability plans for Phase III, cost estimates for maintenance activity in the prototype, and limited production equipments, and the actual cost incurred in the feasibility phase.

Units of measure for maintenance effort will be expressed in man-hours per year. This will be required for both the ground and airborne equipment where the man-hour cost will be different but can be applied to each system to determine total maintenance cost. For the ground systems, the current FAA maintenance policy of repair-on-site to the maximum extent possible should be used to apportion labor hours between on-site and center maintenance points. However, if the contractor can show that significant cost advantages can be gained through alterations of current FAA maintenance policy, then these procedures should be outlined and the maintenance requirements reported. The maintenance policy outlined by the contractor must be supported by the Reliability and Maintainability Plan, the Logistics Support Plan plus the initial and sustaining training requirements. It should be assumed that for configuration "K" equipment, there will be 24 hour per day maintenance service provided. Configuration "D" facilities can be assumed to have 8 hour per day coverage with call back for facility repair.

For the avionics equipment, the maintenance practice currently subscribed to by the major air carriers and corresponding electronic system suppliers can be assumed. That is, for cost estimating purposes, maintenance of the avionics equipment will be provided by the manufacturer for a 1-year period. Thereafter, maintenance and repair of the equipment will be the responsibility of the purchasing air carrier.

Operation cost will be a function of the power budget and will include all electrical power requirements for the subsystem and environmental equipments. Hardware specifications for prototype equipment with the associated power budget plus the shelter and environmental equipment will form the framework for engineering estimates. Units of measure for the operations cost will be expressed in kilowatt hours per year. For the purpose of uniform computations for environmental equipment power consumption, the following standard temperature profile for Kansas City International Airport shall be assumed:

JAN	FEB	MAR	APR	MAY	JUN	JUL	AUG	SEP	OCT	NOV	DEC
36	40	54	72	75	89	90	89	83	75	56	44
19	24	33	48	54	70	67	67	62	53	38	30

1. Preventive maintenance

Any work activity that must be performed by the electronics or electromechanical technicians to maintain the equipment in operational status. This will include any periodic maintenance required to keep the system operating within the reliability constraints imposed by the contractor's plan.

2. Unscheduled maintenance

All work activity that must be performed to restore a failed equipment to operational status. Frequency of occurrence and labor hours to repair will be a direct function of the MTBF and MTTR and must be supported through the Reliability and Maintainability Plans.

D. Logistic Support (Extract)

1. Replacement modules and assemblies (throwaways)

This section of the cost schedule should contain the cost of replacement modules and assemblies for both ground and airborne systems. Costs may be computed on a lot basis. However, cost should be entered in the schedule as $ per year per system for both ground and airborne systems. The supporting rationale for these costs should be consistent with the contractor's maintenance philosophy and reliability estimates and logistic support plan. The supporting rationale should consider the following information:

A. Definition

A replacement module or assembly is one which will not be repaired or re-entered into the supply system. This cost category is associated with the replenishment requirements in a given time period of the replacement modules necessary to maintain the appropriate stocking levels due to deletion by failures.

B. Quantitative requirements

Quantitative annual requirements for a particular module are a function of the failure rate, the number of modules per system, and the usage factor.

C. Usage factors

The number of failures for a particular component will be directly related to the number of hours of normal use for that component (i.e. hours for which the component is receiving its normal stress level).

Equipments perform in three modes of operation. These modes are the operating mode, the standby mode and the emergency mode:

(1) *Operation mode*
   Operational requirements for the "K" configuration ground system provide 24 hours a day operation. For redundant system equipment, this operational time should be equally shared. "D" configuration ground systems provide 16 hours a day operation.
(2) *Standby mode*
   Certain components may be stressed only in the standby mode or in the standby mode in addition to the operating mode. These factors should be used in the estimation of component usage.
(3) *Emergency mode*
   Certain components are stressed only in the emergency mode. Usage for these components should be based upon the expected hours of emergency mode operation.

D. Cost per unit

As an approximation, costs of initial hardware acquisition may be used. The contractor may modify this figure according to the total quantity of modules required for the life of the system (25 years).

E. Assignment by hardware category

Allocation of modules to hardware systems should be to another angle or DME equipment. No further differentiation is necessary.

F. Airborne replacement modules (Category III)

Airborne replacement modules should be computed for Category III systems (configuration "K") in accordance with the maintenance philosophy as previously defined. Configuration "D" airborne systems are not applicable to this cost category. Two modes of operation, operating and standby, are applicable to airborne equipments. For Category III aircraft, 2700 hours annual operating time should be assumed. This time should be divided equally among redundant equipments.

3. Repair parts

This section of the cost schedule should contain the cost of repair parts for both ground and airborne systems. Costs may be computed on a lot basis. However costs should be entered in the schedule as $ per year per system for both ground and airborne systems. The supporting rationale for these costs should be consistent with the contractor's maintenance philosophy and reliability estimates. The supporting rationale should consider the following information:

A. Definition

Repair parts are those parts stocked for the purpose of repairing modules or assemblies which are economically repairable.

B. Quantitative requirement

Quantitative requirement for a particular repair part is a function of:

    (1) Associated level of maintenance (central depot facility or airport).
    (2) Failure rate.
    (3) Number of modules per system.
    (4) Usage factor.

C. Usage factors, cost per unit, assignment by hardware categories, airborne repair parts

These considerations are the same for repair parts and replacement modules.

3. Packaging, handling, and transportation

This section of the cost schedule should contain the cost of packaging, handling, and transportation for logistic elements for both ground and airborne systems. Cost should be based on $ per year per unit system. The supporting rationale should be consistent with the contractor's maintenance philosophy and should consider the following information:

A. Ground

    (1) *Replacement modules*
        Replacement modules are packaged, handled, and transported in bulk lots, consistent with proper protection and replenishment requirements to the central

depot facility located at Oklahoma City, Oklahoma. In addition, a one-way shipment of 1000 miles average should be computed for each group of failed modules, to represent delivery to the requiring airport.

(2) *Repair parts*

Repair parts are packaged, handled, and transported in bulk lots, consistent with proper protection and replenishment requirements to the central depot facility. In addition a round trip of 1000 miles each way should be computed for each module repaired at the central depot facility based on failure rate, number of units and usage. Also, a one-way shipment of 1000 miles should be computed for groups of failed parts representing replenishment at the airport at which repairs are made.

B. Airborne (Category II and III)

Packaging, handling, and transportation costs equivalent to two air freight deliveries per computed failure for a distance of 1000 miles should be computed. Computed failures per year per system should be used and a $ cost per year per system computed.

4. Storage

Since storage costs will be integral to shelter costs at airports, the only costs necessary here are for the central depot facility. The required costs should be represented by the number of square feet required to store the stocking levels of replacement parts, repair parts and test equipment.

Storage costs for Category III (configuration "K") airborne equipment should be treated on the basis of total square feet required without considering individual airlines. Storage cost for configuration "D" airborne equipment is not applicable.

5. General—sustaining

A. Training

This category represents sustaining training costs required by the attrition of trained personnel. No data is required from the contractor under this cost category.

B. Test equipment

This category includes the maintenance and calibration costs of special test and handling equipment for each year group and the prorata costs of general test and handling equipment. Cost in $ per year per system should be entered in the cost breakdown work schedule.

C. Documentation

This category includes the cost of modifications and changes to the initial documentation for each year group. Cost in $ per year per system should be entered in the cost breakdown work schedule.

IV SCOPE OF CONTRACTOR COSTING EFFORT

[No extracts]

ENCLOSURES

[No extracts]

14.3.4 Cost estimates

Cost estimates can be obtained using comparative and parametric techniques. In **comparative** costing the estimate is determined by comparing the material and labor content against a schedule for a similar existing item or activity of known cost. The schedule should allow differences affecting cost (as distinct from performance or function) to be highlighted. The comparative technique is equally applicable to the estimation of both recurring costs (e.g. maintenance) and nonrecurring costs (e.g. acquisition). **Parametric** costing depends on the establishment of relationships between cost and specific characteristics. This may appear self-evident for some factors (related to performance), but the technique is again applicable in other areas, including production, maintenance and personnel. Both techniques can be used to support the stepwise refinement of an LCC cost model.

In estimating the value of each cost item for such a model, a conundrum always presents itself. The further the set of cost items is developed, the less project-specific these items become, and the easier it is to compare them to items in past projects. Consequently, by the time the system design is completed, it is possible to estimate the LCC with great precision. However, the actual cost is by then all but predetermined; and while the estimate can be useful in fine tuning cash flow predictions and financing arrangements, it can no longer have any influence on the system design. As LCC modeling is intended as a design tool, the earlier in the engineering process it can be put to use, the more valuable it should be. So to the conundrum: How can models that will predict the system LCC be developed while the system itself is still being designed (or even analyzed)?

As with all estimation, the estimation of system LCC must necessarily be based on past experience—on **comparison** with previous projects. However, whereas the classical estimation process (e.g. as it is employed by quantity surveyors) is based on comparison with objects of known costs that differ only in quantity (e.g. number, volume, weight), the system-level estimation process needs to be considerably more subtle and complex.

It is unlikely that a previous project could be found that differs from the present one in size only. Instead, it is necessary to determine the group of parameters to which the system cost is particularly **sensitive**. In addition to the quantity of various element types, this group may include such system parameters as mean time to repair (MTTR), mean time before failure (MTBF) and weight (but with low weight corresponding to high cost), and also parameters describing the environment in which the system must operate (e.g. according to the environmental groups defined in Chapter 10). As the partitioning of the system progresses, each element will have its own group of parameters and/or its own set of sensitivities.

This **parametric** development of LCC is particularly suitable for providing a design tool within systems engineering, as it allows the overall effect of a proposed change to a system parameter to be looked at directly. However, it presupposes an availability of (or a willingness to develop) the various parametric **cost models** required. On the one hand, such models are based on experience with other (but not necessarily similar) systems; on the other hand, they express general functional relationships.

As an example of the latter, consider a system (e.g. a communications system) that consists of a number n of identical elements (e.g. transceivers), all of which must be operating for the system to operate. The system MTBF is then simply the element MTBF divided by n. It is now possible to build a model of system cost as a function of the system MTBF. This model must include such cost items as element acquisition cost and cost per repair (on average), but it must also consider the cost of stocking spare elements, assuming that failed elements are replaced while they are sent away for repair. If the time taken to get an element repaired is T, there must be enough elements in store initially to replace those that fail during the first T hours of operation. If the system MTBF is large compared with T, one spare will be adequate; if MTBF is small compared with T, the number of spares required will be inversely proportional to MTBF. Thus the simple but nonlinear relationship shown in Figure 14.2 applies.

However, note that in this example the cost of the elements—and thereby of the spares—also depends on MTBF. As a consequence the true effect of changing MTBF can be ascertained only by estimating all the cost items and then running the LCC calculation.

14.4 Case study
14.4.1 Overview

In Section 6.5.1 a simple economic model of the combustion optimization system was developed. It gave a relationship between the three system parameters A, q and s, where:

A = acquisition cost
q = system performance parameter
s = yearly maintenance costs, as a fraction of A.

The relationship was expressed as

$$Q_{min} = A (s + 0.333)/F(q)$$

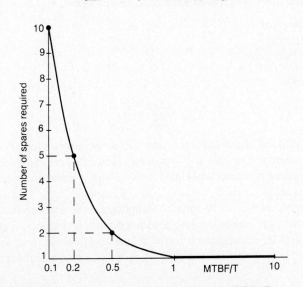

Figure 14.2 Number of spares required as a function of MTBF/T.

where

Q_{min} = minimum value of yearly fuel costs for which it would be economic to install the combustion optimization system

$$F(q) = 1.05 \left(\frac{1}{18.5} - \frac{1}{(20 - 1.5q)} \right)$$

The relationship was illustrated in Figure 6.11.

Such a model is very useful in looking, for example, at the viability of introducing the system into a particular market, but it is not adequate as a design tool once the decision has been made to proceed with the development of the system. Stated somewhat simply, this is because the model gives an expression for a boundary value Q_{min} rather than a criterion for an optimal design. What is needed is a model for the cost-effectiveness of the system.

14.4.2 Definition of cost-effectiveness

The system effectiveness *(SE)* is the degree to which the system reduces the excess flue losses (i.e. flue losses resulting from variation in the combustion process parameters). Using the same simplifying assumptions as in Section 6.5.1,

$$SE = F(q)/F(0) = 13.5 \left(1 - \frac{12.3}{13.3 - q} \right)$$

Note that, for $q > 1$, SE becomes negative!

In Section 9.3.2.1 the global system performance parameter q was further defined by relating it to the three design parameters e, d and u, where:

e = acquisition error
d = deadband error
u = feedback loop gain.

The relationship was

$$q = e \frac{x_0}{\Delta x} + \frac{d}{x} \cdot \frac{1}{u}$$

By inserting this into the above expression for SE, an expression for SE in terms of the three design parameters e, d and u is obtained. However, achieving given values for these three parameters has costs associated with it, and it therefore becomes necessary to look at the cost side.

The costs consist, as well, of two major components: the acquisition cost A, and the yearly maintenance cost (including loss associated with downtime). Taking the latter over the design lifetime, discounting it to the acquisition point in time, and denoting its present worth by S, the total cost is simply $A + S$. The cost-effectiveness can then be defined as

$$CE = SE/(A + S)$$

However, both A and S are functions of e, d and u. Whenever a change in any one (or more) of these parameters is contemplated, it is not sufficient to see how this will influence system performance (or SE). An increase in SE is not good in itself; it is only good if the effect on the cost is such that CE increases.

14.4.3 Discussion

This example is intended to illustrate the following:

1. The usefulness of the system is clearly and very directly related to cost. Irrespective of how well engineered it is, there is a **limiting cost**.
2. On the other hand, the usefulness of the system is related in a simple way to its performance; and no matter how low the cost is, there is a **limiting performance** below which the system is of no benefit.
3. Depending on whether the system is close to the cost limit or to the performance limit, any further engineering effort should be put into **cost reduction** or **performance improvement**, respectively.
4. The only correct way to judge the merit of a design change is to determine the resultant **change in cost-effectiveness**.
5. The relationship between performance, as measured by design parameters, and benefit (i.e. system effectiveness) is usually much too complex to be handled in an ad hoc manner. A **model** needs to be developed.

14.5 Summary

1. **Cost-effectiveness (CE)** is the ultimate or all-encompassing measure of the quality of the engineering of a system. If CE is correctly defined, an increase in CE is **always** desirable.
2. The problem with CE, and what often has limited its usefulness, is that, while it is usually not too difficult to determine or estimate the cost, it is often both complicated and controversial to put a value on **system effectiveness (SE)**. What is the value of increasing MTBF from 10,000 h to 12,000 h? Or what is the value of decreasing noise from 4.5 dB to 4.0 dB?
3. Aside from the above problem, the principles involved in defining CE for a particular system are straightforward. For each system parameter, relate its value to a dollar value, and then combine them so as to obtain a quantitative measure of overall system performance or effectiveness (the latter being the degree to which the system provides the required service). Then determine every cost item related to the creation of the system and its operation over its lifetime, and refer them all to one point in time so as to obtain the **lifecycle cost (LCC)**. CE is generally the ratio of the two, although sometimes it makes more sense to take the difference.
4. A change in any system parameter will result in a change in both SE and LCC. Whether it is a **desirable change** or not can only be determined by considering both changes together (i.e. by looking at whether CE increases or decreases).

14.6 Short questions

1. What is meant by the term "system effectiveness"?
2. What are some of the major system characteristics that determine system effectiveness, and what are some of the factors that influence their importance?
3. Give one reason for why the definition of system effectiveness is generally a difficult task.
4. How is lifecycle cost (LCC) defined?
5. What is a cost item?
6. What is net present (or future) value?
7. What is the internal rate of return?
8. Why does net present value normally decrease with increasing discount rate?

14.7 Problems

MMIC fabrication: A fabrication plant for microwave monolithic integrated circuits (MMICs) incurs direct costs through its requirement for plant, personnel and materials.

Plant consists of dedicated buildings and equipment, currently valued at $2 million and $5 million respectively. Each is subject to depreciation (by 5% and 25% respectively per annum) and to annual outlays for utilization maintenance and development (of 10% and 20% respectively of undepreciated value). Plant capacity is limited to an average of 500 wafers per month.

Personnel costs covering salaries and related on-costs are proportional to the throughput or workload, averaging $700 per processed MMIC wafer of N circuits, $15 per die prior to packaging and $10 per final assembly. Testing is an integral part of this workload, the results of which are measured in terms of process yields.

Material costs cover raw wafers, process consumables and supporting services, averaging $400 per wafer.

Wafer yields y_w are determined by material quality and process control; die yields y_d depend on the complexity of integrated circuits (i.e. the component count n); and package yields y_p are set by quality control. The die yield follows from the average component yield ($\bar{y}_c$) through the equation

$$y_d = (\bar{y}_c)^n$$

(a) Prepare a structure diagram showing all the cost elements of a MMIC successfully produced at the fabrication plant.
(b) Determine the average cost of a successful MMIC in terms of the given variables, with the plant operating at full capacity. (Interest rates of 18% should be assumed if necessary. Any other assumptions should be clearly stated.)
(c) If values of y_w and y_p in the range 70–90% are equally probable, determine the selling price for a high volume MMIC comprising 100 circuit components (of average yield $\bar{y}_c = 99.5\%$) whose dimensions allow 500 dies per wafer. The selling price should be set to ensure a 100% gross margin over direct costs. State clearly the basis of any average cost assumed in the calculation.

(d) If the same circuit function could be achieved with two separate (smaller) chips, each of 75 circuit components requiring 80% of the former surface area per die, would the fully integrated (single-chip) solution be more attractive in terms of cost? (A qualitative answer addressing the inherent issues will suffice.)

14.8 References

Redlien, H.W. and Kelly, R.J., "Microwave Landing System: the new international standard", *Advances in Electronics and Electron Physics*, vol. 57.

Index

Index